Contents

Design and construction of reinforced steel chain wall

Kepha Abongo[1], Makoto Kimura[2], Akihiro Kitamura[3]

[1]Postgraduate student, Department of Urban and Environmental Eng., Kyoto University, Kyoto 615-8520 Japan;abongo.k@ht3.ecs.kyoto-u.ac.jp
[2]Professor, Innovative Collaboration Center, Kyoto University, Kyoto 615-8520 Japan; kimura@iickyoto-u.ac.jp
[3]Engineer, Department of Reinforced Earth, Showa Kikai Shoji Co. Ltd., Suminoue-Ku ,Kitakagaya-1 Chome-3-23, Osaka 559-0011, Japan; kitamura@showa-kikai.co.jp

ABSTRACT: This paper reports the result of insitu pullout test of chain reinforced wall constructed in mountainous region of Japan. The insitu test was performed for chain with and without end bearing plate anchor and horizontal bar. Test results revealed that the end bearing anchor greatly improved the pullout force of chain than the horizontal bar. The horizontal bar was found to be weak in tension and therefore did not greatly enhance the pullout force of chain. Comparison of the measured field values and the design values calculated from existing design methods yielded results that were in good agreement.

INTRODUCTION

Chain has been used as a reinforcement inclusion in the reinforcement of earthfill in Japan. Despite the fact that little is known about the mechanisms that generate pullout force of chain, field construction of reinforced wall using chain has already been done. Flexibility of chain that is its ability to be folded easing its transportation is one of the factors that have contributed in its use for fill slope reinforcement in recent times. It has been though that due to its flexibility, chain can follow the deformation of the soil around it incase of uncertainty such as a consolidation of the ground. Its ability to generate high frictional coefficient is another factor that has given rise to its use in the recent days. Comparative study on chains of different shapes and sizes with other reinforcement materials such as rounds bars and ribbed steel bars revealed that chain generates higher frictional resistance than these materials (Fukuda et al., 2007). In addition, separate studies reveled that pullout force of chain is dependent on the outer width of the chain (Kitamura et al., 2006).

Photo 1. Construction of reinforced chain wall

In this method of constructing retaining wall, the insitu soil is used as the fill material hence it does not pay attention to judicious selection of fill material like other reinforcement methods such as the Terre-Armee making it cheaper in terms of construction cost.

Field construction of chain wall was done in Nara prefecture of Japan. This area was on a mountainous region that isolated two villages. Construction of the wall involved placing the insitu soil into thin layers and compacting using roller as shown in **Photo 1**. At the edges, closer to the frame wall, compaction was achieved by the use diesel operated hand vibrator. One month after completion of the construction, insitu pullout test was conducted on the wall in order to verify the performance of the chain with or without tension bars and end bearing plate anchors attached on its end.

REINFORCED CHAIN WALL

Chain reinforcement method involves integration of steel made frames with chain. The steel frames serve the purpose of the facing panel in which the chain is clipped to using a flexible joint. **FIG.1.** shows the schematic representation of the components of chain wall consisting of the following parts;

(1) Bottom rail which hold the steel frames into position during construction. The rail serves the same purpose as leveling pads in reinforced wall.
(2) Hollow Steel frame spaced at an interval of 50 cm serving as facing panel. The steel frames are joined over each other as the construction height increases.
(3) Top rail that joins the top of the facing panels together after the construction has been completed.
(4) Flexible joint that is able slide onto the frame and hence follows the movement of the chain in case of eventuality such as settlement of the soil around the chain or movement of the chain during compaction process.
(5) Steel made chain used as a reinforcing material.
(6) An end bearing plate anchor fixed at the end of the chain which enhances the pullout resistance of chain.
(7) A woven grid of aperture size 2 cm by 2 cm is laid around the facing panel during construction in order to hold the soil firmly. Once the woven grid has been placed in position, a vegetation mat laced with grass seeds is wrapped from the inside of the wall. Under suitable environmental conditions, these seeds germinate into grass that grows on the wall thereby preventing it against erosion.

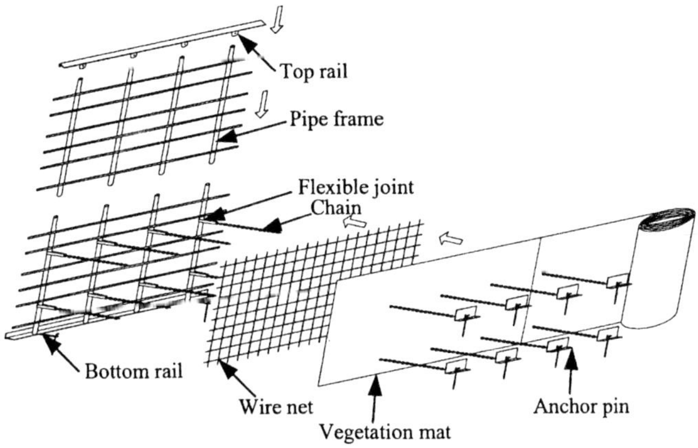

FIG. 1. Components of chain reinforcement method

PULLOUT FORCE OF CHAIN AND DESIGN METHOD

Cylindrical Model

Fukuda et al. (2004) proposed a cylindrical model shown in **FIG.2**. for evaluating the frictional resistance of steel chain. This model assumed that the inner space of chain is entirely filled with the soil. When chain is examined at the point where two links intersect, a cross pattern shape show in **FIG. 2.** can be seen. However, filling of the space is limited when the soil grain becomes large.

During pullout, shearing of the soil occurs and therefore the soil around the chain tends to dilate. The assumption is that as dilatancy occurs, the failure surface is circular and hence the diameter of the slip surface expands by amount β times that of the outer diameter of the chain, B. Based on this assumption, equations (1) to (3) of the cylindrical model were proposed for calculating the pullout force of steel chain.

$$F_f = \pi B \times \beta \times L \times \tan\phi' \times (\frac{1+K_0}{2}) \times \sigma_v' \qquad (1)$$

$$\alpha = \beta(\frac{1+K_0}{2}) \qquad (2)$$

$$F_f = \alpha \times \pi B \times L \times \sigma_v' \times \tan\phi' \qquad (3)$$

Where F_f is the frictional force generated during the pullout of chain, σ'_v is the measured vertical pressure, L ,is the apparent length of the chain embedded inside the soil, B is the diameter of the cylindrical model (equivalent to the outer width of the chain), β is the outer width correction factor that adjusts the outer diameter of the cylindrical model to the assumed slip surface, K_0 is the coefficient of earth pressure at rest, ϕ' is soil's friction angle and α is the frictional correction coefficient.

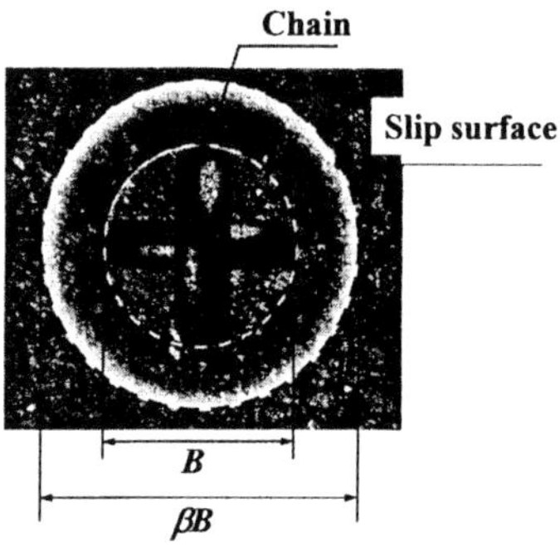

FIG. 2 The cylindrical model of the chain

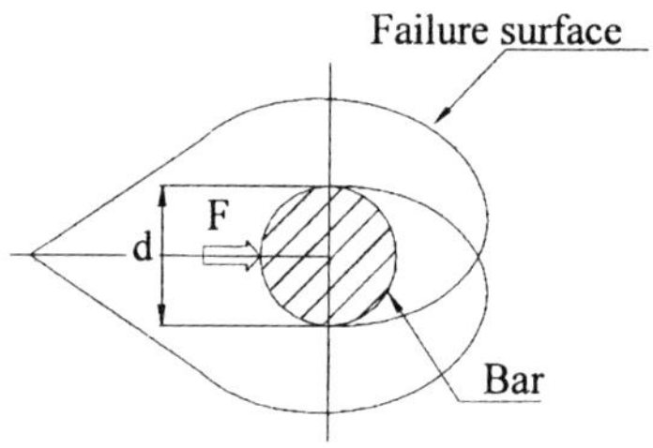

FIG.3. Cross section of a horizontal bar

Design Method

Fukuda (2007) observed through laboratory pullout test that the parameter α depends on the applied pressure. As a matter of fact, α decreased as the applied pressure was increased. In order to obtain a value of α that is independent of the applied pressure, Fukuda et al., 2007 normalized the obtained α at pressure of 100kPa using equation (4) below. The value of n was found to be equal to 0.6. In addition two design equation (equations 5 and 6) for obtaining the value of α_0 was suggested, one based the effect of dilatancy and the other based on angle of internal friction of soil.

$$\alpha_0 = \alpha\left(\frac{\sigma_v'}{100}\right)^n \quad (4)$$

$$\alpha_0 = 3.5\left(-\frac{dv}{d\varepsilon}\right) + 0.6 \quad (5)$$

$$\alpha_0 = 1.8\tan\phi - 0.2 \quad (6)$$

In this paper, equation (6) is used in design due to its simplicity and ease with which soil's friction angle can be obtained.

INSTRUMENTATION OF THE WALL

Three test cases were executed as shown in **Photo2**. The test arrangement involved conducting pullout test for (1) chain alone (2) chain with horizontal bar and (3) chain with an L-type end bearing plate anchor attached to it. The horizontal bars used in test were of 6mm diameter (**FIG.3.**) while the end bearing anchors used in the construction were of length 0.2 m and width 0.1 m. **Photo 2** also shows the set up of the pullout test on the wall. Two meters of the chain was embedded inside the soil, however one meter was enclosed inside the rubber sheath and the other one meter left as the testing length. The purpose of the rubber sheath was to minimize the wall effect and to allow

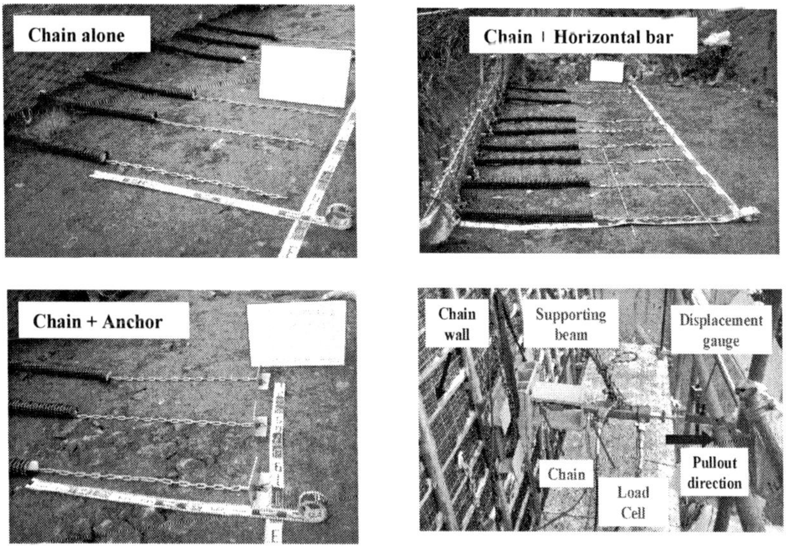

Photo 2. Testing patterns and In-situ test arrangements

Table 1. Soil and chain properties and testing conditions

Chain Type	**Length (m)**	**B (mm)**	**H (m)**	**ϕ (°)**	**γ_t (kN/m^3**	**σ_v (kN/m^2)**	**α_0**
CWC-6.3	1.0	23	4.3	37	20	86	1.2
CWC-8.0	1.0	29.8	8.7	37	20	172	1.2

for only the portion of the wall with constant overburden pressure to be tested. **Table 1** above gives the properties of soil and the wall. Two types of chain were tested for namely CWC-6.3 and CWC-8.0 (wire diameter 6.0 and 8.0 mm respectively) and the heights of the potions of the wall in which they were constructed were 4.3 m (portion A in **Photo 1**) and 8.7 m (portion B in **Photo 1**) respectively. The insitu soil has γ_d of 20kN/m^2 and ϕ of 37°.

RESULTS AND DISCUSSION

Insitu pullout test

FIG.4. shows test results obtained for both CWC-6.3 and CWC-8.0. In both cases, the peak pullout force was observed for chain alone and chain with horizontal bar, however for the case of pulling out chain with anchor, peak pullout force was not observed. In this case, peak pullout force was obtained by drawing tangents on the straight portions of the graphs as shown in **FIG. 4.** The point at which both the lines

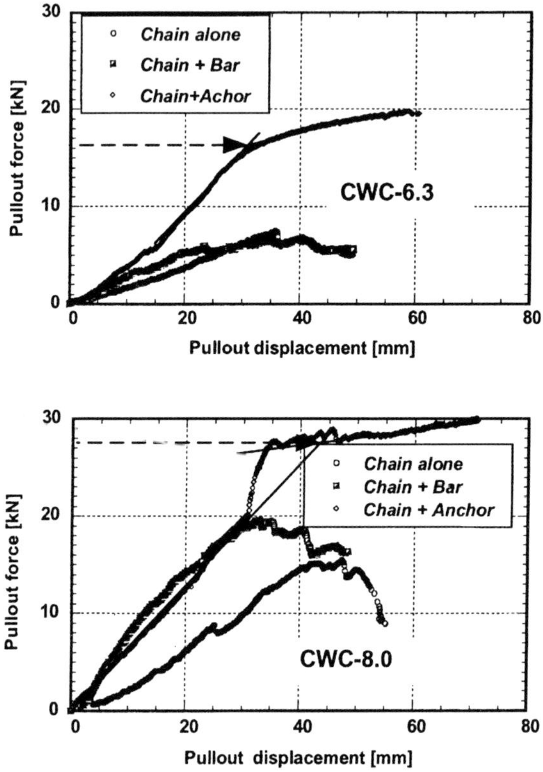

FIG. 4. Pullout – displacement graphs for chains tested under various

intersected was taken as the peak pullout force. Pullout force of chain is greatly enhanced when chain is combined with anchor than when chain is combined with horizontal bar. Comparison of the test result with the result of chain alone indicated that pullout force increases by 21 % and 100% for chain with horizontal bar and chain with anchor respectively. It can be concluded therefore that L-type anchor greatly increases the pullout capacity of chain.

Design of Maximum pullout force of chain

Equation 1 of the cylindrical model was used in the design.

$$\alpha_0 = 1.8 \tan\phi - 0.2 \tag{6}$$

$$\alpha = \alpha_0 \left(\frac{100}{\sigma_v^{'}} \right)^{0.6} \tag{4}$$

$$F_c = \alpha \times \pi B \times L_{eq} \times \tan\phi \times \sigma_v \tag{3}$$

F_c = <u>6.1 kN</u> (CWC-6.3) F_c = <u>9.4 kN</u> (CWC-8.0)

F_c: Calculated pullout force of chain, L_{eq}: Tested length of chain (= 1 m)

Design value of the anchor

Terzhagi- Buisman's equation applied for the design of the anchor. Cohesion of the soil was zero and therefore the first term in equation 7 was neglected.

$$Q_a = A_a\left(\beta.cN_c + q_p.N_q - q_p\right) \tag{7}$$

$$\beta = 1.0 + 0.3\frac{B_a}{L_a} \tag{8}$$

$$q_p = K_a \sigma_v \tag{9}$$

$$K_a = \frac{1 - \sin\phi_1}{1 + \sin\phi_1} \tag{10}$$

$$N_q = \frac{1 + \sin\phi_1}{1 - \sin\phi_1}\exp(\pi \tan\phi_1) \tag{11}$$

$$N_c = \left(N_q - 1\right)\cot\phi_1 \tag{12}$$

$$\tan\phi_1 = \frac{2}{3}\tan\phi \tag{13}$$

Q_a = <u>7.90 kN</u> (CWC-6.3) Q_a = <u>15.80 kN</u> (CWC-8.0)

Design value of round bar

$$F_{rb} = d\left(c \times N_c + \frac{\gamma_d}{2} d \times N + \sigma_v \times N_q \right) \tag{14}$$

Maximum fore generated by the horizontal bars is calculated from equation (14).Since d is very small and cohesion is zero, the first and second terms in equation (14) above are ignored.

$$F_{rb} = M \times d \times N_q \times \sigma_v \tag{15}$$

F_{rb} = <u>6.80 kN</u> (CWC-6.3) F_{rb} = <u>13.60 kN</u> (CWC-8.0)

Table 2 Summary of the test result and design values (design values in bracket)

Chain type	Chain alone (kN)	Chain+ Anchor (kN)	Chain +Bar(kN)
CWC-6.3	7.40 (6.10)	17.00 (14.00)	8.5 (12.90)
CWC-8.0	14.70 (12.10)	28.10 (27.90)	19.00 (13.60)

Comparison of the measured and design values

Table 2 below gives the summary of the test results for the three test cases. From this table, it can be seen that the pullout force of chain can be estimated from the cylindrical model equation with some degree of accuracy. The calculated design value from a combination of chain and anchor was closer to the measured value. However for the case of chain with horizontal bar, the design value was higher than the measured value implying that horizontal bar is not effective in enhancing the pullout force of chain. Weakness of the horizontal bar is tension and the possibility of a wide clearance between the inner space of the chain and bar are though to be the main reasons for this.

CONCLUSIONS

The following conclusions can be drawn from the research;

1) Equation of the cylindrical model together with the design method based on angle of internal friction gave a good estimation of the pullout force of chain.
2) When L-angle is used at the end of the chain, the pullout resistance of the chain is doubled.
3) Test case with horizontal bar yielded lower peak value than the design value, implying that horizontal bar is not effective in enhancing the pullout force of chain.

ACKNOWLEDGEMENTS

The authors would like to convey their gratitude Showa Kikai Shoji Co. Ltd, Osaka for their support and willingness to collaborate in research especially agreeing for pullout test to be performed in the constructed portion of the wall in Nara prefecture.

REFERENCES

Fukuda, M., Kitamura, A., Mochizuki, Y., Hongo, T., Inoue, S., Fujimura, E. and Kimura, M. (2007) "Resistance of steel chain in pullout test with and without sliding box" *New Horizon in Earth reinforcement*, IS-Kyushu, pp. 299-305.

Fukuda, M., Hongo, T. and Konishi, K. (2004) "Frictional force of steel made chain reinforcement applied to fill slope". *Proc. Of Earth reinforcement*, IW-Shiga 2004, pp 152-161.

Kitamura, A., Shigeyoshi, M., Fukuda, M. Hongo, T. Inoue, S. and Fujimura, E. (2006) "Frictional pullout characteristics of chain as a reinforcing material." 41st *Geotechnical Engineering Society Annual conference*, pp. 1843-1844. (In Japanese)

Deformation regularity and simplified calculation method for foundation pit with confined water during excavation and dewatering

DING Chunlin[1] MENG Xiaohong[2]

[1]Associate Professor , [2]Engineer, Key Laboratory of Road and Traffic Engineering of the Ministry of Education, Tongji University, Shanghai 200092, China; chunlinding308@tongji.edu.cn

ABSTRACT□Based on a deep foundation pit of some station in Shanghai Metro without layer 6 (stiff-soil aquiclude), a centrifuge model test was employed to study the deformation regularity of the foundation pit with confined water during excavation and dewatering. According to the test results, the simplified calculation equations for the deformation of the foundation pit with confined water during excavation and dewatering were studied by using the regression analysis theory, and the predicted results of the simplified calculation equations are identical with the test results. In the end, some valuable suggestions are proposed for the design and construction of the foundation pit with confined water.

INTRODUCTION

A subway station in Shanghai is located in a downtown area. There is mainly silt clay in saturation state, and a small amount of thin fine sand and cultch in the range of the excavation depth according to the reconnaissance data. Two layers of confined aquifer exist under the pit bottom, one is the layer 5-2, which is a micro-confined aquifer, and the other is the layer 7 ,which is composed of the layer 7-1(grass yellow sandy silt) and the layer 7-2(gray silt). The micro-confined aquifer has a shallow buried depth, and is close to the pit bottom. The layer 6(stiff-soil aquiclude) between the micro-confined aquifer and the second confined aquifer is absent at most sites. According to the pumping test, the two confined aquifers are connected, which is not conducive to the stability of the foundation pit. In addition, according to the judgment formula of the inrushing stability, the stability of the foundation pit can only be guaranteed by dewatering. However, the geographical location of the station is in heavy traffic section, and it's also an area with densely covered buildings and underground pipelines, so the permit deformation of the pit is small while the requirement on environmental protection is strict (Ding el at, 2004). Therefore, it is very important to know the deformation regularity of the foundation pit with confined water without the layer 6(stiff-soil aquiclude). In this paper, a centrifuge model test is employed to study the deformation regularity of the foundation pit with confined water, and the simplified calculation equations of the deformation are studied.

PRINCIPLE OF CENTRIFUGE MODEL TEST

The basic principle of the centrifuge model test is to put the model made by prototype material in a certain proportion under the centrifugal force field generated by the high-speed rotation of the centrifuge, so the self weight of the soil is increased by the centrifugal acceleration, and the loss of the self weight caused by the model reduction is compensated, so that the model can achieve the same stress state with the prototype, and shows a similar process of deformation and failure with the prototype (Taylor,1995).

The similarity law of the centrifugal model test of the foundation pit with confined water can be obtained by analyzing control equations according to the equilibrium equations, continuity equations, geometric equations, physical equations, boundary conditions and initial conditions in the seepage-stress coupling field (Ding and Meng, 2008), as is shown in Table 1.

Table 1 Similarity Ratio of Physical Parameters between Model and Prototype

physical quantity	geometric dimension	area	acceleration	unit weight	stress	strain	displacement
prototype	1	1	1	1	1	1	1
model	1/n	$1/n^2$	n	n	1	1	1/n
physical quantity	**permeability coefficient**	**time**	**pore water pressure**	**water head**	**flow rate**	**surface force**	**volume force**
prototype	1	1	1	1	1	1	1
model	n	$1/n^2$	1	1/n	n	1	n

DESIGN OF CENTRIFUGE MODEL TEST

The model ratio n is selected as 120 according to the purpose of the test, the excavation dimension, the clearance size of the model box and the working conditions of the geotechnical centrifuge.

The soil tested is disturbed soil taken from the site, and the section of the selected soil is shown in Fig. 1. The physical and mechanical properties and other indicators of the soil can be seen in Ref. (Ding el at, 2004).

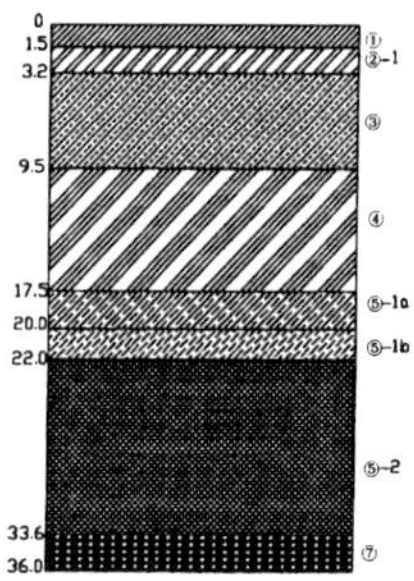

Fig.1 Cross-section of the soil

The equivalence and similarity principle of bending stiffness and compressive strength of the material is mainly considered in the simulation of the diaphragm wall and the support. For the actual diaphragm wall structure, the physical dimension of the model is 1/120 of the prototype, accounting that the model is simulated by reinforced concrete, and the thickness of the model is only a few millimeters, which is inconvenient to produce; therefore, an aluminum plate is used as a substitution. For the alternative model, the thickness of the plate is determined by the principle that the bending stiffness of the model is similar to the raw material; finally, the model thickness is determined to 4.8mm. The Φ609 mm steel support on site is simulated by an aluminum bar, according to the principle of equivalent compressive strength, the diameter and the wall thickness of the aluminum bar in the simulation model test can be determined to 5mm and 1mm respectively.

The retaining structure of the foundation pit simulated in the test is internal-shoring diaphragm wall, with an excavation width 19.0m and a diaphragm wall depth 30.0m. According to the excavation depth of the pit standard section and the shield working well, the excavation depth simulated is designed for three conditions with excavation depths 14.7, 16.7 and 18.0m respectively. When the excavation depth is 14.7, 16.7 and 18.0m, 4, 5, 6 Φ609mm steel supports are located along the depth direction, as shown in Fig.2. The average horizontal distance of the support is 3.0m. In each case of the excavation depth, three conditions are also simulated according to the confined water head before dewatering on-site and the water head without inrushing, which are the conditions with confined water head 20.0, 24.0 and 28.0m respectively.

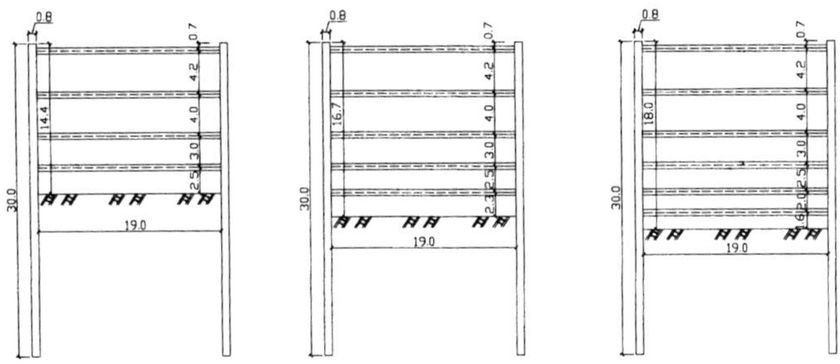

(a) case 1 (b) case 2 (c) case 3

Fig.2 Section of foundation pit in test cases(unit: m)

ANALYSIS OF TEST RESULTS

Fig. 3 is the pictures for the centrifuge test with the confined water head 20.0m and different excavation depths. Fig. 4 is the deformation diagram of the foundation pit based on Figure 3.

From Fig. 4, it can be concluded that: for different confined water heads and excavation depths, the distance between the point with maximum ground settlement behind the wall and the foundation pit is 0.4—0.5 times of the excavation depth, and the deeper the excavation depth, the longer the distance. The maximum horizontal displacement of the diaphragm wall emerges 2. 0—3.0m under the excavation face.

(a) excavation depth 14.4 m (b) excavation depth 16.7 m (c) excavation depth 18.0 m

Fig.3 Pictures of centrifuge test for foundation pit with confined water head 20.0 m

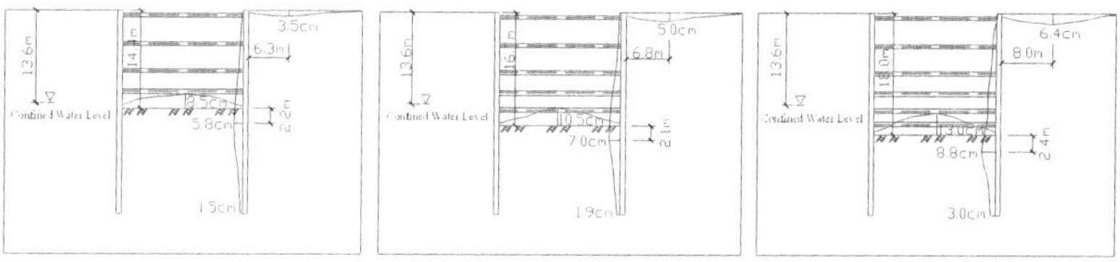

(a) excavation depth 14.4 m (b) excavation depth 16.7 m (c) excavation depth 18.0 m

Fig.4 Diagram of deformation for foundation pit with confined water head 20.0 m

The relation between the deformation and the confined water head in a certain excavation depth is shown in Fig. 5. It can be seen: when the excavation depth is smaller than 16.7 m, as the confined water head decreases, the maximum ground settlement around the foundation pit has a linear reduction, so do the maximum horizontal displacement of the wall, the maximum horizontal displacement at the bottom of the wall and the maximum bottom heave; however, when the excavation depth is larger than 16.7 m, as the confined water head decreases, the deformation decreases nonlinearly. When the excavation depth is 14.4, 16.7 and 18.0m respectively, and the confined water head drops from 28.0m to 20.0m, the maximum ground settlement around the foundation pit, the maximum horizontal displacement of the wall, the maximum horizontal displacement at the bottom of the wall and the maximum bottom heave decrease 22.0%, 16.0%, 33.0% and 32.0% respectively.

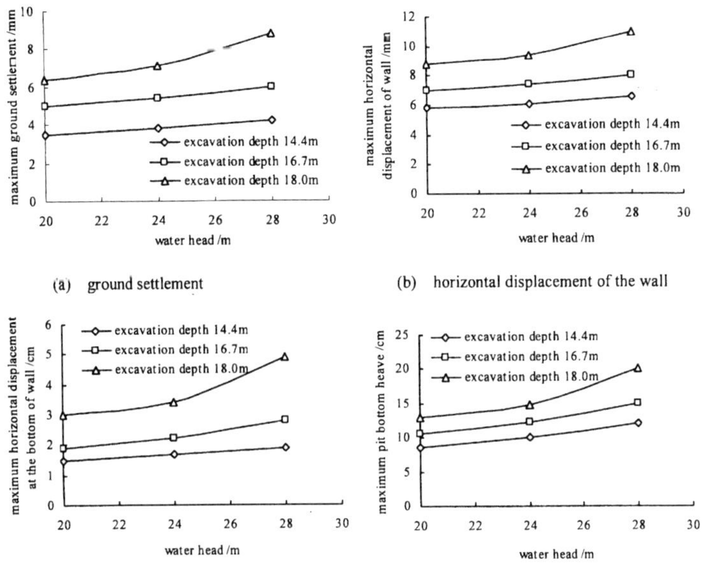

(a) ground settlement (b) horizontal displacement of the wall

(c) horizontal displacement at the bottom of the wall (d) bottom heave of the foundation pit

Fig.5 Relation between deformation and confined water head of the foundation pit

The relation between the deformation and the excavation depth for a certain confined water head is shown in Fig. 6. It can be seen: when the confined water head is determined, as the excavation depth decreases, the maximum ground settlement around the foundation pit has a nonlinear reduction, so do the maximum horizontal displacement of the wall, the maximum horizontal displacement at the bottom of the wall and the maximum bottom heave, and the higher the confined water head, the more obvious the decrease trend is. When the confined water head is 20.0, 24.0 and 28.0m respectively, and the excavation depth decreases from 18.0m to 14.4.0m, the maximum ground settlement around the foundation pit, the maximum horizontal displacement of the wall, the maximum horizontal displacement at the bottom of the wall and the maximum bottom heave decrease 48.0%, 37.0%, 55.0% and 36.0% respectively.

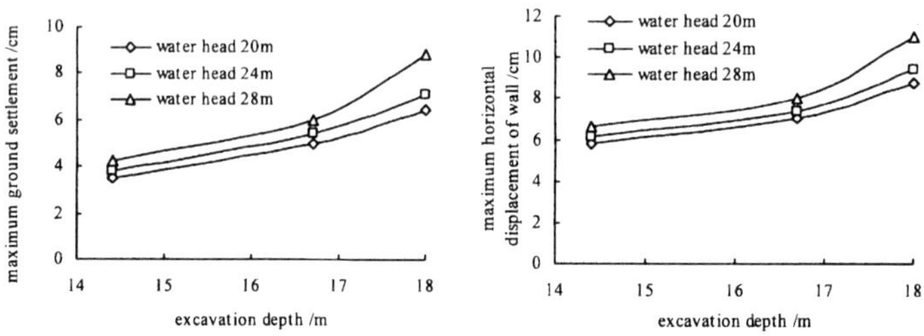

(a) ground settlement (b) horizontal displacement of the wall

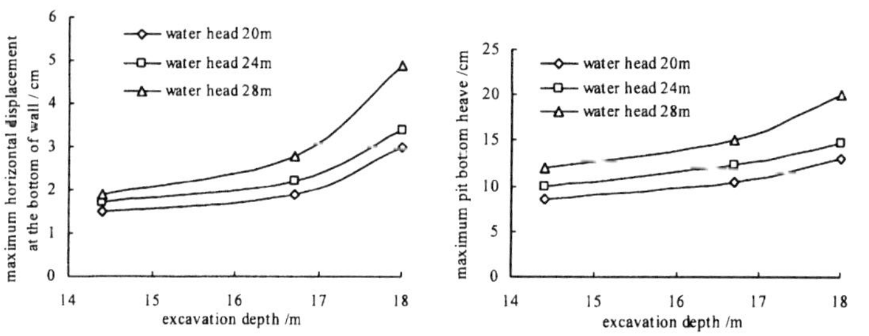

(c) horizontal displacement at the bottom of the wall (d) bottom heave of the foundation pit

Fig.6 Relation between deformation and excavation depth of the foundation pit

In order to explore the relation between the deformation and the excavation depth & the confined water head so as to use it conveniently in the project, a study is made by using regression analysis according to the test results in this paper (He el at,2002; Gui el at,1999). In order to set up the relation between the deformation and the excavation depth & the confined water head, based on univariate regression analysis, it is assumed that the multiple regression model of the deformation Y and the excavation depth and the confined water head is as follows:

$$Y = a_0 + a_1X_1 + a_2X_2 + a_3X_3 + a_4X_4 + \varepsilon \quad (1)$$

Where, the independent variables $X_1 = h$ (h is the excavation depth), $X_2 = h^2$, $X_3 = h_w$ (h_w is the confined water head), $X_4 = h_w^2$, the dependent variable Y is the maximum ground settlement δ_{max} (or the maximum horizontal displacement of the wall, the maximum horizontal displacement at the bottom of the wall, and the maximum bottom heave), ε is the error term, which follows normal distribution $N(0,\sigma^2)$, $a_0, a_1, \cdots, a_4$ are the regression coefficients.

A self-developed multiple regression analysis program MLRAP is applied based on multiple regression analysis theory in this paper, and a regression calculation analysis is carried out on the samples, finally, the regression equations for the deformation of the foundation pit are obtained as follows:

$$\left.\begin{aligned}
\delta_g &= 0.2904 - 0.61552h + 0.047674h^2 + 0.03523h_w + 0.002842h_w^2 \\
\delta_w &= 0.6735 - 0.40020h + 0.038816h^2 + 0.11095h_w + 0.000959h_w^2 \\
\delta_{wf} &= 0.4320 - 0.32608h + 0.027030h^2 - 0.11053h_w + 0.004866h_w^2 \\
\delta_{fb} &= -1.0527 - 0.86440h + 0.070640h^2 + 0.15446h_w + 0.009197h_w^2
\end{aligned}\right\} \quad (2)$$

Where δ_g、δ_w、δ_{wf}、δ_{fb} are the maximum ground settlement around the foundation pit, the maximum horizontal displacement of the wall, the maximum horizontal displacement at the bottom of the wall and the maximum bottom heave, h is the excavation depth(m), h_w is the confined water head(m). And the applicable conditions for the formulas mentioned above are as follows: $h \geq 5m$, $h_w \geq 5m$.

The multiple correlation coefficients R for the regression equations above are 0.9847, 0.9770, 0.9585 and 0.9874 respectively, which are close to 1, indicating the significant regression effects of the equations. A contrast analysis diagram of the predicted values by regression and the test values for the samples is showed in Fig.7, and the results show that the two coincide with each other.

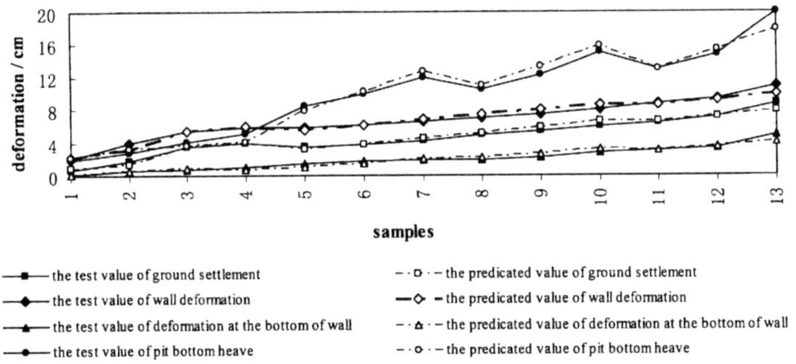

Fig.7 Comparison on predicted values by regression and test values for samples

CONCLUSIONS

(1) For the foundation pit without layer 6(stiff-soil aquiclude) at different confined water head and excavation depth, the distance between the point with maximum ground settlement behind the wall and the foundation pit is 0.4—0.5 times of the excavation depth, and the deeper the excavation depth, the longer the

distance. The maximum horizontal displacement of the diaphragm wall emerges 2—3m under the excavation face.

(2) As the confined water head decreases, the deformation of the foundation pit has a reduction. What the confined water affects most are the deformation at the bottom of the wall and the bottom heave. So the main measure should be dewatering when the bottom heave of the foundation pit with confined water cannot be guaranteed.

(3) As the excavation depth decreases, the deformation of the foundation pit has a nonlinear reduction, and the higher the confined water head, the more obvious the decrease trend is. What the excavation depth affects most are the ground settlement and the horizontal displacement at the bottom of the wall. Therefore, in order to reduce the deformation around the foundation pit caused by excavation, the excavation depth should be reduced, and the designed buried depth of excavation should be reduced if possible.

(4) According to the results of the centrifuge tests and the regression coefficients of the regression equations, it can be seen that: for the foundation pit without layer 6 stiff soil (aquiclude), the effect of the excavation depth on deformation is more sensitive than the confined water head. Therefore, the excavation procedures and excavation parameters should be mainly considered in order to reduce the deformation of the foundation pit with confined water.

(5) The predicted values of the simplified calculation equations obtained by regression analysis coincide with the test values, which can provide a reference for the design and construction of foundation pit with confined water.

REFERENCES

Ding Chunlin, Gong Quanmei, Meng Xiaohong (2004). A study on treatment technology of confined water for deep foundation pit of metro station during excavation[R]. Tongji University.

Taylor R. N.(1995).Geotechnical centrifuge technology[M]. London: Chapman and Hall.

Ding Chunlin, Meng Xiaohong(2008). Centrifuge model test and field measurement analysis for foundation pit with confined water [J], *Journal of Tongji University (Natural Science)* , 36(1):22—26.

He Shaohua, Wen Zhuqing, Lou Tao(2002). Experiment design and date procession[M]. Beijing: Defence Science Technology University Press.

Gui Liangjin, Wang Jun, Dong Bo(1999). Use and programming of fortran powerstation[M]. Beijing: Beijing University of Aeronautics & Astronautics press.

Model tests and numerical simulations on pile-soil interaction of passive piles

ZHOU Jian[1,2], CHEN Xiao-liang[1,2],ZENG Qing-you[3], WU Biao[1,2]

[1] Department of Geotechnical Engineering, Tongji University, Shanghai 200092, China;
[2] Key Laboratory of Geotechnical and Underground Engineering of Ministry of Education, Tongji University, Shanghai 200092, China;
[3] Fujian Communication Planning & Design Institute, Fuzhou 350004, China.

ABSTRACT: This paper presents the experimental and numerical studies on mechanisms of pile-soil interaction of a single free-head short pile subjected to horizontal soil movement. The earth pressures on upstream and downstream face were measured during test and the load transfer mechanism of passive pile were revealed from the model test results. Moreover, the behavior of soil surrounding the pile and the interaction between pile and soil were analyzed via a two-dimensional discrete element method, namely PFC. It has been found that the numerical experiment results were generally in good agreement with the physical test results. The study would be helpful to deeply reveal the mechanism of pile-soil interaction and facilitate the design process of passive laterally loaded piles.

KEY WORDS: Passive pile; Model test; Numerical simulation; PFC; Pile-soil interaction

INTRODUCTION

The majority of piles are designed to support "active" loads, that is, loads from superstructure are directly transferred to the pile foundation by the cap. However, in many cases, piles are designed to withstand "passive" loads, which are created by the deformation and movement of soil surrounding the piles due to the weight of soil and the surcharge. Several empirical and numerical methods have been proposed for analyzing the response of single pile and pile group subjected to lateral loading from horizontal soil movements. A comprehensive review on these methods has been made by Stewart et al.

Based on the summary of the published papers, the latest model tests and numerical analysis results on passive piles, it is noticed that almost all of the research work focused on the pile shaft itself, whereas little attention is paid to the soil surrounding the pile. However, it is known that pile-soil interaction depends on the behavior of the surrounding soil to a large extent. So the behavior of surrounding soil should be concerned when the pile-soil interaction is involved.

In this study, an experimental apparatus was carefully designed to analyze the behavior of a single passive pile subjected to lateral soil movement. A model test was performed using loose sand. Based on the model test results, a two-dimensional discrete element method, namely PFC^{2D}(Particle Flow Code in two dimensions), was used to analyze the mechanism of pile-soil interaction. Benefiting from the advantages of PFC^{2D}, more details of the pile-soil interaction were revealed and the mechanisms of pile-soil interaction were analyzed more comprehensively.

PHYSICAL MODEL TEST

The experimental apparatus is briefly described below and more details can be found in ZENG, Q.Y. (2005). The testing container consisted of some steel sheets and a transparent glass side through which photos can be taken by high-resolution camera. Two instrumented model piles were made of wood with 30 mm length, 30 mm width and 270 mm height, and each one was instrumented with five earth pressure cells along the pile shaft for measuring earth pressure. In this study, the sand samples was prepared with void ratio 0.823, relative density 0.076 and density 1450 kg/ m^3. The layout of the test is shown in Fig.1.

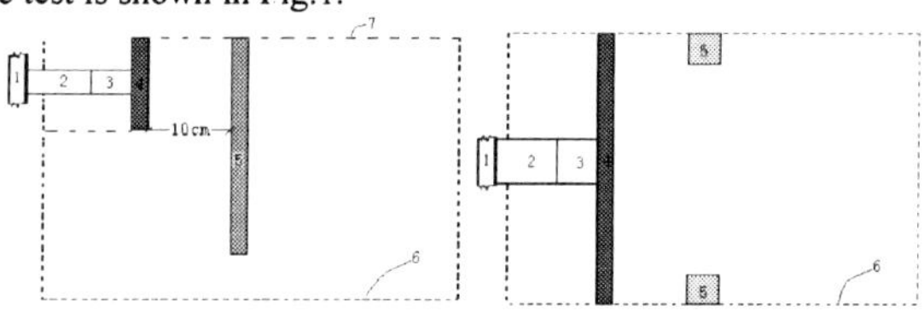

(a) Front view (b) Top view

1 reaction frame ; 2 jack ; 3 load sensor ; 4 retaining wall ; 5 pile ; 6 testing container. 7 sand surface

FIG. 1. Layout of model test device of passive pile

Experimental results

The earth pressures on upstream and downstream face at different positions along pile shaft are plotted in Fig.2. It should be noted that the downstream face is the side away from the retaining wall, while upstream face is the side near the retaining wall. The total earth pressures along pile shaft are the combination of the pressure on downstream and upstream face. From Fig.2, some meaningful conclusions can be summarized as follows: (1) Above the turning point, the curves of earth pressure on downstream face are seen to have a more-or-less parabolic shape. Below the turning point, the soil in front of the pile is in a state of active earth pressure due to the rotation of pile, and, apparently, the active earth pressures are too small; (2) There is a critical point in the pile shaft at eighty percent of the embedded length from the pile top. The curves of earth pressures on upstream face, above the critical point, are seen to be parabolic shape. However, beneath the point, those increase linearly with the increase of depth; and (3) It can be seen that the curves of total earth pressure have two critical points along pile shaft: one located at 40 percent of the embedded length of pile, which is consistent with the location of retaining wall bottom and the other located at 90 percent, which is at the vicinity of the turning point. The curves could be divided into three sections by two critical points: parabolic distribution on the upstream face, parabolic distribution on the downstream face and then increasing linearly with the increase of depth. The mechanisms of pile-soil interaction were revealed by the distribution of total earth pressure, that is, in the upper unstable layer, the pile was subjected to the movement of sand, and the lower pile transferred the loads to the surrounding sand.

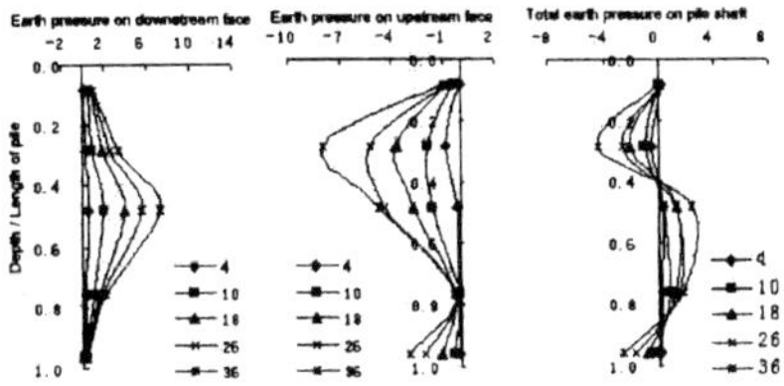

FIG. 2. Earth pressure distribution of pile shaft in loose sand(kPa)

NUMERICAL SIMULATIONS

In order to facilitate the description of the process of experiments, the height of retaining wall is referred to hereafter as H, likewise, distance between retaining wall and pile as S and the length of pile as L. It is very obvious that the mechanisms of pile-soil interaction are directly related to the ratios of $H:S:L$. PFC^{2D} was used to analyze the behavior of the pile subjected to soil horizontal movement under different heights of retaining wall. A series of simulations were performed under different situations, that is, the ratios equaling to 1:2:5, 2:2:5, 3:2:5 and 4:2:5, respectively, and the displacement of pile shaft and the earth pressure on the interface between pile and soil were computed. Then, by means of the numerical simulations, the characteristics of passive pile under different retaining wall heights were analyzed and the results of numerical simulations with the ratios of $H:S:L$ equaling to 2:2:5 could be directly compared with the physical model tests results.

Numerical model

In this work, the height of numerical simulation model box is 0.65 m, the width is 1.26 m and the rest parameters of model are shown in Table 1. Four models were generated according to the requirements as mentioned above. Due to space limitations, the numerical model with the ratios of $H:S:L$ equaling to 1:2:5 is shown in Fig.3.

Table 1. Parameters of PFC^{2D} model

	Particle density (kg/m3)	Radius of particle (m)	Shearing stiffness (N/m)	Normal stiffness (N/m)	Coefficient of friction
Sand particle	2650	0.7 ~2.0	2.0×10^6	2.0×10^6	2
Model pile	1000	0.03	1.0×10^6	1.0×10^6	1

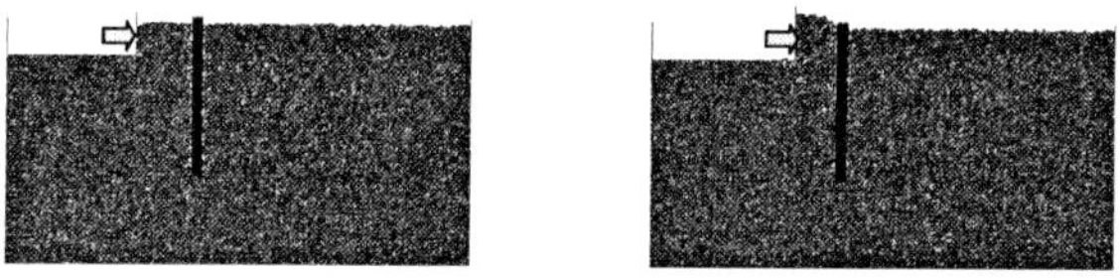

$H:S:L=1:2:5$, y = 0mm $\qquad$ $H:S:L=1:2:5$, y = 20mm

FIG. 3. Numerical model of PFC^{2D}

The relationship between displacement and load of retaining wall

Fig.4 shows the curves of load-displacement of retaining wall with different heights in PFC^{2D}. It could be seen that all the curves were quite similar in shape, but different in magnitude each other. The ultimate loads of retaining wall for the four tests were 200, 2130, 5523 and 8470 N, respectively, which increased sharply with the retaining wall height increasing. Besides, the corresponding displacement of retaining wall to ultimate load was different from each other, that is, the higher the retaining wall was, the larger the corresponding displacement to ultimate load was.

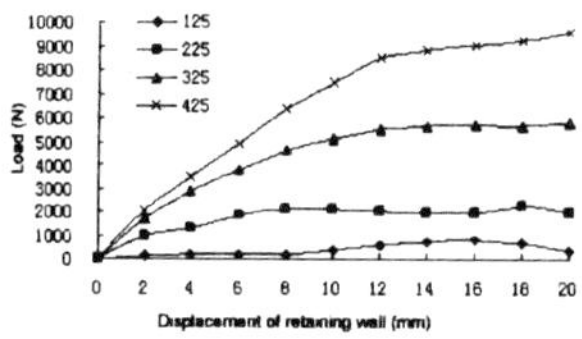

FIG. 4. Curves of load-displacement of retaining wall in PFC^{2D}

The reinforced effect on sand due to pile insertion in different conditions is analyzed as follows:

According to Rankine theory, the passive earth pressure of sand without pile is,

$$E_p = \frac{\gamma h^2}{2}\tan^2\left(45+\frac{\varphi}{2}\right) \tag{1}$$

From the equation above, it could be found that the Rankine passive pressure is proportional to the square of the height of retaining wall(h) with the unit weight(r) and internal friction angle(φ) keeping constant. In order to analyze the reinforced effect without the interference of retaining wall height, the ultimate load were divided by the square of the height. Therefore, the ratios of the reinforcement effect on sand for pile insertion were 200: 532: 614: 529, namely, 1: 2.66: 3.07: 2.65. It is apparent that the reinforced effect was most significant when the ratios equal to 3:2:5. There existed an interface between unstable and stable layer. Due to the interface, the pile could be divided into two portions, passive and active pile, as demonstrated above. The position of the interface was determined by the height of retaining wall and the distance between the wall and pile. Therefore, the reinforced effect on sand for pile is directly related to the ratios of $H:S:L$.

Distribution of earth pressure along pile shaft

The distributions of earth pressure along pile shaft in PFC^{2D} with the ratios of $H:S:L$ equaling to 2:2:5 are shown in Fig.5. It could be seen that the turning point of pile is almost fixed and the depth is about ninety percent of the embedded length of the pile. A fairly good agreement can be seen between numerical simulation and physical model test in the shape.

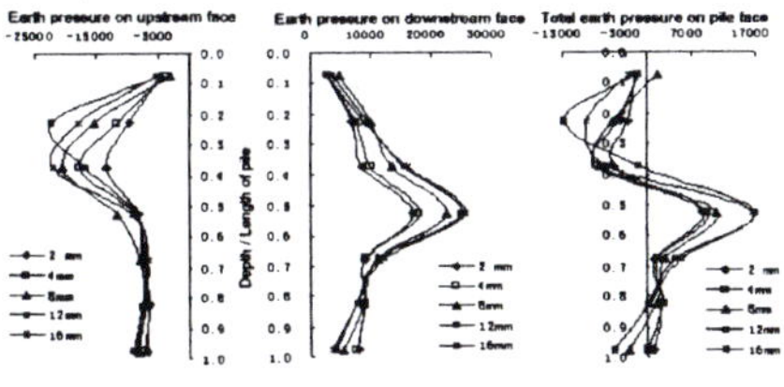

FIG. 5. Earth pressure distribution with $H:S:L=2:2:5$ in PFC^{2D}(kPa)

The earth pressure distributions along pile shaft with four kinds of values of $H:S:L$ when the displacement of retaining wall was 12 mm were plotted in Fig.6. The earth pressure distributions are significantly different from each other with different ratios of $H:S:L$. There are a few differences in the magnitude of earth pressure on the downstream face at the upper section of pile, while those at the lower section of pile increased with the retaining wall height increasing. That is because the displacement of the lower section of pile increase gradually with the height of retaining wall increasing. And the distribution of earth pressure on the pile upstream face is seen to be parabolic shape, with the maximum pressure being at the vicinity of the middle center of the retaining wall. However, the distribution of the total earth pressure along pile shaft is very complicated.

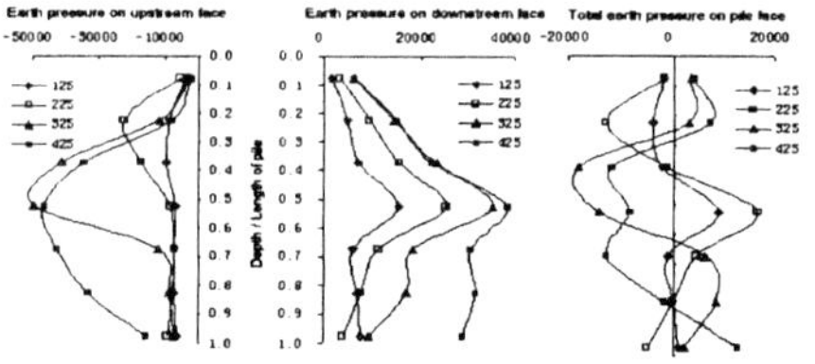

FIG. 6. Earth pressure distribution of piles with different values of $H:S:L$ in PFC^{2D} (kPa)

CONCLUSIONS

In this study, a physical model test with a single pile embedded in loose sand subjected to horizontal movement was performed and the earth pressures on upstream and downstream face were measured during test. The curves of total earth pressure could be divided into three sections by two critical points and the mechanisms of pile-soil interaction were revealed by the distribution of total earth pressure, that is, in the upper unstable layer, the pile was subjected to the movement of sand and the lower section transferred the loads to the surrounding soil. Based on the model test results, PFC^{2D} was used to simulate the interaction between pile and soil for lateral movement of soil. From the numerical simulation results under four different ratios of $H:S:L$, When the ratios equal to 3:2:5, the reinforced effect was most significant. Therefore, the reinforced effect on sand for pile is directly related to the value of $H:S:L$. This would be helpful to the design of passive laterally loaded piles. Moreover, when the value of $H:S:L$ equal to $2:2:5$, the distributions of earth pressure computed by PFC^{2D} agree fairly with those measured in physical model test in the shape. Therefore, it menifests that PFC^{2D} is feasible to analyze the behavior of pile subjected to soil horizontal movement.

ACKNOWLEDGMENTS

This study was primarily supported by the National Natural Science Foundation of China (Grant No,50578122) and Shanghai Leading Academic Discipline Project.

REFERENCES

Stewart, D.P. and Jewell, R.J. and Randolph, M.F. (1994). "Design of piled bridge abutments on soft clay for loading from lateral soil movements." *J. Geotechnique.*, Vol. 44 (2): 277–296.

Poulos, H.G. and Chen, L.T. and Hull, T.S. (1995). "Model tests on single piles subjected to lateral soil movement." *J. Soil and foundations.*, Vol. 35 (4): 85–92.

Leung, C.F and Chow, Y.K. and Shen, R.F. (2001). "Behavior of pile subjected to excavation-induced soil movement." J. Journal of Geotechnical and Geoenviromental Engineering., Vol. 126 (11): 947–954.

Pan, J.L and Goh, A.T.C and Wong, K.S. (2002). "Ultimate soil pressures for piles subjected to lateral soil movements." *J. Journal of Geotechnical and Geoenvironmental Engineering.*, Vol. 128 (6): 530–535.

Poulos, H.G. and Chen, L.T. (1997). "piles subjected to lateral soil movement." *J. Journal of Geotechnical and Geoenvironmental Engineering.*, Vol. 123(9): 802-811.

Chia-Cheng Fan and James H. Long. (2005). "Assessment of existing methods for predicting soil response of laterally loaded piles in sand." *J. Computers and Geotechnics.*, Vol. 32 (3): 274-289.

Zhou, J, Chi, Y. (2004). "Simulation soil properties by particle flow code." *J. Acta Mechanica Solida Sinica.*, Vol. 25 (4):377 – 382. (in Chinese)

Zhou, J and Chi, Y.W. (2000). "Simulation of biaxial test on sand by particle flow code." *J. Chinese Journal of Geotechnical Engineering.*, Vol. 22 (6): 701–704. (in Chinese)

Zeng, Q.Y. (2005). "Model tests and PFC simulation for laterally loaded piles." PhD thesis. Shanghai, Tongji University. (in Chinese)

Li, Y.H. (2003). "Study on digital photography-based deformation measurement technique and its application in geo-physical model experiment." PhD thesis, Shanghai: Tongji University. (in Chinese)

Design and Research of Circular Diaphragm Wall for Deep Excavation of Shanghai Tower

Jian Jia XiaoLin Xie ChuanPing Liu JieQun Zhai Yu Zhang

Architectural Design & Research Institute of Tongji University, 1239 Siping Road, Shanghai 200092, China; PH: (86-21)6598-2775; FAX: (86-21)65985288; E-mail: jiajian@tongji.edu.cn

ABSTRACT: Shanghai Tower, the tallest skyscraper in China with 632m building height is under construction. For such height of building, time for construction completion of the tower is the most critical issue in planning the project. It would be the best beneficial to the whole project if the excavation of basements could be commenced as early as possible and the period of construction could be as short as possible. A best solution came up after numerous discussion and comparison - conventional open excavation for basements in tower area and top-down method in podium area. Moreover, a cylinder-shaped diaphragm with outer diameter 123.4m, 1.2m thick and 50m in height was adopted as a lateral retaining structure for basement excavation in tower area with 31.1m excavation depth. This scheme provides a faster program by means of no extra lateral supporting strut installation and demolition during excavation and basement construction. Although the design scheme of cylinder-shaped diaphragm without lateral supporting strut has been applied in some projects domestically and oversea, in view of the project scale (excavation depth and excavation area), design difficulty (excavation in soft clay in extensive depth and diameter), it would be the first project in China mainland. This paper serves to summarize the discussion and design calculation, analysis and control measure on deep and large diameter cylinder-shaped diaphragm wall in conventional open excavation, and also intends to provide as a reference for future similar projects.

INTRODUCTION

Shanghai tower, located in the center of Lujiazui financial district, neighbor of Jinmao Tower and Shanghai World Financial Center, has been under construction. The tower will be mainly office and hotel floors combined with exhibition and entertainment halls, as well as retail stores in the podiums floors. The building height 632m will make this tower not only the highest building in China, but the most important landmark building in the area.

Planning on the master construction program of a 632m high building is a challenging task. The construction period should be kept as short as possible in order to generate the revenue as early as possible. Passing through numerous comparisons, conventional open excavation for basements in tower area and top-down method in podium area was finally adopted. The construction site will be separated into two phases. Firstly, start open excavation in the tower area to the polygonal pile cap. In order to speed up the construction cycle, a large diameter (123.4m) cylinder-shaped diaphragm with no lateral supporting strut was adopted. It will not only shorten the excavation time, but also shorten the basement construction time because of avoidance of demolition of lateral supporting strut. Secondly, until completing the basements to ground level in tower area, start the construction of basements in podium area by top-down method.

The cylinder-shaped diaphragm, by fully utilizing the benefit of high compressive capacity of concrete and the cylindrical effect, by transferring the lateral loads from underground soil and water pressure into lateral centripetal forces, has been recently used in the projects domestically and oversea. However, in view of the project scale (excavation depth and excavation area), design difficulty (excavation in soft clay in extensive depth and diameter), it would be the first project in China mainland.

This paper serves to summarize the discussion and design calculation, analysis and control measure on deep and large diameter cylinder-shaped diaphragm wall in conventional open excavation, and also intends to provide as a reference for future similar projects.

PROJECT DESCRIPTION AND THE DESIGN SCHEME

Project Description

Figure 1. Prospective View of Shanghai Tower

Figure 2. Bird's Eye View of the Site

The project – Shanghai Tower, is located south of Huayuanshiqiao Road (next to Jinmao Tower), east of Yinchengzhong Road, north of Lujiazuihuan Road and west of Dongtai Road (next to Shanghai World Financial Center). The scale of the tower is shown as bellows: 632-m building height, 580-m structural height, 121-number of usable floors, 5-number of basement floors, 34960-sq. m excavation area, 31.1-m excavation depth in tower area and 26.7-m excavation depth in podium area.

Geological Condition. Shanghai is located southeast of the entry of Yangtze River Delta, it is a delta alluvial plain. The compositions of the geology are mainly the saturated clayey soil, silt and sand, which are generally stratified distributed. In this construction site, layer ⑧ is absent. The soil layers distributed over 24m, are mainly soft silty clay, with the unfavorable geological features of high moisture content, high void ratio, high sensitivity, low-intensity and high-compressive. Shallow groundwater is diving in this area, and the groundwater level generally is 1.0 ~ 1.7-m below ground. The first confined water is distributed in layer ⑦ sandy soil, which is 27m below ground. The second confined water is distributed in layer ⑨ sandy soil. The two layers are connected with each other and full of water.

Distinction of the Project. The major consideration of this project is to keep the construction progress of the tower, to satisfying the demand of the construction working space, and to protect the surrounding existing buildings and utilities.

Design of Excavation and Lateral Support. We have preliminary considered 4 design schemes for the excavation and lateral support of this particular project:

1. Conventional open excavation for the entire area;
2. Top-down method for the entire area;
3. Divided into two zones - tower and podium, both adopted conventional open excavation;
4. Conventional open excavation for the tower area and top-down method for the podium area

All schemes above are technical feasible. However, after the detail discussion and comparison, the scheme 4 – conventional open excavation for the tower area and top-down method for the podium area was adopted.

In this scheme, the tower area and the podium area will be divided into two separated excavation zones. Firstly open excavation in the tower area, and secondly worked the podium area by top-down method. In order to speed up the construction cycle, and considering the feature of the tower foundation, which is a polygonal pile cap, the scheme of a 123.4-m external diameter (121-m internal diameter) cylinder-shaped diaphragm wall without lateral supporting struts was adopted. Until the tower area going back to ground level and completed the ground floor structure, the podium area will then be commenced. The advantages of adopting this scheme is listed below:

1. Minimizing the excavation area in the tower area and kick off the construction of the tower basement first in order to speed up the progress.
2. By fully utilizing the benefit of cylindrical effect, the lateral loads from underground soil and water will be transferred into lateral centripetal forces. Therefore, no lateral supporting strut is required to be installed during the excavation. As well as no need of demolition of the struts during basement construction in order to shorten the construction program.
3. The podium area can be used as a working site during the excavation of tower area.
4. The ground floor slab can be used as a working site during the top-down excavation of the podium area in order to resolve the problem of limited working spaces.
5. Cost and time saving for using the ground floor structure as a lateral support to the diaphragm wall during the top-down excavation of the podium area.
6. The stiff basement floor structures (slabs and beams) are used as lateral support in order to minimize the deformation of the diaphragm wall, to protect the surrounding existing utilities.
7. The top-down method can minimize the construction noise and dust during the excavation of podium area, which can create an environmental friendly working place.

For the above advantages, this design scheme can satisfy the need for this particular project.

Nevertheless, some problems are required to be considered and solved during the design:

1. In the saturated soft clay area, for the huge project scale and construction difficulty, there are little local successful experiences. Some detail research and analysis are required to do for the design.
2. Top-down method in podium requires closely coordination in the design between the excavation, lateral and the permanent main structure. It makes the design more difficult. In addition, it requires better construction workmanship.

DESIGN AND CONTROL FOR CONVENTIONAL OPEN EXCAVATION IN CIRCULAR TRENCH

This paper summarizes the design of large diameter cylinder-shaped trench in the project of Shanghai Tower. It is focused on the design and calculation.

Application and Load Transfer Characteristic of circular Trench. Cylinder-shaped diaphragm has been applied in some projects domestically and oversea. The following Table 1 shows the current projects adopted the circular retaining structure for deep excavation.

Table 1. Application of large circular retaining structure for deep excavation

Projects	Diameter (m)	Depth of excavation (m)	Thickness of diaphragm wall (m)	Depth of diaphrag m wall (m)	Embedment ratio	Number of the ring beam	Construction method
Shanghai Tower	123.4 (outer diameter)	31.1	1.2	50	0.60	6	Open Excavation
Shanghai World Financial Center	100 (outer diameter)	18.3	1.0	31.6~ 33.6	0.72	4	Open Excavation
Shanghai World Expo 500kv underground transmission substation	130 (outer diameter)	33.7	1.2	57.5	0.71	4 floors of structure slab and 3 floors of ring beam	Top-down excavation
HongKong International Financial Center	61.5 (inner diameter)	35	1.5	38	Diaphragm wall embed into the base rock	4	Open Excavation
500kV underground Shin-Toyosu substation	144 (inner diameter)	29	2.0 (inner wall involved)	/	/	/	Open Excavation

As reference from the above projects completed, cylinder-shaped retaining structure can transfer the lateral loads from underground soil and water pressure into axial compression forces, in order to reduce the lateral supporting struts as well as to lower the project cost. In addition, a certain reduction on soil pressure to the cylindrical diaphragm wall can be achieved due to the arching effect [1]. Also, certain reduction in embedment length of diaphragm wall can also be reached, as the stability of soil improved due to cylindrical space effect.

Application of larger diameter cylinder-shaped diaphragm wall without lateral supporting strut is not only for economy, but more workable and safe in construction.

Design Calculation Method. In excavation projects, load transfer of soil and water pressure is related to the stiffness of retaining structure. The soil and water pressure during excavation are undertaken by the horizontal arch and vertical bending action of the cylinder-shaped diaphragm wall. The stiffness ratio of horizontal arch action and vertical bending action determines the ratio of load distribution from soil and water pressure. Increase of excavation trench diameter will gradually decrease the stiffness of horizontal arch action, keeping the stiffness of vertical bending unchanged, the ratio of the load undertaken by horizontal arch action will be gradually reduced. Therefore, calculation on the lateral forces acting on the circular trench and the diameter of the trench are closely related.

For the Shanghai Tower large deep circular trench, the diaphragm wall is simulated as the revised vertical curved structure and spatial arching structure respectively. The comprehensive result are being compared and applied to the design.

Calculation Method 1 - Amended plane elastic ground beam "m" Method. In design of excavation and lateral support, usually the conventional vertical plane beam on elastic foundation method (plane "m" method for short) is used for mechanical analysis for cylinder-shaped retaining structure. The diaphragm wall is simulated as a simple vertical moment resisting members, ignoring its arching effect. The result of plane "m" method is not reasonable. Therefore, plane "m" method requires to be amended.

Cylindrical diaphragm wall can be simulated as an axis-symmetry structure. The model is based on the wall structure with a unit width, and is simulated as a vertical beam on elastic foundation. The horizontal arching effect is considered as a series of the spring. The equivalent spring stiffness of the ring beam (lateral support) and the diaphragm wall is shown as bellows [2]:

$$K_d = E_d b / R_0^2 \qquad \text{(formula 1)}$$

$$K_h = E_h A_h / R_{h0}^2 \qquad \text{(formula 2)}$$

K_d——equivalent elasticity coefficient of cylindrical wall (kN/m^3) ;

b——thickness of the cylindrical wall (m) ;

$E_d=\alpha E$——modulus of circumferential compressibility of the cylindrical wall, E is the elastic modulus of the concrete of the cylindrical wall, the value of α varies 0.5~0.7;

R_0——initial radius of the cylindrical wall;

K_h——equivalent elasticity coefficient of cylindrical beam (kN/m^3) ;

E_h——elastic modulus of the concrete of the cylindrical beam (kN/m^2) ;

A_h——section area of the cylindrical beam;

R_h——initial radius of the cylindrical beam.

The structure internal force is calculated in accordance with the construction procedure, and the envelope diagram of internal force and displacement of the cylindrical wall could be got. By the results of the displacement of the cylindrical wall and cylindrical beam, the maximum compression stress could be calculated.

Calculation Method 2 - three-dimensional "m" method of the slab on elastic foundation.

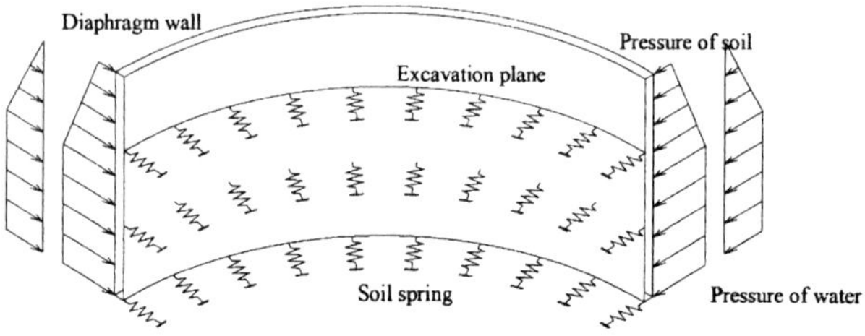

Figure 3 Calculation diagram of three-dimensional "m" method

At present, the three-dimensional "m" method of the slab on elastic foundation (three-dimensional "m" method for short) is widely adopted in the analysis of the large circular retaining structure for deep excavation [3]. First, the cylindrical effect of the cylindrical retaining structure could be considered in the three-dimensional model. Second, this method is derived from the two-dimensional model, the value choose of the parameter is reasonable. Third, the comparison of the results got from the two methods could make the design more reasonable. The calculation diagram is showing in the following Fig.3.

The soil under the excavation level is simulated by the spring, of which the stiffness is calculated by the following formula:

$K_H=k_h \cdot b \cdot h=m \cdot z \cdot b \cdot h$..(formula 3)

In which, K_H——stiffness of the soil spring;

k_h——horizontal coefficient of the ground foundation;

m——proportionality constant of k_h with depth;

z——distance between soil spring and the excavation level;

b, h——horizontal, vertical distance of the soil spring.

To make the model reasonable, the cylindrical wall is considered as a polygon with 2m-4m polygonal line, and the intensity reduction of the joint of the cylindrical wall is considered.

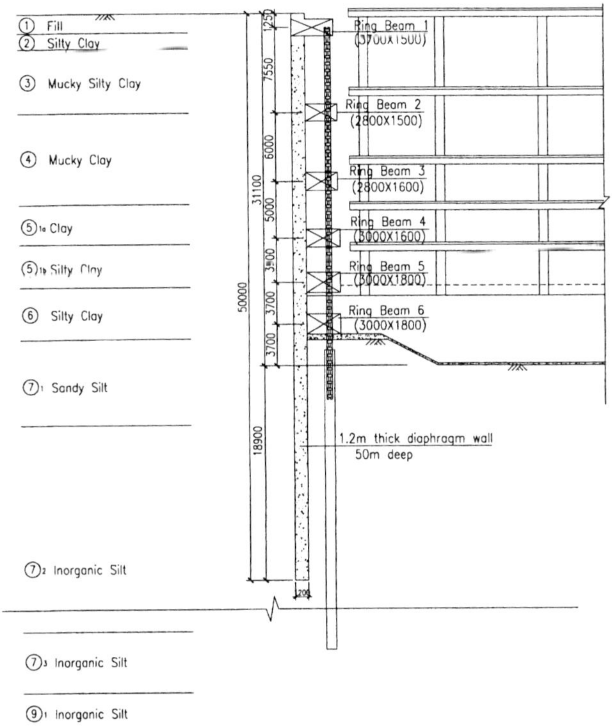

Figure 4. Section of Retaining Structure

Result of Design Calculation. After the trial calculations by the two methods mentioned above, a safe and reasonable design of retaining structure is finally determined and its specifications are: 50-m deep, 1.2-m thick walls, equipped with six ring beam (lateral support), the section is shown in Figure 4.

The corresponding calculation results of the above design are as bellows:

Calculation Method 1(plane "m" method). The corresponding calculation results shown in Figure 5.

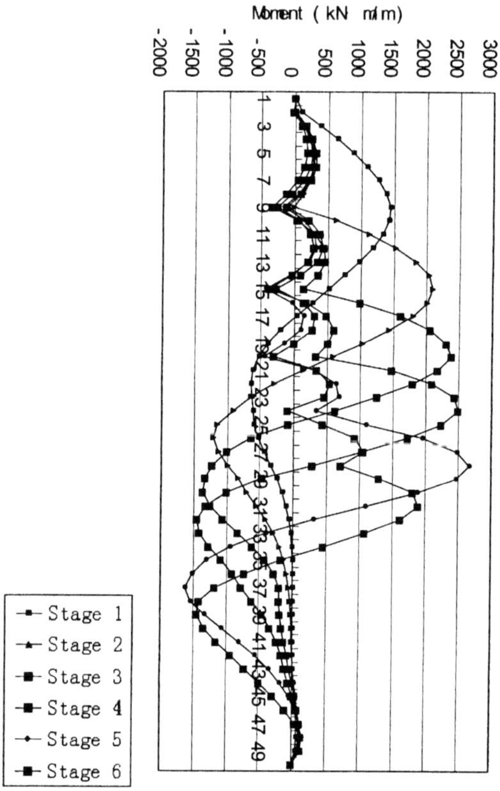

Figure 5. Calculation result of diaphragm wall bending moment

Calculation method 2 (three-dimensional "m" method). In the application of three-dimensional "m" method, both uniform soil and water pressure and non-uniform pressure is considered, in order to simulate the uneven distribution of the soil layers and the uneven unloading due to the excavation.

Calculation models and calculation results shown in Figure 6, Figure 7a, Figure 7b, Figure 8a, Figure 8b

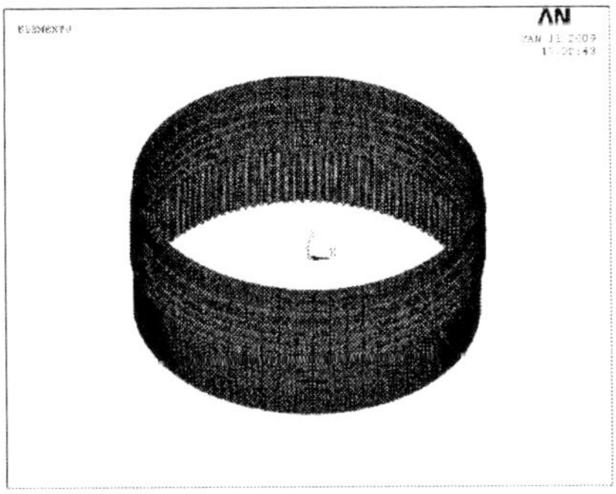

Figure 6. Calculation model

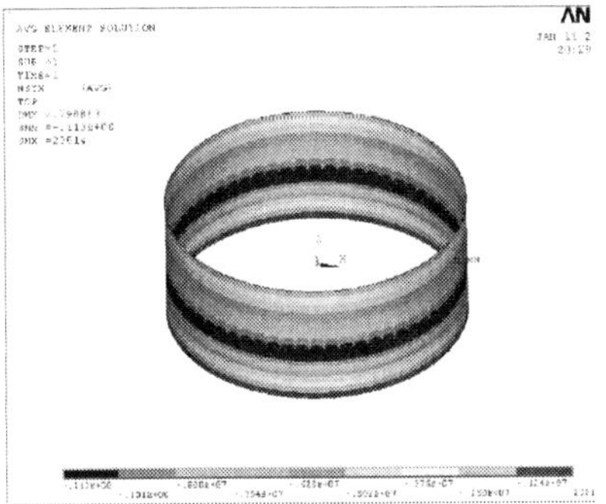

Figure 7a. Axis force of diaphragm wall under uniform pressure（N）

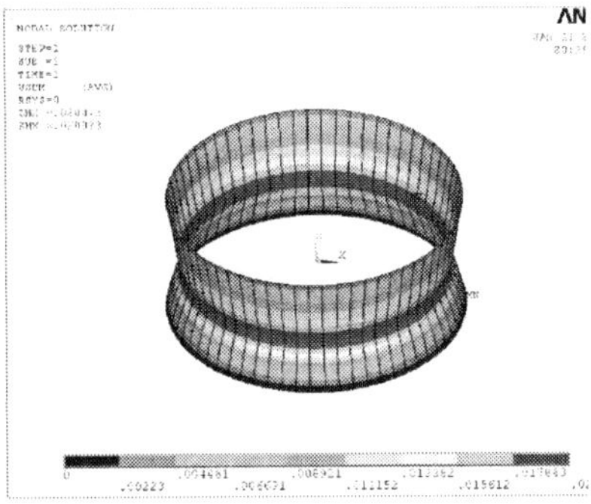

Figure 7b. Displacement of diaphragm wall under uniform pressure（m）

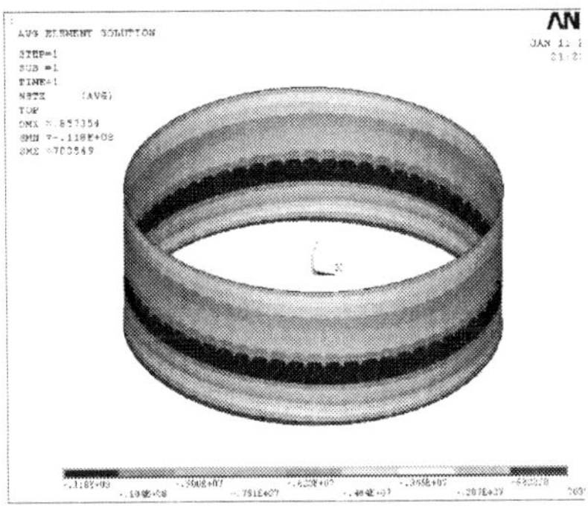

Figure 8a. Axis force of diaphragm wall under non-uniform pressure（N）

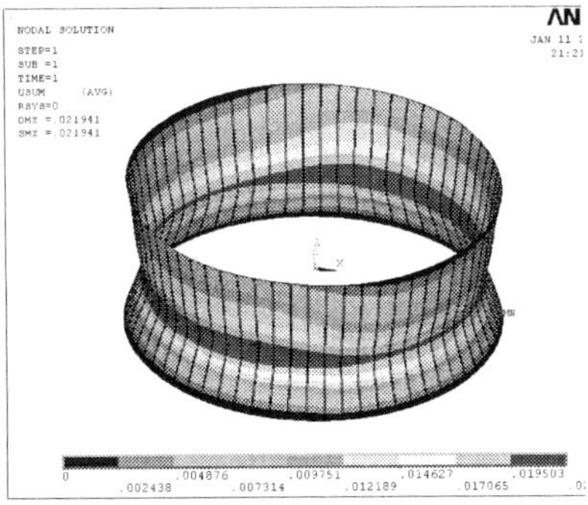

Figure 8b. Displacement of diaphragm wall under non-uniform pressure（m）

Design Control and Measure. For this project, to keep the roundness of the diaphragm wall is the major key issue. Each piece of wall sectors should be close to another. Symmetrical and balanced excavation and construction should be kept in all time, so the arching effect can work effectively by transferring the soil and water pressure into axial force. To ensure the overall balance of excavation, the following design control and measures are established:

Control on Sector of Cylinder-shaped Diaphragm. In order to keep the roundness, the diaphragm wall is designed to be an equilateral polygon, each sector of wall to be cylinder-shape, to ensure the load could transfer reliable, shown in Figure 9.

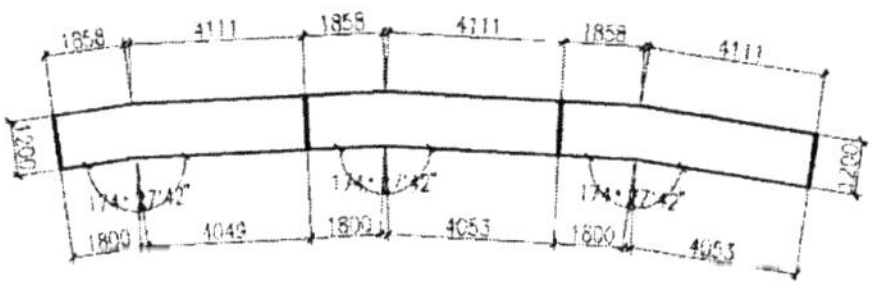

Figure 9. Sector control of diaphragm wall

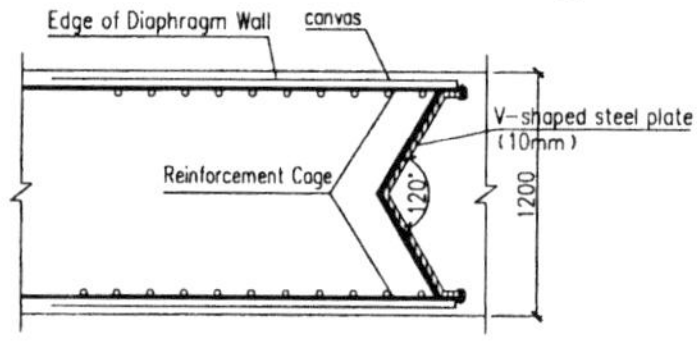

Figure 10. Measurement in Joint

Measurement in joint of Cylinder-shaped Diaphragm wall. In order to ensure the quality of construction joints in diaphragm wall, V-shaped thin steel plate is placed in the end and the canvas is wrapped outside the reinforcement cage to smoothing the pulling of locking tube and to prevent unwanted concrete flow (see figure 10). In addition, milling joint is used in some of the diaphragm walls construction to improve the integrity of the retaining structure. Further ensure that the loads in the diaphragm wall can be effective transferred in order to well perform the arching effect. This is the first project to apply milling joint in Shanghai. (see Figure 11).

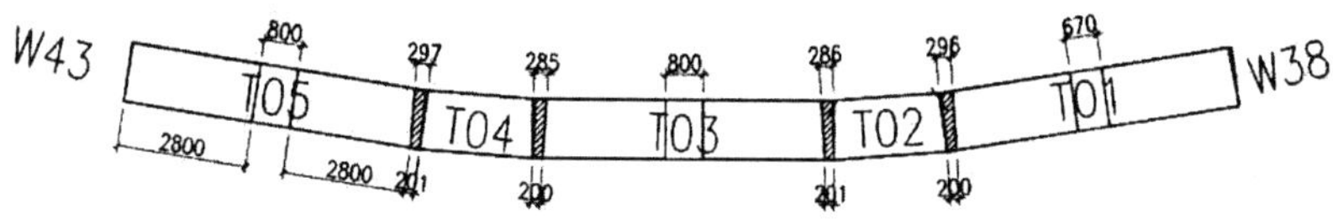

Figure 11. Sector of diaphragm wall with milling joint

Workmanship Requirement for Construction of cylinder-shaped Diaphragm wall and ring beam. To keep the roundness of the walls, each piece of sector has to control the distance (radius of the circle) between the outer turning point to the center of the circle. Also a series of accuracy requirement of the sitting and dimension is set.

Workmanship Requirement for Excavation. To ensure the stability and balance of the circular trench, the excavation depth and area per time should be controlled, and the soil should be excavated symmetrically. The island-excavation should be carried. The detailed requirement is as bellows: First, the soil near the diaphragm wall should be excavated symmetrically and rapidly, and the ring beam should be constructed in time, in order to reduce the exposure time without lateral support and control the deformation of the wall. Second, the rest soil can't be excavated until the lateral support is completed, the soil could serve as the ballast to keep the deformation of the wall.

Monitoring in the stage of Construction. Considering the huge project scale, difficulty in construction and high protection requirement of surrounding utilities, the real time monitoring is required in this project, in order to ensure safety and stability of excavation trench and the surroundings. A real-time monitor system is developed, which can analyses the data of monitor in time, the construction workmanship and scheme can be adjusted and optimized if necessary.

CONCLUSION

Project of Shanghai Tower is currently the tallest building under construction. Passing through comparison of several scheme, and numerous discussion and analysis, a circular trench with open excavation in tower area and top-down method in podium area is adopted in order to fulfill the tight construction schedule. Although the circular trench is considerably deep and large, and it's very difficult in view of design and construction, the reasonable design was got after theoretical study and detail calculation analysis. Moreover, a real time monitor system is developed to ensure that the deep excavation in tower area can be carried out safety and successfully.

Figure 12. Current site photo

Upon finishing this paper, the excavation is under construction. Current site photographs are shown in Figure 12. When the project completed, all the data and result will provide a reference for future similar projects.

REFERENCES

[1].Liu fa-qian, Wang Jian-hua. Theoretical study on active earth pressure of circular foundation pits. Chinese Journal of Geotechnical Engineering,2008,Oct,Vol.30 Supp:5~8.(in Chinese)

[2].JTJ 303-2003 Design and Construction Technical Code for Diaphragm Wall Structure of Port Engineering [S]. Beijing: China Communications Press, 2004.

[3].Shen Jian, Wang Jian-hua, Gao Shao-wu.3-D analysis method of retaining structure of deep excavation based on "m" method[J]. Chinese Journal of Underground Space and Engineering,2005,1(4):530-533.(in Chinese).

Analysis of case histories on deep excavations in marine clay

S.Y. Lam[1], X. Ma[2] and M.D. Bolton[3]

[1]PhD Student, University of Cambridge; syl33@eng.cam.ac.uk
[2]Associate Professor, University of Tongji; xf.ma@tongji.edu.cn
[3]Professor, University of Cambridge; mdb@eng.cam.ac.uk

ABSTRACT: Two case histories on deep excavation of marine clay are used to study the use of a decision-making tool based on a new deign method called the Mobilized Strength Design (MSD) method which allows the designer to use a simple method of predicting ground displacements during deep excavation. This application can approximately satisfy both safety and serviceability requirements by predicting stresses and displacements under working conditions by introducing the concept of "Mobilizable soil strength". The new method accommodates a number of features which are important to design of underground construction between retaining walls, including different deformation mechanism in different stages of excavation. The influence of wall depth, wall flexibility and stratified ground are the major focus of this paper. These developments should make it possible for a design engineer to take informed decisions on the influence of wall stiffness, or on the need for a jet-grouted base slab, for example, without having to conduct project-specific Finite Element Analysis.

INTRODUCTION

The Mobilizable Strength Design (MSD) method is a general, unified design methodology, which aims to satisfy both safety and serviceability requirements in a single calculation procedure, contrasting with conventional design methodology which treats stability problems and serviceability problems separately. In the method, actual stress-strain data is used to select a design strength that limits ground deformations and also the energy dissipation in the ground while taking plastic soil analysis. Thus solution can be acquired by satisfying the energy conservation which also satisfies equilibrium conditions without the use of empirical safety factors.

APPLICATION OF MSD FOR DEEP EXCAVATION PROBLEM

For an in-situ wall supporting a deep excavation, the total deformation can be approximated as the sum of the cantilever movement prior to propping, and the subsequent bulging movement that accretes incrementally with every sequence of propping and excavation.

A method for estimating the cantilever movement had been suggested earlier by Osman and Bolton (2004). It begins by considering the lateral earth pressure distribution for a smooth, rigid, cantilever wall rotating about a point some way above its toe, in undrained conditions. A simple mobilized strength ratio is introduced to characterize the average degree of mobilization of undrained shear strength throughout the soil. By using horizontal force and moment equilibrium equations, the two unknowns, the position of the pivot point and the mobilized strength ratio, are obtained. Then, a mobilized strain value is read off from the shear stress-strain curve of a soil element appropriate to the representative depth of the mechanism at the mid-depth of the wall Accordingly, for the initial cantilever phase, the wall rotation is estimated as one half of the shear strain required to induce the degree of mobilization of shear strength necessary to hold the wall in equilibrium.

A typical increment of bulging, on the other hand, was calculated in Bolton et al. (2008) by considering an admissible plastic mechanism of base heave for both very wide and narrow excavations. In this case, the mobilized shear strength was deduced from the kinematically admissible mechanism itself, using energy conservation principles. The energy dissipated by shearing balances the loss of potential energy due to the simultaneous formation of a subsidence trough on the retained soil surface and a matching volume of heave inside the excavation in addition to structural energy stored in the supporting structures for each increment of excavation. Progressively incorporating elastic strain energy requires the calculation procedure to be fully incremental. A fully incremental solution, admitting layered ground with different stress strain curves, will permit the accumulation of different mobilized shear strengths, and shear strains, at different depths in the ground, thereby improving accuracy. The deformation is estimated using the relationship between the boundary displacements and the average mobilized shear strain, in accordance with the original mechanism.

FORMULATIONS

Consider now the deformations of a multi-propped wall supporting a deep excavation in soft, undrained clay. At each stage of excavation the incremental displacement profile of the ground and the wall below the lowest prop can be assumed to be a cosine function (O'Rourke,1993) as follows:

$$\delta w = \frac{\delta w_{\max}}{2}\left(1 - \cos(\frac{2\pi y}{\lambda})\right) \tag{1}$$

Here δw is the incremental wall displacement at any distance y below the lowest support, δw_{max} is its maximum value, and λ is the wavelength of the deformation.

An incremental plastic deformation mechanism conforming to Equation 1 was proposed by Osman and Bolton (2006) for an infinitely wide multi-propped excavation in clay. In this mechanism, the wall is assumed to be fixed incrementally in position and direction at the lowest prop, implying that the wall has sufficient strength to avoid the formation of a plastic hinge. The wall and soil are deforming compatibly and the soil deformation also follows the cosine function of Equation 1. The dimensions of this mechanism depend on the wavelength λ. The solution includes four zones of distributed shear which consist of a column of soil adjoining the excavation above the level of the lowest prop, a circular fan zone centred at the lowest prop, another circular fan zone with its apex at the junction of the wall and the excavation surface and a 45 degree isosceles wedge below the excavation surface. It is required that the soil shears compatibly and continuously with no relative sliding at the boundaries of each zone. The dotted lines with arrows show the direction of the flow. Along each of these lines the displacement is constant and is given by the cosine function of Equation 1. It is assumed that the zone outside the deformation zones is rigid. This mechanism is simple and neat, but it only applies to very wide excavations. In the case of a narrow excavation, the width of the triangular wedge could be bigger than the actual width of the excavation. In view of this, a new mechanism for narrow excavations is proposed in Figure 1(b). The mechanism in the passive zone (zone EFHI) is replaced. The new mechanism meets the condition for undrained shearing, which means that the volumetric strain remains zero throughout the zone.

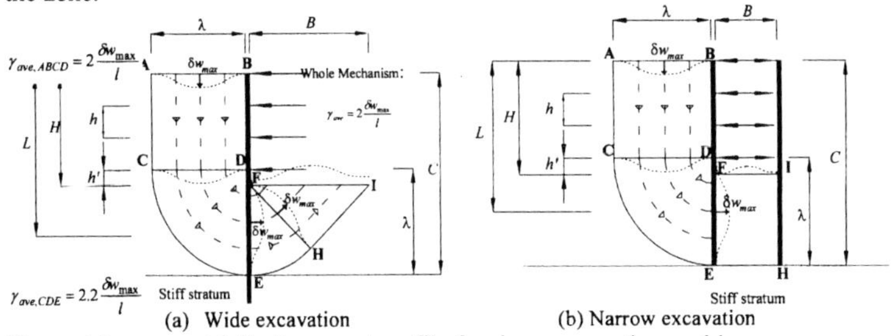

Figure 1 Incremental displacement profile for deep excavation problem

INCREMENTAL ENERGY BALANCE

By conservation of energy, the total loss of potential energy of the soil (ΔP) must balance the total dissipated energy due to plastic shearing of the soil (ΔD) and the total stored elastic strain energy in bending the wall (ΔU).

$$\Delta P = \Delta D + \Delta U \tag{2}$$

The potential energy loss on the active side of the wall and the potential energy gain of soil on the passive side can be estimated easily. The net change of gravitational potential energy (ΔP) is given by the sum of the potential energy changes in each layer:

$$\Delta P = \sum_{i=1}^{n}\left[\int_{volume} \gamma_{sat}(m,i)\,dw_y(m,i)\,dVol\right] \tag{3}$$

where $dw_y(m,i)$ is the vertical component of displacement of soil in the ith layer for the mth construction; $\gamma_{sat}(m,i)$ is the saturated unit weight of soil in the ith layer for the mth construction.

Since there are no displacement discontinuities, the total plastic work done by shearing of soil is given by summing the internal dissipation in each layer:

$$\Delta D = \sum_{i=1}^{n}\left[\int_{Volume} \beta(m,i)\,c_u(m,i)\left|d\gamma(m,i)\right|dVol\right] \tag{4}$$

where $c_u(m,i)$ is the undrained shear strength of soil in the ith layer for the mth construction; $d\gamma(m,i)$ is the shear strain increment of soil in the ith layer for the mth

construction; and the corresponding mobilized strength ratio $\beta(m,i)$ is given by the ratio between mobilized shear strength and :undrained shear strength.

The total elastic strain energy stored in the wall, ΔU, can be evaluated by repeatedly updating the deflected shape of the wall. It is necessary to do this since ΔU is a quadratic function of displacement:

$$\Delta U = \frac{EI}{2}\int_0^s \left[\frac{d^2 w_x}{dy^2}\right]^2 dx \tag{5}$$

where E and I are the elastic modulus and the second moment of area per unit length of wall, and s is the length of the wall in bending. L can be the length of wall s below the lowest prop.

By solving the energy balance equation, designers can easily obtain maximum incremental boundary displacement and hence maximum accumulated lateral wall deflections. Other details of the formulations can be found in Bolton et al. (2008).

ANALYSIS OF CASE HISTORIES

The extended MSD method is verified against two case histories of excavations in Singapore marine clay. The new features of layered soil and bending stiffness of the wall would be implemented. Reasonable predictions can be obtained. This demonstrates the practicality of application of MSD method in the deep excavation problem.

The typical non-linear stress-strain behaviour of Singapore marine clay is obtained by fitting hyperbolic curve over triaxial tress-strain data using FE. The relationship between stress and strains is given by the hyperbolic formulation as follows:

$$\sigma_1 - \sigma_3 = \frac{\varepsilon_1}{a + b\varepsilon_1} \tag{6}$$

,where σ_1 and σ_3 are the major and minor principal stresses respectively and ε_1 is the major principal strain. The parameter a and b are estimated by plotting the triaxial data in $(\sigma_1\text{-}\sigma_3)$ vs ε_a space. The initial Young's modulus $E_i=1/a$ and at large strain the deviator stress reaches $(\sigma_1\text{-}\sigma_3)_{ultimate} = 1/b$. The practical failure deviator stress is a fraction of the ultimate stress, $(\sigma_1\text{-}\sigma_3)_{ultimate}$. The expression for Young's modulus E_t can be expressed as:

$$E_t = \left[1 - \frac{R_f(\sigma_1 - \sigma_3)}{2c_u}\right]^2 E_i \tag{7}$$

,where c_u is the undrained shear strength.

Case 1: Braced sheet-pile excavation of 2-level basement in Singapore marine clay

A narrow excavation of a two-level basement of a sub-structure in soft marine clay in Singapore is investigated. The excavation area measured 110m by 70m is supported by relatively flexible sheet pile wall (Type YSPIII) with a sectional modulus of 1310 cm^3/m. The depth of excavation varied from 6.4m to 7.5m. The sheet pile wall was supported by three levels of bolted struts. The vertical spacing varied from 1.4m to 1.8m. The length of the sheet pile ranged from 12m to 24m. The subsoil profile consists of two layers of soft marine clay deposit of a considerable thickness. The detailed description of construction sequence and instrumentations are given in Tan *et al.* (1985).

In this analysis, two cases of the sheet pile wall with length of 12m and 24m in layered soils are considered. The measured profiles of lateral displacement are shown in figure 2(a) & (b) at various excavation stages at two inclinometers (I-2 and I-9) are chosen for comparison. From measured results from inclinometer I-2, a relatively large lateral wall displacement of 330mm is observed at an excavation depth of 6.8m due to the considerable extent of the depth of soft clay. While at inclinometer of I-9, the maximum lateral wall displacements were significantly smaller due to the relatively shallow depth of soft clay. Owing to the lack of published actual stress strain data from the site, the simulated stress-strain behavior of Wong and Broms(1989) is used in the MSD calculation. Wong and Broms (1989) assumed R_f= 0.7, E_i= 200c_u and c_u= 15kPa. The undrained shear strength of soft clay is assigned to layers at different depths accordingly. The calculated MSD deflection profiles are plotted in the same figure for comparison. Comparisons of maximum lateral deflections between measured data and predicted results are made. A less than twenty percents difference between estimated and computed wall displacement can be obtained. In general, a reasonable estimation of wall displacement can be obtained from stress strain data without the use of FE.

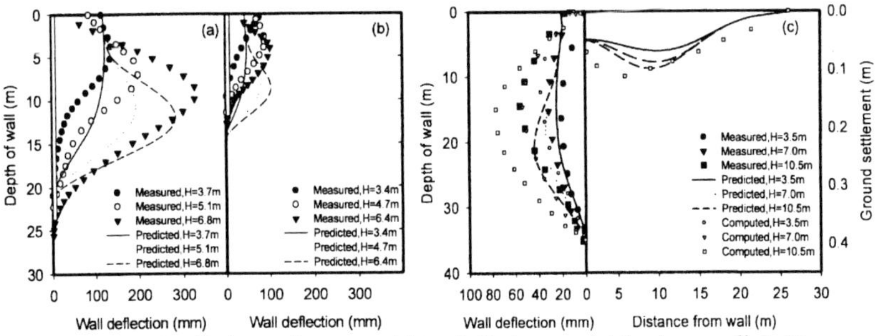

Figure 2 MSD Prediction and measured lateral displacement (a) a long wall L=25m (b) a short wall L=12.5m for *Case 1* (c) Case 2

Case 2: Excavation of a deep basement in Singapore marine clay supported by diaphragm wall

The construction of the new headquarters of the United Overseas Bank in Singapore, UOB Plaza, included excavation of a 3 level basement. The site is underlain by up to 30m of soft marine clay over hard bouldery clay. The sides of the excavation were supported by diaphragm walls up to 1.2m thick and 35m deep. The basement was constructed using three level of propping with an average prop spacing of 3.5m. The maximum excavation depth is 10 to 12m.

The soil conditions were investigated by boreholes, in-situ vane shear and piezocone tests to establish the shear strength of the soft clay. A desiccated clay layer with a c_u of above 60kPa is sandwiched by a very soft layer of lower marine clay with a c_u of 22kPa and a layer of lower marine clay with a c_u of 50kPa. The water level was located close to the ground surface. For details of the construction method, readers may refer to Wallace *et al.* (1992).

The lateral movement of the diaphragm wall was monitored by inclinometer I111. The inclinometer was installed to the toe of the diaphragm wall which was embedded about 7m into residual soil and hard bouldery clay. Thus, it is assumed that movement of the bottom of the wall toe was zero. The measured movements for a typical diaphragm wall section are shown in figure 2(c). The relatively small movement is primarily due to the high stiffness of the retaining wall system. The construction sequence was clearly defined and closely supervised with care control of excavation, prompt insertion of prop and preloading. Numerical simulation by FREW is also included for comparisons. In general, the computed values over-predicted the lateral wall displacement, which is primarily due to the choice of soil stiffness. On the other hand, MSD prediction results slightly under-predicted the lateral movement at the last stage of excavation owing to the rigid prop assumption. The predicted and computed ground settlement toughs are shown in the figure for comparison. In general, the immediate settlement tough right after the excavation predicted by MSD method agrees well with the computed results.

CONCLUSIONS

An extended MSD method is introduced to calculate the maximum wall displacement profile of a multi-propped wall retaining an excavation in soft clay. The back-analysis of excavation in Singapore marine clay allows the authors to validate the capability of the new method in dealing with excavation with different wall stiffness, wall length, various soil strata and also different excavation

ACKNOWLEDGEMENTS

The authors would like to acknowledge the Platform Grant (GR/T18660/01) awarded by the UK Engineering and Physical Sciences Research Council and the Research Grant (618006) provided by Research Grant Council of the HKSAR government.

REFERENCES

Bolton, M. D., Lam, S.Y. and Osman, A.S.(2008). Supporting excavations in clay-from analysis to decision-making. Special Lecture, 6th Int. Symp. of Geotech. Asp. of Underground Const. in Soft Ground. April, Shanghai.

Osman, A.S. and Bolton, M.D. (2004) A new design method for retaining walls in clay. Canadian Geotechnical Journal, 41 (3), 451-466.

Osman, A.S. and Bolton, M.D. (2006) Ground movement predictions for braced excavations in undrained clay. ASCE J. of Geotech. and Geoenviron. Engrg, 132 (4), 465-477.

O'Rourke, T.D. (1993). Base stability and ground movement prediction for excavations in soft clay. Retaining Structures, Thomas Telford, London,131-139.

Tan, S.B., Tan, S. L. and Chin, Y.K. (1985) A braced sheetpile excavation in soft Singapore marine clay, 11th Int. Conf. Soil Mech. Fdn. Engrg, Vol.3, pp.1671-1674.

Wallace, J.C. and Long, M.M.(1992) Retaining wall behavior for a deep basement in Singapore marine clay, Proc. Conf. Retain. Structures, Cambridge, pp.195-204.

Wong, K. S., Brom, B. B. (1989). Lateral wall deflections of braced excavations in clay. ASCE Journal of Geotechnical Engineering, Vol. 115, No. 6, June, 1989.

Analytic Method of Load-displacement Curve for Tension Anchors Based on Hyperbolic Load-transfer Function

Wei LIU[1,2], Long-zhu CHEN[1], Xiao-zhou XI[1],

[1]Institute of Engineering Safety and Disaster Prevention, Dept. of Civil Engineering, Shanghai Jiao Tong University, Shanghai 200240, China
[2]Dept. of Civil Engineering, Ningbo Institute of Technology, Zhejiang University, Ningbo 315100, China

ABSTRACT: Analysis of the mechanism of anchors requires consideration of nonlinear soil behavior and relative motions at the anchor-soil interface. Hyperbolic load-transfer function model, which can allow for these factors, is used to simulate the mechanical behavior of this interface. For tensile type anchors, a set of analytical equations for axial load-displacement relationships is established. Based on the solution, initial anchor pullout stiffness is determined to deduce an approximate formula for critical anchorage length which indicated the efficiency of anchors. Further study shows that the anchorage length has a very important influence on both frictional resistance distribution and ultimate bearing capacity of anchors. Thus, advices for design of anchors are given. The analytical solution provides satisfactory agreement with observed measurements in field test, and thus represents that the analytical solution is reasonable in theory and feasible in practice.

INTRODUCTION

Ground anchors, which are installed into deep soils, are used to mobilize and improve abilities of self-intensity and self-stability. Various types of ground anchors are successively applied to provide uplift or lateral resistance for slope, foundation pit, tunnel, earth retaining structure and excavation support, etc. Based on the different load transfer mechanisms of the grout-strand interface, grouting bolts can be divided into three kinds: tension anchors, compression anchors, and shear-type anchors (Cheng, 1996). At present, tension anchors are widely used in engineering, compared with other types. For tension anchors, ultimate bearing capacity is determined by both tensile strength of strand and pull-out resistance of anchorage body.

Large amounts of experimental data show that the bond stress distribution is very uneven along bonded length especially along long anchors. Many researchers studied the load transfer behavior of ground anchors, and some practical bond stress distribution along soil-grout and grout-strand interfaces have been proposed. In various kinds of research approaches, the load-transfer method, which employs load-transfer function to model interface behavior of anchoring system, is simple and practical. Some of unit side friction versus local bolt displacement relations (q_s-w curves) have been put forward such as linear elastic model (Zhang, 2002), ideal elastic-plastic model (Kim, 2003) and bilinear hardening model (Zhang, 2005a).

Analysis of the mechanism of anchors requires consideration of nonlinear soil behavior and relative motions at the anchor-soil interface. Hyperbolic load-transfer function model, which can allow for these factors, is used to simulate the mechanical behavior of the interface in this paper. Based on the proposed model, a set of analytical solution is derived for the axial load-displacement relationship for tension anchors. The results obtained from the proposed analytical solution have been compared with the field test results. It is found that the calculated results are in good agreement with the measured data.

Meanwhile, based on the developed analytical solution, initial anchor pullout stiffness is determined to deduce an approximate formula for critical anchorage length. The approximate formula also shows a trend of more uniform distribution of the frictional resistance when the anchorage is less than critical anchorage length. Thus, an appropriate designed anchorage length of bolt is suggested.

AN ANALYTICAL SOLUTION FOR THE LOAD-DISPLACEMENT RELATIONSHIP FOR TENSION ANCHORS

Modeling of anchor-soil interface

Under uplift loading, a mechanical model for anchor-soil system is established as shown in Fig.1. For the relationship between unit side friction $q_s(z)$ and local bolt displacement $w(z)$, the following hyperbolic curve is expressed as

$$q_s(z) = \frac{w(z)}{a + bw(z)} \tag{1}$$

where, $1/a$ is the initial tangent slope of the load-transfer function and is called initial shear stiffness, and $1/b$ is the asymptotic value of the load-transfer function. Determination of the constants a and b is the key to the successful fitting of anchor-soil interaction. If there have measured load-transfer function curves, constants a and b could be obtained through data fitting. Otherwise, constants can only be obtained by empirical formulas (Kim et al., 1999).

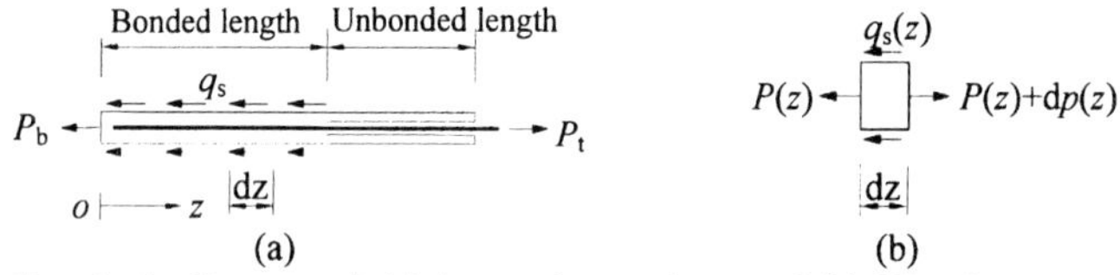

FIG.1. Free body diagram of: (a) An tension anchor; and (b) Slice dz

Solution procedure

The free body diagram of slice dz at location z along the bonded length is shown in Fig.1(b), and the force balance of the anchor-soil interface of slice dz is given by the following equilibrium equation:

$$\mathrm{d}P(z)/\mathrm{d}z = uq_s(z) \tag{2}$$

where u is the perimeter of the anchor section. Assuming that the anchor acts as an elastic bar and from Hooke's law, it follows

$$\frac{\mathrm{d}w(z)}{\mathrm{d}z} = \frac{P(z)}{EA} \tag{3}$$

where E = elastic modulus of anchor, and A is the cross-sectional area of the anchor. Combining Eq.(2) with Eq.(3) gives

$$\frac{\mathrm{d}P(z)}{\mathrm{d}w(z)} = \frac{EAu}{P(z)} q_s(z) \tag{4}$$

By substituting Eq.(1) into Eq.(4) and integrating, Eq.(4) becomes

$$P(z) = \frac{\sqrt{2EAu}}{b}\sqrt{bw(z) - a\ln(1 + \frac{b}{a}w(z)) + c_1} \tag{5}$$

where c_1 is the constant to be determined from the boundary conditions that are $P(z)|_{z=0}=P_b$ and $w(z)|_{z=0}=w_b$, then Eq.(5) can be presented as

$$P(z) = \sqrt{P_b^2 + \frac{2EAu}{b^2}\left\{b(w(z)-w_b) - a\ln[1 + \frac{b(w(z)-w_b)}{a+bw_b}]\right\}} \tag{6}$$

By substituting Eq.(6) into Eq.(3) and integrating, Eq.(3) becomes

$$z = \int_{w_b}^{w(z)} \frac{EA}{\sqrt{P_b^2 + \frac{2EAu}{b^2}\left\{b(w(z)-w_b) - a\ln[1 + \frac{b(w(z)-w_b)}{a+bw_b}]\right\}}} \mathrm{d}w(z) \tag{7}$$

Given that the anchor has a zero force at the anchor tip (i.e., P_b=0) and has an initial upward tip displacement w_b, Eq.(7) has only two unknown values: z and $w(z)$. Therefore, substituting a series of assumed values $w(z)$ into Eq.(7), a series of values z can be calculated. When the calculated z is just equal to the bonded length l_a, corresponding $w(z)$ is the displacement at the top of the bond zone, S_a. Subsequently, applied load Q (i.e., $p(z)|_{z=l_a}$) can be obtained from Eq.(6). Thereupon, for an initial tip displacement value, the relevant load and displacement at the top of the bond zone, Q and S_a, are obtained. Meanwhile, values of axial force $P(z)$ are acquired at any value of z to give the force-depth curve.

Considering the deformation of the bonded portion, total displacement of anchor, S, can be obtained as follows:

$$S = S_a + \frac{Ql_f}{E_s A_s} \tag{8}$$

where l_f is the unbonded length, E_s is the elastic modulus of the strand, and A_s is the cross-sectional area of the strand. Then for a series of initial tip displacement w_b, the relevant load and displacement at the anchor head could be calculated by Eqs.(6) , (7) and (8). Therefore the load-displacement relationship for tension anchor is obtained in a single-layered homogeneous soil. This method can also be used in layered soils.

It is noteworthy that for Eq.(7), singular integral is needed to be treated as the integrand function tends to infinity in the vicinity of z=0, where $w(z)$=w_b and P_b=0. Denoting this integrand function to be $F(w)$, Eq.(7) may be written as

$$z = \int_{w_b}^{w_b+\varepsilon} F(w)\,\mathrm{d}\,w(z) + \int_{w_b+\varepsilon}^{w(z)} F(w)\,\mathrm{d}\,w(z) \tag{7a}$$

where ε is an undetermined positive number which tend to zero. Among them, the second part $\int_{w_b+\varepsilon}^{w(z)} F(w)\mathrm{d}w$ could be obtained by numerical integration easily. Based on the Taylor Formula $\ln(1+x) = x - \frac{1}{2!}\frac{x^2}{(1+\theta x)^2}$, where 0<θ<1, the first part is expressed as

$$\int_{w_b}^{w_b+\varepsilon} F(w)\,\mathrm{d}\,w(z) < \int_{w_b}^{w_b+\varepsilon} \frac{k_1}{\sqrt{k_2\varepsilon}}\mathrm{d}\,w(z) = 2k_1\sqrt{\varepsilon / k_2} \tag{9}$$

where $k_1 = b\sqrt{2EAu}/2u$, and $k_2 = b - ab/(a+bw_b)$. Denoting the permissible integration error of equation Eq.(7) be δ, which could be a determined value (for instance, 0.01mm), ε is given by

$$2k_1\sqrt{\varepsilon / k_2} = \delta \tag{10}$$

It seems that the first part $\int_{w_b}^{w_b+\varepsilon} F(w)\mathrm{d}w$ could be neglected. Eq.(7) can then be simplified as

$$z \approx \int_{w_b+\varepsilon}^{w(z)} F(w)\,\mathrm{d}\,w(z) \tag{7b}$$

The procedure mentioned above can be implemented by computers, and a general calculation program has been developed.

DETERMINATION OF CRITICAL ANCHORAGE LENGTH

It is generally considered that when the bonded length is over a certain value, any further increase in length will only marginally increase the bearing capacity of anchor. Therefore, suitable anchorage length is stipulated by the technical codes. According to the Technical specification for ground anchors, anchorage length should take a value from 3~8m for rock bolts, while 6~12m for soil anchors (CECS 22: 2005, 2005). Although the approximate scope of anchorage length has been specified, the critical anchorage length needs to be investigated to better understand the behavior of ground anchors.

For simplicity, the stiffness of the anchor here is defined as the ratio of the applied load and the displacement at the top of the bond zone, Q/S_a. Denoting the initial stiffness to be K_0, it can be calculated as follows:

$$K_0 = EA\sqrt{\frac{ku}{EA}}\,\mathrm{th}\left(\sqrt{\frac{ku}{EA}}l_a\right) \tag{11}$$

where E , A and u are the elastic modulus, the cross-sectional area, and section perimeter of anchor respectively, l_a is the bonded length, and k is equal to constant $1/a$ (i.e., the initial tangent slope of the load-transfer function). Using a non-dimensionalized initial stiffness $\overline{K}_0$, Eq.(11) may be written as

$$\overline{K}_0 = \frac{K_0}{EA/l_a} = 2\sqrt{n}\frac{l_a}{d}\,\mathrm{th}\left(2\sqrt{n}\frac{l_a}{d}\right) \tag{12}$$

where l_a/d is the ratio of bond length and diameter, and $n=kd/E$ is the ratio between the modulus of the anchor-soil interface, kd, and the elastic modulus of anchor, E.

However for $2\sqrt{n}\frac{l_a}{d} > 2$, the value of $th(2\sqrt{n}\frac{l_a}{d})$ is larger than 0.964 and is approach to the function limit of 1. Obviously, contribution of bonded length to the initial anchor stiffness becomes limited when the bonded length beyond a certain value, which is called the critical anchorage length. Denoting the critical anchorage length to be l_{ac}, it can be estimated as:

$$l_{ac} = d / \sqrt{n} = \sqrt{Ed / k} \quad (13)$$

Eq.(13) needs to be verified. Zhang (2005b) reported that the whole friction resistance was mainly focused on the anterior two-thirds segment of bonded length. Let ξ denote the ratio of friction resistance provided by the posterior one-third segment and the whole friction, it can be expressed by

$$\xi = \int_0^{\frac{1}{3}l_a} \tau(z)\,dz \Big/ \int_0^{l_a} \tau(z)\,dz \quad (14)$$

The smaller the ξ value, the less uniform the friction distribution along the bonded length. Fig.2 gives the ratio ξ for various anchorage lengths. As shown, ξ is equal to 0.175 at length ratio of 0.5; ξ is equal to 0.091 at length ratio of 0.75; and ξ is equal to 0.052 at length ratio of 0.9. This results shows the former given Eq.(13) is reasonable to estimate the critical anchorage length.

However, objective to obtain the critical anchorage length is not to use this value to get the maximum anchoring resistance, but to find out a suitable anchorage length which makes anchor design more economic and effective. So the actual anchorage length is suggested not to exceed $0.5l_{ac}$. Furthermore, according to the Eq.(13), it should be paid more attention that the softer the soil, the smaller n is and the larger the critical anchorage length is. In addition, the initial stiffness of anchor can be efficiently increased by enlarging bolt diameter and suitable bond length as well.

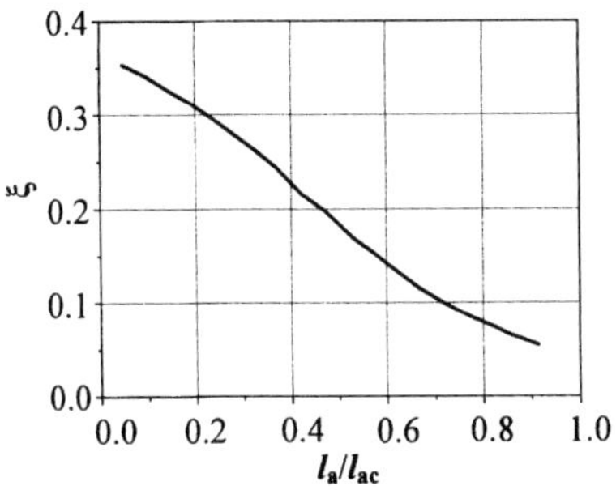

FIG. 2. ξ-l_a/l_{ac} **curve**

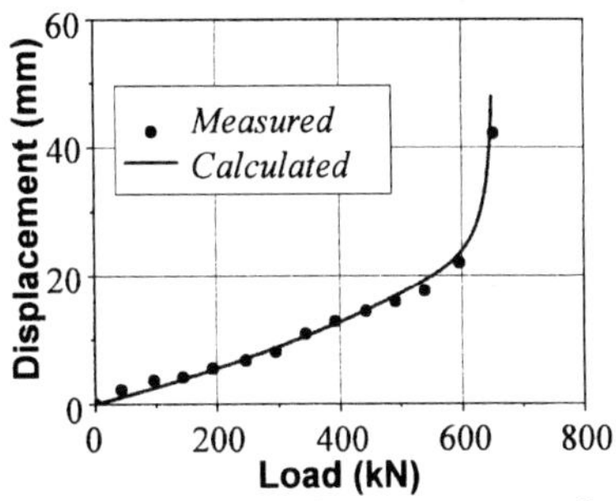

FIG.3. Comparison between measured and calculated curves

COMPARISON WITH CASE HISTORIES

Zhang Ji-ru presented the pull-out test of anchor M3 designed for a deep foundation pit (Zhang, 2002). The bonded length is 17.5m, and the unbonded length is 5.5m. The grout has a cross section area of 11,800 mm^2 and an elastic modulus of

2.0×10^7 kN/m^2, while the strand has a cross section area of 1470mm^2 and an elastic modulus of 2.0×10^8 kN/m^2. So the anchor has a synthetic area of 13,270 mm^2 and a synthetic elastic modulus of 4.0×10^7 kN/m^2.

Constants used in hyperbolic transfer functions are obtained through data fitting of the measured load-displacement of anchor M3. It is found that constants are: $1/a$=175.44MN/m^3, $1/b$=92.16kPa. Fig.3 gives the measured and calculated load-displacement curves for anchor M3. The calculated curve approaches the measured value at most points, except in the initial part of curves. Encouragingly close agreement are obtained between measured and calculated curves.

CONCLUSIONS

A mathematical model developed in this paper provides a practical method to the analysis of tension anchors. Based on this model, analytical equations for anchor load-displacement relationships are obtained. To validate the proposed method, case studies are carried out. It is found that the calculated results from the analytical equations are in good agreement with the measured data.

Meanwhile, initial anchor pullout stiffness is determined to deduce an approximate formula for critical anchorage length, l_{ac}. To make anchor design more economic and effective, the actual anchorage length is suggested not to exceed $0.5l_{ac}$.

ACKNOWLEDGEMENTS

The study reported herein is supported by the National Natural Science Foundation of China with the granted number 50779034.

REFERENCES

CECS 22: 2005. Technical specification for ground anchors. Beijing: China Planning Press, 2005.

Cheng, L.K. (2001). "Present status and development of ground anchorages." *China Civil Engineering Journal*, 34(3): 7-12 (in Chinese).

Kim, N.K. (2003). "Performance of tension and compression anchors in weathered soil." *J. Geotech Geoenviron Eng, ASCE*, 129(12):1138-50.

Kim, S., Jeong, S., Cho, S., and Park, I. (1999). "Shear load transfer characteristics of drilled shafts in weathered rocks." *Journal of geotechnical and geoenvironmental engineering*, 11: 999-1010.

Zhang, J. Shang, Y.Q. and Ye, B. (2005a). "Analytic calculation of P-S curve of bolts based on bilinear load-transfer function." *Chinese Journal of Rock Mechanics and Engineering*, 24(6): 1072—1076. (in Chinese)

Zhang, J., Shang, Y.Q., and Ye, B. (2005b). "Analytical calculations of critical anchorage length of bolts." *Chinese Journal of Rock Mechanics and Engineering*, 24(7): 1134-1138(in Chinese)

Zhang, J.R., Tang, B.F. (2002). "Hyperbolic function model to analyze load transfer mechanism on bolts." *Chinese Journal of Geotechnical Engineering*, 24(2): 188-192.(in Chinese)

Experimental research of jet-grouting parameters in different soil conditions

Alexey Malinin[1], Ilya Gladkov[2], Dmitriy Malinin[3]

[1]Technical director, "InzhProektStroy", Perm, Russia; info-ips@yandex.ru
[2]Leading engineer, "InzhProektStroy", Perm, Russia; info@geo-soft.ru
[3]Leading engineer, "InzhProektStroy", Perm, Russia; cct-info@yandex.ru

ABSTRACT: Results of experimental research of soil condition and jet-grouting parameter influence on diameter value of jet-grouting columns are described in this article. Three series of experimental work in soils of different cohesion were made. After construction of jet-grouting columns a new device for diameter measurement was tested. Dependences of jet-grouting column diameter value on soil treatment time, cement grout pressure and monitor rotation speed were received.

INTRODUCTION

Jet-grouting physics is formed presently in separate scientific trend, which allows solving main design tasks relating to underground construction. All characteristics of underground structures created using jet-grouting technology depend on soil properties, cement quantity and features of technological process, principles of which are determined by multifactor dependences. Experimental research of jet-grouting column diameter dependence on soil properties and jet-grouting parameters is presented in this article.

Jet-grouting is scientific and technical trend in modern geotechnology.

This technology got the main practical application at task solution in the field of underground construction. Such tasks as soil strengthening at construction of different tunnels, mines and underground openings are successfully solved in many countries using jet-grouting technology. This technology is widely used at construction of sheet pile walls, waterproofing membranes, pile foundations etc.

Diameter of jet-grouting columns is the most important parameter which defines efficiency of jet-grouting results almost in all fields of its application. Especially it concerns waterproofing membrane construction, because crossing of columns at specified interval determines quality of waterproofing works.

Now most jet-grouting characteristics are specified empirically, on the basis of expensive experimental works. In most cases at designing of jet-grouting structures and works, data from similar jobsites are used.

In order to get more accurate assessment of jet-grouting column diameter, experimental works are made in conditions of real jobsite. Choice of operating mode is made on the basis of received data.

Column diameter depends on soil type, lifting speed and rotation frequency of monitor, as well as injected grout volume, nozzle diameter and pressure. Two main designed factors – time of soil treatment and pressure may vary till specified limits depending on natural soil type.

If time of soil erosion is not enough, jet-grouting structure can be made of spiral form. It is reasonably to use such structures, e.g. at construction of highways or foundation plates, where it is necessary to increase soil properties. Long time of soil erosion will require big cement consumption, which is not reasonable from economic point of view. So fixing of optimal soil erosion time is extremely important factor which specifies technical and economic effects of jet-grouting.

Three jobsites constructed by "InzhProektStroy" (Russia) were chosen for experimental research. On the first jobsite there was cohesive soil, on the second jobsite – non-cohesive soil and on the third one – soil with low cohesion.

EXPERIMENTAL WORKS IN COHESIVE SOILS

Cement plant in Novotroitsk (Russia) is under construction. At building of multisection cement silo designers decided to make a pile-raft foundation: circular foundation plate with a 42 m diameter, 3,4 m thick, and 199 jet-grouting piles 20 m long. The main bearing layer of foundation plate is light clay with gruss and interlayers of light pulverescent clay. During working procedure lots of limestone interlayers were detected, which considerably complicated construction of jet-grouting columns.

Now the main method of jet-grouting column diameter determination is digging out of already hardened columns and their subsequent measurement. This process is quite difficult and long-term. On this jobsite diameter was measured immediately after construction, using special patented measuring device (fig. 1).

After pile construction the measuring device was lowered into the column body (before cement grout setting). This measuring device is in the form of two crossing bars, which were moved apart to pile's wall. Then in the upper part of device a measurement was made. It should be noted that edges of erosive zone were evident, because it was hard clay. Diameter measurement was made at a depth of 1 metre.

Several similar measurements were made, during them one of main jet-grouting parameters was changed – time of soil treatment. Other parameters such as pressure, lifting stroke and rotation speed of monitor remained the same. Time of monitor lifting for 1 meter was changing from 110 to 230 seconds. At the same time diameter was changing from 370 to 600 mm.

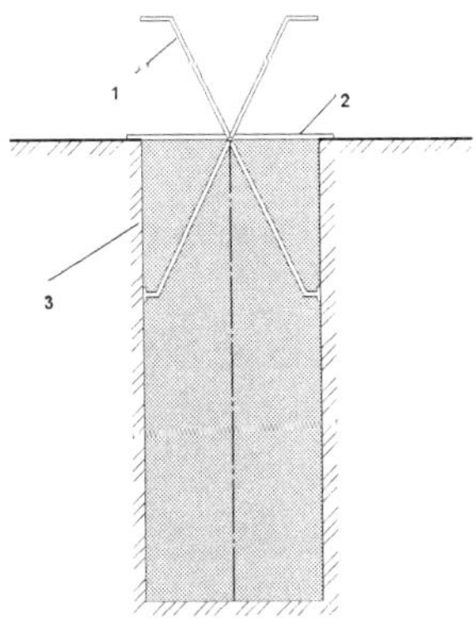

FIG. 1. Scheme of column diameter measurement
1 – measuring device; 2 – holder; 3 – jet-grouting column wall.

Dependence of jet-grouting column diameter on soil erosion time is shown on figure 2(1). Wide spread of values is conditioned by the fact that strengthening soil was heterogeneous, consisting from different clays. That was confirmed by colour of coming out pulp. Its colour was changing from dark to bright red. Moreover clay also had interlayers of limestone. At the same time this diagram shows increase of pile diameter value depending on period of soil treatment.

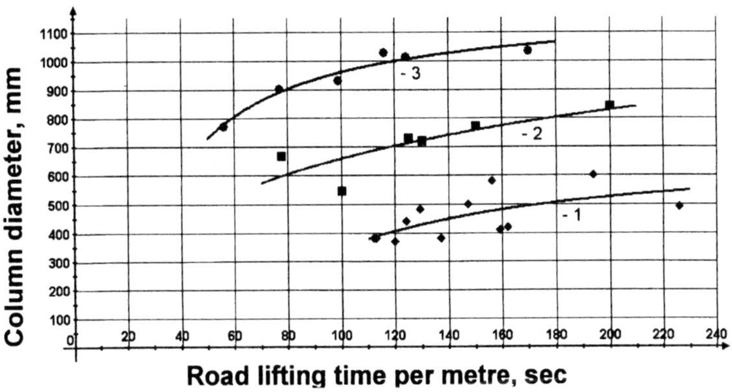

FIG. 2. Dependences of jet-grouting column diameter on soil treatment time
1 - cohesive soil (c = 47 kPa); 2 – soil with low cohesion (c = 7 kPa);
3 - non-cohesive soil (c = 1 kPa).

Besides several piles were constructed using preliminary soil erosion (PSE) by water (pressure 40 MPa). In was known that preliminary soil erosion increased pile diameter on the average for 15-30%.

For control of soil-concrete quality, cores with a 10-cm diameter were drilled out from columns (fig. 3). As shown on this figure there are pieces of not mixed clay in the sample. That can be caused by insufficient time of soil treatment in some parts of jet-grouting column.

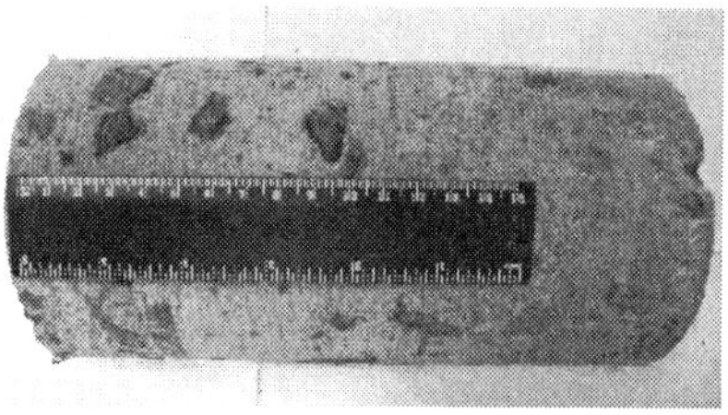

FIG. 3. Core sample drilled out from jet-grouting column in clay.

EXPERIMENTAL WORKS IN NON-COHESIVE SOILS AND SOILS WITH LOW COHESION

The second jobsite where experimental measurements were made was situated in Pushkino, Moscow region (Russia). The soil was homogeneous, consisting of fine sand of average consistency. There was construction of sheet pile wall. As apposed to the previous jobsite here columns were constructed in non-cohesive soil. In order to limit groundwater entry into the pit, sheet pile wall was made by solid wall.

Check measurements of constructed columns were made during the process using above-mentioned measuring device in order to provide wall solidity. Received values of jet-grouting column diameters exceeded designed values and equaled to 703-885 mm.

In addition to column diameter measurements, experimental pile field from 16 piles was constructed. Piles were made using different parameters. The main parameters of technology such as: injection pressure, lifting speed and rotation frequency of monitor were changed.

Column diameters were measured after their construction and then all columns were digged out. Results received by measuring device almost coincide with data received at pile digging out (disarrangements over the range 5%). Dependences of column diameter on injection pressure, soil treatment time and monitor rotation frequency are graphed according to received results.

Injection pressure was varying over the range 10 to 50 MPa. This range is caused by technological limits – parameters of injecting cement pump and some features of Jet1 technology. As shown on figure 4 (1) visible increase of column diameter caused by increase of pressure.

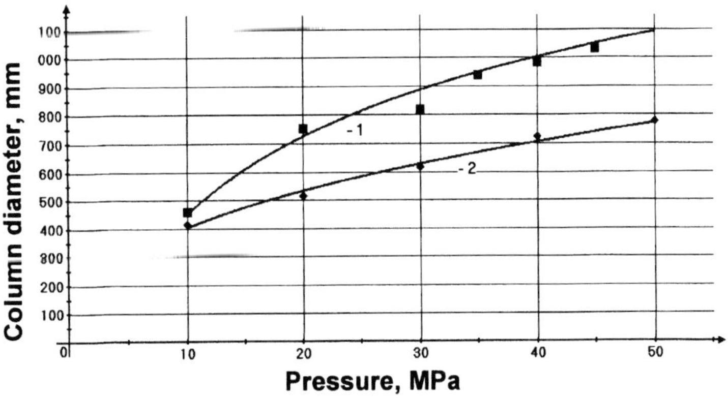

FIG. 4. Dependences of jet-grouting column diameter on soil treatment time
1 – non-cohesive soil (c = 1 kPa); 2 – soil with low cohesion (c = 7 kPa).

Dependence of column diameter on soil treatment time is indicated on figure 2 (3). Time of treatment for 1 m was changing over the range 56 to 176 seconds.

Results show that the largest increase of diameter happens during initial time interval, and subsequent increase of treatment time doesn't change diameter considerably.

Several columns were constructed using different rotation speeds - from 11 to 35 rpm (revolutions per minute). The optimal speed value is 20 rpm, under such speed column diameter is the largest one (fig. 5).

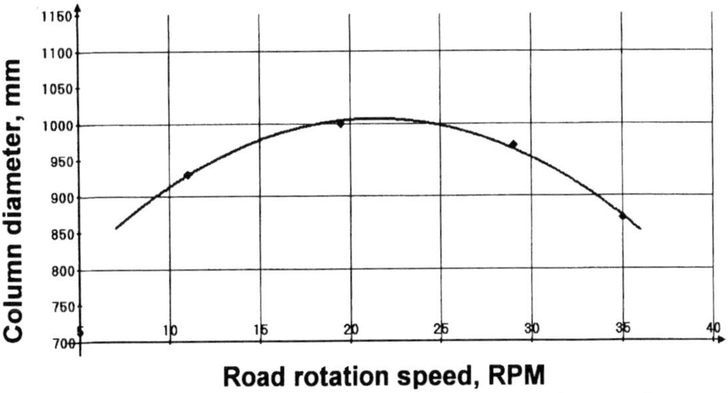

FIG. 5. Dependence of jet-grouting column diameter on monitor rotation speed.

For the third series of experimental works at jobsite in Perm (Russia) experimental pile field was made (fig. 6).

FIG. 6. Experimental jet-grouting columns.

In this case the soil was homogeneous, consisting of fine compact sand, with low cohesion (c = 7 kPa).

The main technological parameters such as injection pressure and monitor lifting speed were also changed like it was made in the previous jobsite. Dependence of column diameter on injection pressure is given on figure 4 (2). At pressure changing from 10 to 50 MPa, pile diameter was varying from 400 to 800 mm. Dependence of jet-grouting column diameter on monitor lifting speed is given on figure 2 (2).

CONCLUSIONS

1. Physical-mechanical soil properties are the determining factors at construction of jet-grouting columns. When choosing a technological mode first of all it is necessary to define cohesion of soil. Type of soil has influence on the column diameter variation which can reach up to 50%.
2. Soil heterogeneity has significant influence on forecasting of jet-grouting column diameter.
3. In cohesive soils at double increase of treatment time a 70% increase of column diameter was shown. In non-cohesive soils an increase of treatment time doesn't result in considerable diameter increase. The main increase is shown during the initial stage.
4. In cohesive soils high speed of monitor rotation and short time of soil treatment can result in bad soil mixing – creation of soil pieces in jet-grouting column.

REFERENCES

Malinin A. (2007). "Jet-grouting.", Presstime, Perm/Russia.

Mitigation of Sheet Pile Movements during DDC by Open Trenches

Yong Tan[1], Fangle Peng[2], and Shaoming Liao[3]

[1]Key Laboratory of Geotechnical and Underground Engineering of Ministry of Education, Department of Geotechnical Engineering, Tongji University, 1239 Siping Road, Shanghai 200092, PRC; Email: tanyong21th@hotmail.com

[2]Key Laboratory of Geotechnical and Underground Engineering of Ministry of Education, Department of Geotechnical Engineering, Tongji University, 1239 Siping Road, Shanghai 200092, PRC; Email: pengfangle@gmail.com

[3]Key Laboratory of Geotechnical and Underground Engineering of Ministry of Education, Department of Geotechnical Engineering, Tongji University, 1239 Siping Road, Shanghai 200092, PRC; liaoshaoming@gmail.com

ABSTRACT: This paper investigated the effectiveness of open-trench wave barriers in reducing the movements of sheet pile walls during deep dynamic compaction (DDC), by isolating (screening) the energy waves generated in DDC impact. This investigation was achieved by a series of finite element analyses. The open trenches with different cross-section shapes were investigated. The dimensional effects of trenches were analyzed and discussed, and then the isolation efficiencies of the different trenches were compared. The numerical results indicated that excavating an open trench in front of the retaining walls is an effective measure to reduce wall movements during DDC. For the investigated open trenches, increasing trench depth or trench width can efficiently improve their isolation efficiencies, while the cross-section shapes of the trenches have no apparent influences.

INTRODUCTION

As a flexible retaining system, sheet pile walls are widely used as waterfront structures in roadway, port, or coastal projects to retain loose granular fills as foundation soils. Commonly, these kinds of sites are required to be densified to mitigate potential liquefaction and/or settlement hazards, especially for those sites located in seismic zones. Deep dynamic compaction (DDC) has been widely used due to its notable densification effects, easy application, and economical effectiveness. DDC is carried out by repeatedly dropping a heavy tamper onto the soil surface from a certain height, following a prescribed grid-pattern. Limitations of DDC include significant vibration and subsequent soil movements, which will cause adverse effects to the adjacent structures and facilities. Tan (2005) reported a failure case of sheet pile walls due to intolerable wall deflections caused by DDC impact. The deflection of a sheet pile wall during DDC is the result of transient dynamic lateral earth pressures against it, which are caused by DDC energy waves propagating to the wall. Consequently, a countermeasure, which can effectively isolate (screen) the energy waves propagating to the walls during DDC and then reduce the wall deflections, is desirable here.

Via numerical or experimental methods, many studies had been carried out to examine the isolation of ground-borne vibration by wave barriers (e.g., Woods 1968; Haupt 1981; Ahmad and Al-Hussaini 1991; Haupt 1995; Massarsch 2005). All of these studies were focused on investigating the isolation effectiveness of wave barriers on reducing ground-borne vibrations caused by time-harmonic loading. However, no work was known to study the isolation effects of wave barriers in reducing impulse-loading (e.g. DDC) induced vibrations, especially to study the isolation of DDC impact on movements of retaining walls. The purpose of this study was to examine the isolation (screening) effectiveness of open trenches on reducing the DDC-induced deflections of sheet pile walls as waterfront structures for retaining loose saturated sands. This was achieved by a series of finite element (FE) analyses. The accuracy and reliability of the FE methodology were verified by the laboratory test results from available literature. The reduction effects of various geometrical parameters (i.e., depth, width, and shape) of the open trenches were investigated, and the key parameters that govern the isolation efficiency of an open trench were identified.

PROBLEM DEFINITION

A typical harbor or coastal project is considered in this study. Sheet pile walls are constructed in water away from shoreline, and the areas between the sheet pile walls and shoreline are then filled with sandy soils to serve as loading and handling zones or storage zones or other facilities. Filling is performed by dumping and pushing fill materials from shoreline. Therefore, the dumped fills are in a loose saturated state, which poses risk of settlement and liquefaction hazards when subjected to earthquake shaking. DDC is adopted to densify the fill sands. Figure 1 presents the geometries and soil profiles of the problem studied. The investigated sheet pile wall has a height of 16 m with a penetration depth of 9 m into the in-situ sands below the dredge line. The retained fills consist of 7 m thick loose sands, underlain by the in-situ sandy soils. The water table level is about 3 m below the top of the retaining walls. One pass of DDC is applied to densify the fills, following the prescribed grid-pattern. Compaction is conducted following the order of the tamping points 1 to 5. The distance between two adjacent tamping points is 3 m. Compaction is completed using a 1.5 m diameter tamper with an impact energy of 200 ton-m. Each grid point is repeatedly impacted by 6 drops. An open trench is excavated in the fills near the sheet pile walls to screen (isolate) the DDC energy waves propagating to the waterfront structures to reduce the wall deflections. Three types of open trenches, designated as I, II, and III, with a depth, d, and a width, w, are considered. Trench I is rectangular in cross-section; trench II is trapezoidal in cross-section with a side slope (H:V) of 1:1; trench III is also trapezoidal in cross-section, but with a side slope (H:V) of 1:2. In order to avoid collapse of the open trenches, the trench sides are paved with 10 cm thick concrete. The isolation (screening) effect on reducing wall deflections is evaluated in terms of amplitude reduction ratio, A_r, i.e.,

$$A_r = \frac{\delta_i}{\delta_0} \tag{1}$$

Where, δ_i = deflection at the top of the sheet pile wall with an open trench; δ_0 = deflections at the top of the sheet pile wall without an open trench. To emphasize the reduction in wall deflections due to the existence of an open trench, another parameter called the percentage reduction ratio, R_f, is defined as:

$$R_f(\%) = (1 - A_r) \times 100\% \tag{2}$$

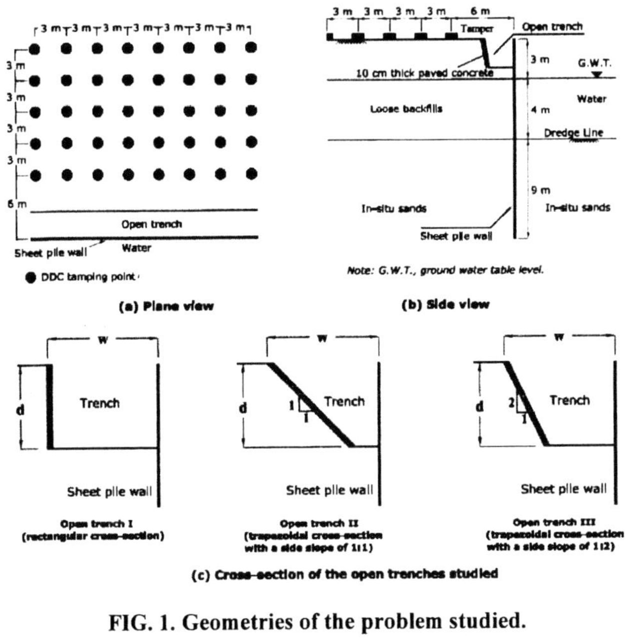

FIG. 1. Geometries of the problem studied.

FE VALIDATION OF ISOLATION EFFECTS BY OPEN TRENCHES

The problem defined in Figure 1 was examined by performing a series of finite element (FE) analyses using the 2D FE code Plaxis 8.2 (Brinkgreve et al. 2002), in which geometrical parameters of the open trenches were varied to study their isolation effects. To check the validity of PLAXIS 8.2, two numerical analyses on vibration reduction ratio of two open trenches were compared with two experimental data available from the published literature. Detailed information about these field model tests can be found in Woods (1968) and will not be repeated here. Figure 2 presents the comparison of the FE analysis results with Woods' test data of PF-8-250 and PF-16-250. It was observed that the FE simulated results were in reasonable agreement with Woods' test data and the maximum differences are less than 25%.

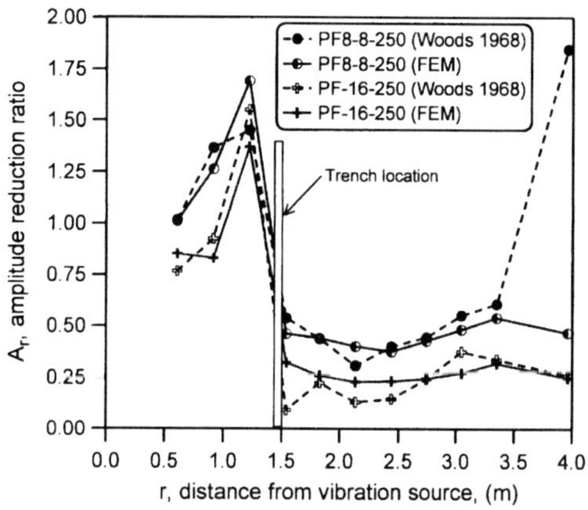

FIG. 2. Comparative study on isolation effects by open trenches.

PARAMETRIC STUDIES OF THE PROBLEM DEFINED

The problem defined in Figure 1 was investigated by a series of parametric studies, in which different geometrical parameters (depth, width, and cross-section shape) of the open trenches were varied. During a parametric study, when a certain geometrical parameter was investigated, the other parameters were kept constant. First, dimensional factors, i.e. depth and width, were examined for each type of the trenches. Then, the reduction efficiencies of different open trench types were compared to investigate the effects of cross-sectional shapes. Since the aim of this study was to investigate the effectiveness of open trenches in reducing the DDC-induced deflection of sheet pile walls, only the numerical analysis results in terms of amplitude reduction ratio, A_r, and percentage reduction ratio, R_f, will be presented and discussed. Prior to investigation of the isolation effects by open trenches, one FE analysis had been performed to investigate the sheet pile wall deflections due to DDC without trenches. The FE-predicted maximum wall deflections without trenches were around 120 cm.

In order to investigate the effects of trench dimensions on reducing deflections of sheet pile walls caused by DDC, three parametric studies were performed. For all three trenches, the trench depth, d, was varied between 0.5 m and 3.0 m, and the width, w, was varied between 0.5 m and 3.0 m.

Figure 3 presents a FE mesh made of 15-node triangular elements along with the constitutive models used in the FE analyses. The left and right vertical boundaries were fixed horizontally, and the horizontal boundary at the bottom was fixed in both horizontal and vertical directions. Absorbent boundaries which can absorb the energy waves were applied to both the vertical and horizontal boundaries to prevent reflection of waves back into the soil bodies. The fills and in-situ sands directly below were simulated by an undrained Mohr-Coulomb model; the paved concrete and steel tamper were simulated by a non-porous linear-elastic material model; the sheet pile wall was simulated by an elastic plate element.

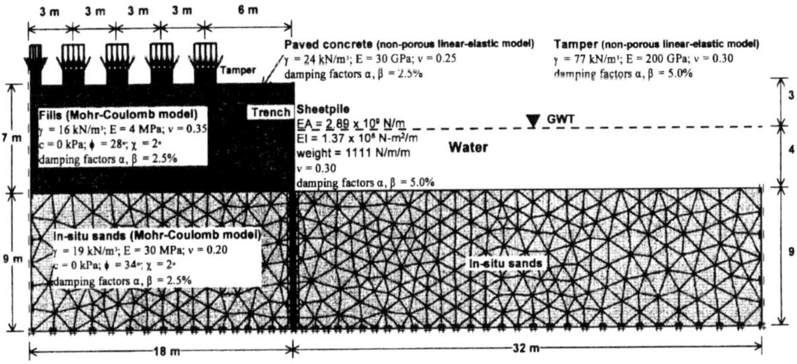

FIG. 3. FE-mesh with an open trench of rectangular cross-section.

Analyses of FE Modeling Results

Figure 4 summarizes the reduction efficiencies of the three investigated open trenches with different cross-section shapes in reducing deflections of the sheet pile wall due to DDC impact. The three parametric studies on one rectangular open trench and two trapezoidal open trenches indicated that for the investigated three trenches, increasing the depth, d, or the width, w, can increase the reduction efficiency of the trenches. Both the depth and the width appeared to be of equal importance in governing the reduction efficiency. This observation was in agreement with the findings of Fuyuki and Matsumoto (1980) and Ahmad and Al-Hussaini (1991).

Segol et al. (1978) pointed out that the geometries (cross-section shape) of shallow open trenches had no significant influences on reduction effects. The comparison of isolation efficiencies for the three investigated trenches presented in Figure 4 also proved this. Based on the field cone penetration tests (CPT) performed before and after DDC, Tan et al. (2007) observed that the disturbance depth in the sands due to surface waves generated in DDC impact was around 0.5 m to 1.5 m below the ground surface. From the standpoint of densification effects, the depth of the open trench should not be too much deeper below the disturbance layers. Otherwise, the excavated open trenches might reduce the lateral confining pressures against the sands below the disturbance layers and then could not provide sufficient confinements during DDC. As a result, densification effects would be reduced. In this case, the DDC-induced sheet pile wall deflections could be reduced by 45~60% by excavating a 2 m deep and 3 m wide rectangular or trapezoidal open trench.

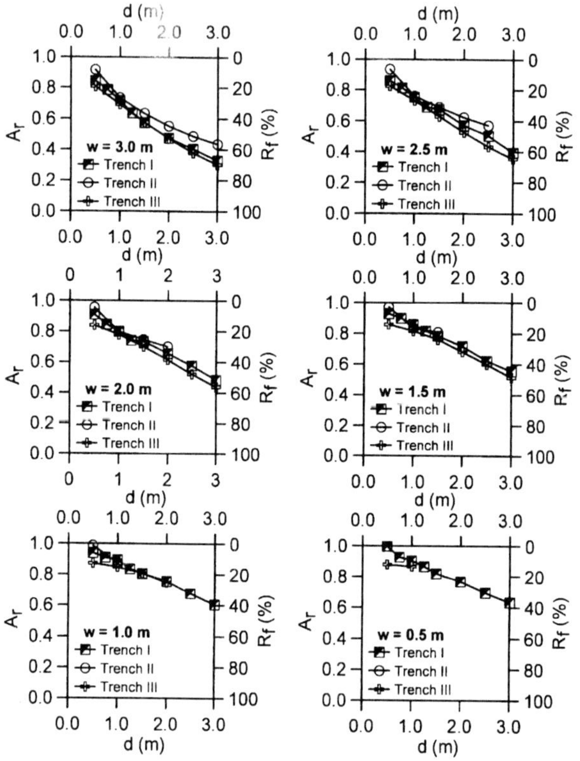

FIG. 4. Summary of reduction efficiencies of the open trenches I, II, and III.

CONCLUSIONS

Based on the findings in the numerical analyses, the following conclusions can be reached:

1. Excavating an open trench in front of the sheet pile walls to be protected is an efficient measure to reduce the DDC-induced wall deflection. The numerical analyses indicated that for an open trench with a depth of 2 m and a width of 3.0 m, around 45-60% of DDC-induced wall deflection may be reduced.
2. For the three investigated open trenches, increasing the depth or the width of the trenches can significantly improve their efficiencies of reducing wall deflections. At the depth $d \leq 3.0$ m, cross-section shapes of the open trenches have no significant effects on reducing wall deflections.
3. Although an open trench deeper than the top disturbance layer might reduce the DDC densification effects, a trench deeper than 3 m can be adopted in practice if there were very strict limitations on wall deflections.

ACKNOWLEDGMENTS

The authors acknowledge gratefully the support provided by the National Natural Science Foundation of China (NSFC Grant. 50908172, 50679056), China National Key Technology R&D Program (2006BAJ27B02-02), SRF for ROCS(SEM), Shuguang Project (No. 05SG25), NCET (06-0378), and Shanghai Leading Academic Discipline Project (B308).

REFERENCES

Ahmad, S., and Al-Hussaini, T.M. (1991). "Simplified design for vibration screening by open and infilled trenches." *Journal of Geotechnical Engineering*, ASCE, 117(1): 67-88.

Brinkgreve, R.B.J., Al-Khoury, R., Bakker, K.J., Bonnier, P.G., Brand, P.J.W., Broere, W., Burd, H.J., Soltys, G., Vermeer, P.A., and Haag, D.D. (2002). Plaxis Finite Element Code for Soil and Rock Analyses. Published and distributed by A. A. Balkema Publisher, The Netherlands.

Fuyuki, M., and Matsumoto, Y. (1980). "Finite difference analysis of Rayleigh wave scattering at a trench." *Bulletin of the Seismological Society of America*, 70(6): 2051-2069.

Haupt, W.A. (1981). "Model tests on screening of surface waves." *In: Proceedings of the 10th International Conference on Soil Mechanics and Foundation Engineering*, Stockholm, Vol. 3, pp.215-222.

Haupt, W.A. (1995). "Wave propagation in the ground and isolation measures." *Proceedings: Third International Conference on Recent Advances in Geotechnical Earthquake Engineering and Soil Dynamics*, Vol. 2, St. Louis, Missouri, pp. 985-1016.

Massarsch, K.R. (2005). "Vibration reduction using gas-filled cushions." *Geotechnical Special Publication: Part of Soil Dynamics Symposium in Honor of Professor Richard D. Woods, GSP 134*, ASCE, pp.1-18.

Segol, G., Lee, P.C.Y., and Abel, J.R. (1978). "Amplitude reduction of surface waves by trenches." *Journal of the Engineering Mechanics Division*, ASCE, 104(3): 621-641.

Tan, Y. (2005). Sheet pile wall design and performance in peat. Doctorial Thesis, University of Massachusetts, Mass.

Tan, Y., Lin, G., Paikowsky, S.G., and Fang, J. (2007). "Evaluation of compaction effects on granular backfill using CPT." *Geotechnical Special Publication: Soil Improvement, GSP 172*, ASCE, pp. 1-10.

Woods, R.D. (1968). "Screening of surface waves in soil." *Journal of the Soil Mechanics and Foundation Division*, ASCE, 94(SM4): 951-979.

Deep excavation induced pile movement in Bangkok subsoil – A numerical investigation

Pornkasem Jongpradist[1],Tanapong Rukdeechuai[2],Sompote Youwai[1], Warat Kongkitkul[3], Attasit Sawatpanich[4] and Jutha Sunitsakul[4]

[1] Assistant Professor, Department of Civil Engineering, King Mongkut's University of Technology Thonburi, Bangkok 10140, Thailand ; pornkasem.jon@kmutt.ac.th
[2] Graduate Student, ditto
[3] Lecturer, ditto
[4] Engineer, Department of Highways, Ministry of Transport, Thailand

ABSTRACT: Simi-empirical equations for predicting lateral pile movement due to adjacent braced excavation in Bangkok subsoil are developed in this paper. The equations are derived from multiple-linear regression analysis of the artificial data, which are the relation between lateral pile deflection and influenced parameters. These data are generated from a large number of finite element (FE) analyzes of pile adjacent to deep excavation using the reasonable soil models and parameters of Bangkok subsoil condition. The soil model in the analyses is specially paid attention to soil stiffness at small strain range. The results of prediction of maximum pile deflection by developed equations are in good agreement with FE analysis data. However, this study deals with only one soil profile, more data with varying soil properties are necessary in further study for more general equations.

INTRODUCTION

In dense urban environments where land is scarce and buildings are closely spaced, cut-and-cover excavations are widely used for basement construction and development of underground transit facilities. One of the main design constraints in these projects is to prevent or minimize damage to adjacent structures. To date, many of the researches have focused on the lateral movements of the retaining wall system and predictions of ground movements. The ground movement during excavation must be lower than allowable limits in order to avoid adjacent structures damaged or destroyed. The magnitude of the settlement and lateral wall movement and their distribution depend on a large number of factors, e.g., soil profile and its geotechnical engineering properties, stiffness of wall and support system, the construction procedure, techniques and workmanship and duration of excavations, boundary conditions of excavation area. There have been many studies on lateral wall movements and surface settlements in excavations with empirical, analytical, and numerical approaches (e.g., O'Rourke,

1981; Peck, 1969; Clough and O'Rourke, 1990; Hashash and Whittle, 1996). In recent years, to mitigate the large amount of lateral wall movements and surface settlements, concrete diaphragm walls as retaining structures, preloading of struts, the top-down construction method, and soil improvement are often implemented. Thus, it is meaningful and desired to estimate the approximate wall movements in deep excavations with the implementations of these mitigating construction methods.

Since many buildings are supported on deep foundations, there is a concern that lateral ground movements resulting from the soil excavation can damage the piles. Although an excavation will cause both vertical and lateral soil movements, the latter component is considered to be more critical as piles are usually designed to sustain significant vertical loads. The lateral loads imposed by soil movements induce bending moments and deflections on the pile, which may lead to structural distress or even failure.

Some researchers have demonstrated that lateral soil movements from excavation activities can be detrimental to nearby existing piles (Finno and Lawrence, 1991; Poulos and Chen, 1997). Similar findings were obtained by Leung et al. (2000) from centrifuge model tests on unstrutted deep excavation in dense sand. Few theoretical methods have been developed (e.g., Poulos and Chen 1997) to evaluate the pile response to these movements. Nevertheless, these methods were derived for typical clay soils and do not consider the realistic soil behavior, such as, one at small strain levels which should be taken into account in analysis of excavation using rigid retaining wall as reported by many previous studies (e.g., Whittle et al., 1993). Thus, the improved method that considers those shortcomings for estimating the pile response is needed. The prediction method of wall deflection by constructing a simplified model from regression analysis of artificial data generated from FEA proposed by Kung et al. (2007) might be an efficient alternative.

In this study, semi-empirical equations to predict the pile movement from nearby deep excavation in Bangkok subsoil using diaphragm wall as support are developed based on reliably artificial data. These data are generated from FEM analyses with a reasonable model of nonlinear, stress-strain behavior of soils at small strain levels verified by measurement data.

NUMERICAL ANALYSES

Evaluation of the effects of excavation adjacent to existing pile foundation is accomplished with the finite element program named PLAXIS. Two dimensional plane strain analysis is employed to model the excavation project as shown in Figure 1. The soil profile used in the analyses is typical Bangkok soil profile consisting of approximately 1.5 m of sedimentary crust underlain by 22.5-m thick clay layer. This total clay layer including 9.5 m thick soft clay, 4 m thick medium clay and 9 m thick first stiff clay, overlies the first sand layer. The soil element composes of triangular element. The sand layer is modeled with Mohr-Coulomb (MC) model with ν', ϕ', c' and E' = 0.3, 36°, 0 and 8×10^4 kPa, respectively. The Hardening Soil model with Small Strain Stiffness (HSsmall) is used with clay layers. The soil parameters used are listed in Table 1 based on the comprehensive site investigation by Prust et al. (2005). The validation of using HSsmall model for clays, MC model for sand layer and the soils parameters in FE analyses of excavation was done by well-documented, high-quality excavation case histories in the previous works (i.e., Wonglert et al., 2008 and Rukdeechuai et al., 2009). Only one half of excavation width has been modeled. The diaphragm wall, basement slab and bore pile are modeled with plate element. The soil element below the final excavation depth, 11.0 to 36.0 m., composes of bore pile which is modeled by elastic model. The undrained condition is assumed in the analysis.

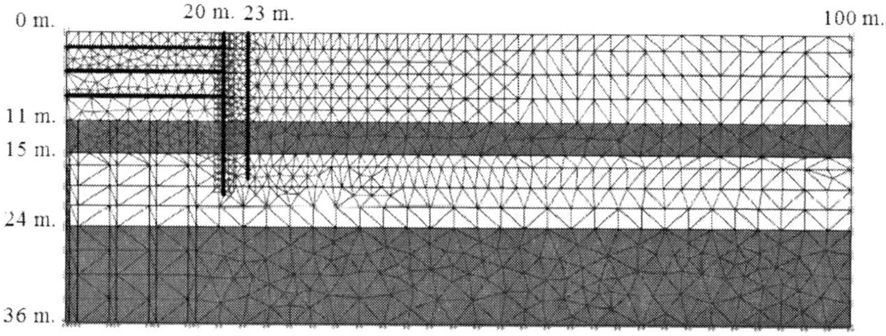

FIG. 1. FE mesh and geometry used in the analysis

The stages of excavation and construction sequence of hypothetical cases are shown in Figure 2a)–2c), the numbers of excavation stages are assumed to be four stages at 2, 5, 8 and 11 m, respectively. The locations of floor slab stages are assumed to be three stages at 1.85, 4.85 and 7.85 m, respectively.

Table 1. HSsmall soil model parameters for Bangkok clays used in this study.

Soil layer	Wea. Crust	Soft Clay	Med. Clay	Stiff Clay	Sand
E_{oed}^{ref} [kPa]	-	5000	20000	60000	-
E_{50}^{ref} [kPa]	-	5000	20000	60000	-
E_{ur}^{ref} [kPa]	-	15000	100000	180000	-
G_0^{ref} [kPa]	-	14200	42000	80000	-
$\gamma_{0.7}$ [-]	-	$1\text{x}10^{-4}$	$1\text{x}10^{-4}$	$1\text{x}10^{-4}$	-
m [-]	-	1	1	1	-
p_{ref} [kPa]	-	100	65	95	-

The analyzed results reveal that the key parameters influencing the response of pile include excavation dimension, excavation support conditions, pile dimension and clearance between pile and wall. The entire analyzed results are reported in Rukdeechuai (2009). Only few of them are presented in this paper. Figure 3 shows the simulated wall movement and pile movement for analysis cases of 40 m excavation width, 20 m diaphragm wall depth, 0.6 m diaphragm wall thickness, 0.25 m pile diameter and 18 m pile length at excavation depth of 11 m for various clearances between wall and pile. It is seen that, for clearance less than 10 m, the shapes of the pile movements are similar to those of diaphragm wall movements having the maximum lateral movement at the depth of about 10 m. The lateral pile movement decreases with increasing clearance between pile and wall. The magnitude of lateral pile and wall movements are almost the same for the clearance less than or equal 1 m. For clearance between pile and wall that is larger than 10 m, the maximum lateral pile movement occurs at the top of pile.

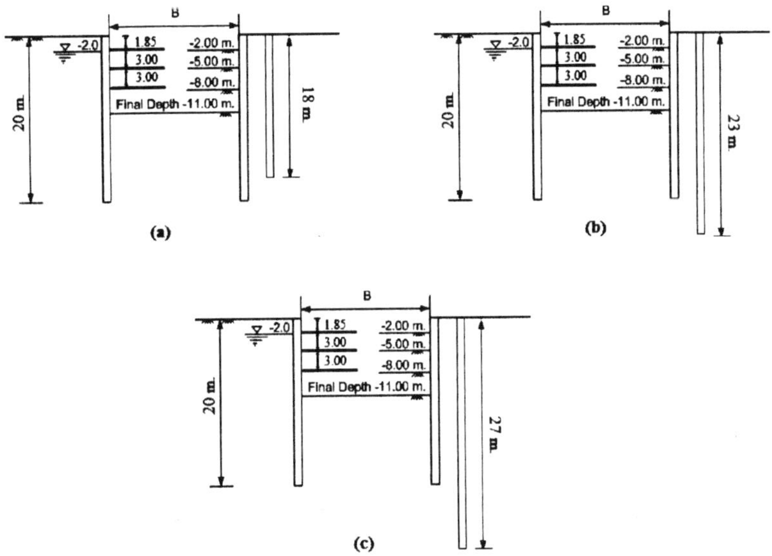

FIG. 2. Reference hypothetical cases for generating artificial data

DEVELOPMENT OF SEMI-EMPIRICAL EQUATION FOR PREDICTION

Table 2. Summary of input variable and range to incorporate in the 2D FEM numerical experiment

Variable	Applicable range
Excavation depth, H_e	5-11 (m)
Excavation width, B	20-60 (m)
Thickness of Diaphragm Wall, t	0.6-1.0 (m)
Depth of Diaphragm Wall	20 (m)
Thickness of Bore pile	0.25-1.0 (m)
Depth of Bore pile	18, 23, 27
Clearance between pile and excavation	0.3-20
Constitutive soil model	Clay: (HSsm

There are six factors including excavation depth, excavation half width, thickness of diaphragm wall, thickness of bore pile, depth of bore pile and clearance between pile and excavation that are found to influence the maximum pile deflection. All these parameters are included in the FEA to generate the artificial data of maximum pile deflection.

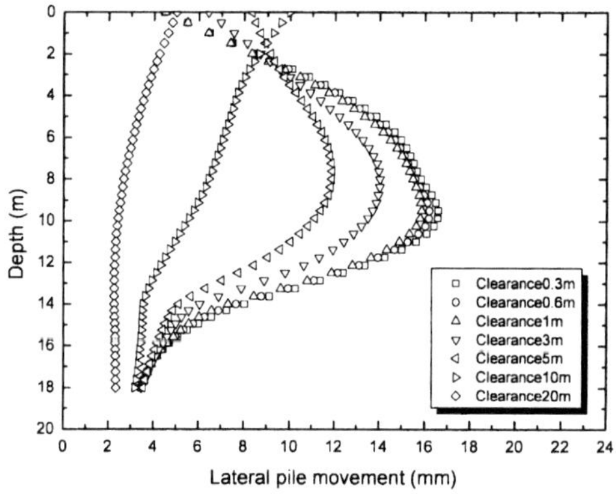

a) pile deflection

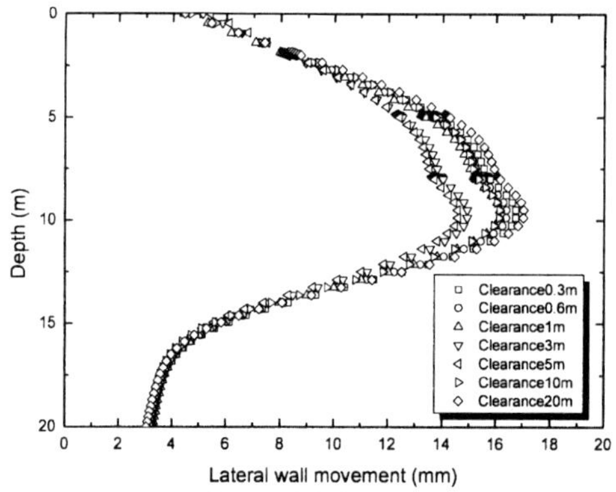

b) wall deflection

FIG. 3. Influence of clearance between pile and wall on (a) lateral pile movement and (b) lateral wall movement due to deep excavation

The ranges of parameters incorporated in the FEM numerical analyses are summarized in Table 2. All analyzed values of lateral pile deflection from FEA of otal 1659 analysis cases are used to establish a semi-empirical equation for estimating e maximum lateral pile movement (δ_{hm}) using multiple linear regression analysis.

Before performing regression analysis, each of the six input variables is sformed to have the linear relationship with the maximum pile deflection. It is observed that all relations between the maximum lateral pile movement and each influencing parameters have a parabolic relation. Therefore, the parabolic function as shown in equation (1) is suitable for transforming the variables to have linear relation with maximum lateral pile movement.

$$X = t(x) = ax^2 + bx + c \quad (1)$$

where x=each of the input variables ($B/2$, H_e, $\ln(EI/h^4_{ave}.\gamma_w)_{wall}$, $\ln(EI)_{pile}$, depth of bore pile, clearance between pile and excavation); and X=transformed variable. The coefficients a, b, and c for each variable are obtained through error minimization and listed in Table 3.

Table 3. Coefficients for Linear Transformation of Six Variables

Variables (x)	Applicable range	Coefficients of Eq. (1) a	b	c
B/2 (m)	10-30	-0.0051	0.3319	6.1779
He (m)	5-11	0.0066	0.3171	7.4529
$\ln(EI/h^4.\gamma_w)$ wall	6.11-7.65	-0.2932	1.7530	12.4683
ln(EI) pile	10.16-14.33	0.2210	-5.1208	39.3827
Depth of Bore pile	18,23,27	0.0267	-1.1070	21.4505
Clearance between pile and excavation	0.3-20	0.0043	-0.5061	13.0157

The developed regression equation is shown in equation (2). Parameters of semi-empirical from the regression are shown in Table 4.

$$\delta_{hm}\,(mm) = b_0 + b_1X_1 + b_2X_2 + b_3X_3 + b_4X_4 + b_5X_5 + b_6X_6 \quad (2)$$

Table 4. Coefficients for Eq. (2) from the regression

Number	X	b
0	-	-51.0520
1	t(B/2)	1.0140
2	$t(H_e)$	0.9120
3	$t[\ln(EI/h^4_{ave}.\gamma_w)]$	0.9120
4	t[(ln(EI)]	1.0430
5	t(Depth of pile)	1.0210
6	t(clearance between pile and excavation)	0.9750

Figure 4 shows maximum pile deflections computed using equation (2) versus respective values obtained from FE analysis. From this figure, equations (1) and (2) are shown to be efficient for estimating the maximum pile deflection giving high coefficient of determination (R^2) and low coefficient of variation (COV). However, for the range of maximum pile deflection less than 4 mm, the equation under-predicts the values from FEA.

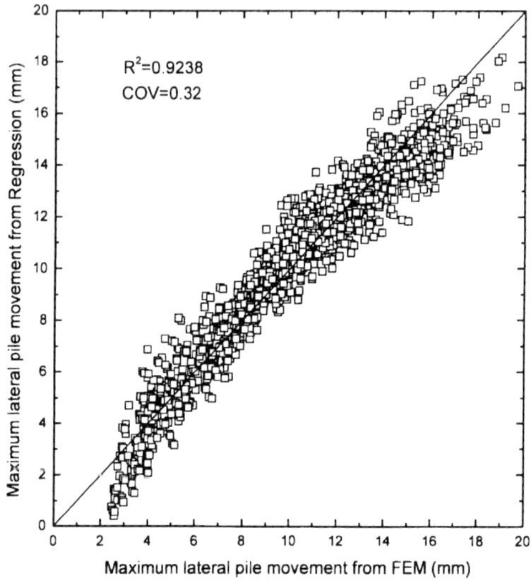

FIG. 4. Performance of prediction by Equations (1) and (2) compared to the FEA data

CONCLUSIONS

The analysis of deep excavation using diaphragm wall adjacent to existing pile in Bangkok subsoil is carried out by means of FEM. The soil model in the analyses is specially paid attention to soil stiffness at small strain range. The key parameters influencing the pile response caused by brace excavation are used in FEA to generate artificial data of the maximum pile deflection. The semi-empirical equation for predicting lateral pile movement in 2D plane strain can be obtained from regression analysis with high coefficient of determination (R^2) of 0.92. This suggests that the method used in this study is effective to develop an empirical equation to estimate the maximum pile deflection from nearby excavation. However, this study deals with only one soil profile, more data with varying soil properties are necessary in further study for more general equations.

ACKNOWLEDGMENTS

The authors appreciate the financial support of the Department of Highway, Ministry of Transport, Thailand and permission of using Plaxis programs from PLAXIS ASIA.

REFERENCES

Clough, G.W. and O'Rourke, T.D. (1990) "Construction-induced movements of in situ Walls", *Proc, Design and Performance of Earth Retaining Structures*, Geotechnical special publication (25) : 439-470.

Finno, R.J. and Lawrence, S.A. (1991). "Analysis of performance of pile groups adjacent to deep excavation", *J. of the Soil Mech. and Found. Div.*, Vol.117 (6) : 934-955.

Hashash,Y.M.A. and Whittle, A.J. (1996) "Ground movement prediction for deep excavation in soft clay." *J. of the Soil Mech. and Found. Div.*, Vol. 122 (6): 474-486.

Kung, G.T.C., Juang, C.H., Hsiao, E.C.L. and Hashash, Y.M.A. (2007). "Simplified model for wall deflection and ground surface settlement caused by braced excavation in clays", *J. of Geotechnical and Geoenv. Engrg.*, Vol.133(6): 1-17.

Leung, C.F., Chow, Y.K. and Shen, R.F. (2000). "Behavior of pile subject to excavation-induced soil movement", *J. of Geotechnical and Geoenv. Engrg.*, Vol. 126 () : 947-954.

O'Rourke, T.D. (1981). "Ground movements caused by braced excavations" *J. of Geotechnical Engrg. Div.*, Vol.107 (6): 1159-1177.

Peck, R.B. (1969). "Deep excavation and tunneling in soft ground" *Proc. of the 7" Int. Conf. on Soil Mech. and Found. Engrg.*, Mexico, State of the Art Volume: 225-290.

Poulos, H.G. and Chen, L.T. (1997). "Pile response due to excavation induced lateral soil movement" *J. of Geotechnical and Geoenv. Engrg.*, Vol. 123 (2): 94-99.

Prust, R.E., Davies, J. and Hu, S. (2005) "Pressuremeter investigate for mass rapid transit in Bangkok, Thailand", *J. of the Transportation Research Record*, No. 1928, : 207 -217.

Rukdeechuai, T. (2009). "Simplified model for predicting lateral pile movement caused by braced excavation in Bangkok subsoil" *Master Thesis, King Mongkut's University of Technology Thonburi (KMUTT), Thailand.*

Rukdeechuai, T., Jongpradist, P., Wonglert, A. and Kaewsri, T. (2009). "Influence of soil models on numerical simulation of geotechnical works in Bangkok subsoil" *EIT Research and Development J.*, Vol. 20 (3) : 17-28.

Whittle, A.J., Y.M.A. Hashash, and R.V. Whitman. (1993) "Analysis of deep excavation in Boston", *J. of Geotechnical Engrg*, Vol.119(1) : 69-90.

Wonglert, A., Jongpradist, P. and Kalasin, T. (2008) "Wall Movement Analysis of Deep Excavations in Bangkok Subsoil considering Small Strain Stiffness" *J. of Res. in Engnr.and Tech.* Vol.5 (4): 393-406.

A Case Study of Retaining Wall with Soil-cement Mixing Reinforcement for Korean Urban Site

YoungSeok KIM[1] and YongSang CHO[2]

[1]Senior Researcher, Korea Institute of Construction Technology, 1190, Simindae-Ro, Ilsanseo-Gu, Goyang-Si, Gyonggi-Do, 411-712, Republic of Korea; kimys@kict.re.kr
[2]Manager, Samsung Construction & Technology, 1321-20, Seocho-2Dong, Seocho-Gu, Seoul, Republic of Korea; yscho.cho@samsung.com

ABSTRACT: This study introduces a novel retaining wall system for sites where the wall is exceptionally close to existing structures. The retaining wall system employs the soil-cement mixing technique, which is commonly used in the deep-cement mixing method to install large diameter reinforcing blocks. The proposed earth retaining wall was applied in an excavation at a Korean urban site to evaluate its practical application. The evaluation results showed that the proposed earth retaining system using the soil-cement mixing method can be applied at urban excavation sites.

INTRODUCTION

In an urban area, constructions near the existing structures have been continuously increased recently to utilize vacant properties and remodel old. In addition, the sizes of urban excavation works have been also increased to meet the demand for huge underground spaces. Consequently, constructions should be conducted where exsiting structures are exceptionally close. In these cases, existing earth retaining systems such as the conventional earth anchor and soil nailing cannot be applied because of the large reinforcement length requirements.

This study introduces a novel earth retaining wall system for sites where the retaining wall is exceptionally close to existing structures. The retaining wall system employs the soil-cement mixing technique, which is commonly used in deep-cement mixing method to install large diameter reinforcing blocks. By installing the large diameter reinforcing blocks (usually Φ300-500 mm), the need for reinforcing depth and the required amount of reinforcement can be reduced.

The proposed earth retaining wall was applied to an excavation at a Korean urban site to evaluate its practical application. The behavior of the retaining wall with soil-cement reinforcing blocks installed into a 10 m depth excavation wall with 20° inclination was evaluated by field measurements. The evaluation result showed that, the proposed earth retaining system using the soil-cement mixing method can be applied at urban excavation sites.

RETAINING WALL WITH SOIL-CEMENT MIXING REINFORCEMENT

General Characteristics

The novel retaining wall system with soil-cement mixing reinforcement applies the slurry mixing technique of deep cement mixing method to the bracing of conventional soil nailing method. For the reinforcement of the wall, this method installs 300-500 mm large diameter soil-cement mixtures in the ground at a certain angle. Because the drilling diameter of conventional soil nailing is small as 100 mm, the proposed method mobilizes more friction, as illustrated in Fig. 1. By installing large diameter soil-cement mixtures, the proposed system maximizes the friction resistance between the ground and soil cement, and therefore efficiently constrains the excavation-caused strain on the wall. Furthermore, the large soil-cement mixtures significantly reduce the depth and the number of reinforcement compared to the conventional earth anchor or nailing method. Rebar or anticorrosive FRP rod is located in the center of the soil cement as tensioning material. Cement milk of high compressive strength is injected around the rod, a 3-layer structure comprising of rod, cement milk and soil cement, which together ensure the sufficient adhesive strength of the rod (Tateyama and Taniguchi 1993; Mimura et al. 1998).

Construction procedure

A retaining wall system with soil-cement mixing reinforcement is designed to drive a pile at a certain angle upon excavation (Japan Advanced Construction Technology Center 1997). The application method of the proposed retaining wall system is given as follows, and the construction procedure of soil-cement reinforcement is shown in Fig. 2.

□ H-pile, S.C.W or steel pile shall be driven in front of the excavation site (it may not be necessary depending on the excavation depth or ground condition).

□ Tensioning material (rebar) shall be injected to the center of the mixer.

□ Cement shall be injected by a mixer while boring to form a 300-500 mm large diameter soil cement.

□ When a large soil-cement reaches the target depth, the mixer shall be pulled out by reverse rotating while the mixing speed and cement injection are controlled so as to form a cement milk.

□ The toe (tensioning material) of the soil-cement shall be fixed with a bolt and nut using wale or plate.

□ The work sequence □-□ shall be repeated while the excavation for the next stage continues.

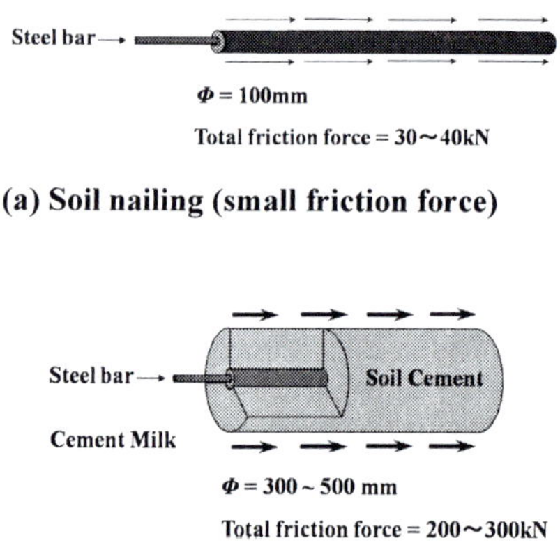

(a) Soil nailing (small friction force)

(b) Soil cement mixing reinforcement (large friction force)

Fig. 1. Comparison between soil nailing and soil-cement mixing reinforcement

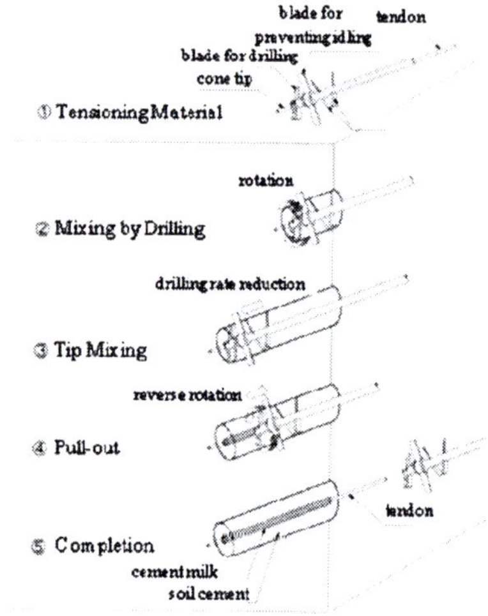

Fig. 2. Construction procedure

EVALUATION OF FIELD APPLICABILITY TO KOREAN SITE

Trial Site Condition

The retaining wall with soil-cement mixing reinforcement was applied at a field trial site in an urban area in order to evaluate its applicability. The trial site was a redevelopment construction site located at northern Seoul. The excavation depth was 10 m and ground water level was 6 m deep from the surface. The ground consisted of fill, deposited soil (silty sand), weathered soil and weathered rock from top to bottom. According to site investigation, SPT N values ranges from 4 to 10 in the deposited soil layer and from 20 to 50 in the residual soil layer (N values from the proved section were 4 to 14). The retaining wall of the trial site was originally designed with Soil Cement Wall (S.C.W) with earth anchors as shown in Fig. 3.

In order to evaluate the applicability, some of the earth anchors were changed to the large diameter soil-cement mixing reinforcement (400 mm diameter). Five benches of the large diameter soil cement mixture were applied to the wall of 10 m • 9.84 m (excavation depth) as shown in Fig. 4. Fig. 5 shows the comparison between two design sections of the conventional earth anchor method and the modified section using the soil-cement mixing reinforcement method. As shown in Fig 5, the required length of the large diameter soil-cement mixing reinforcement was 3.8 m, whereas the required length of conventional earth anchor system was about 10 m from neighboring structure. Therefore, the retaining wall system with soil-cement mixing reinforcement can be a suitable method for urban sites when the construction site is exceptionally close to existing structures (Kim et al. 2009).

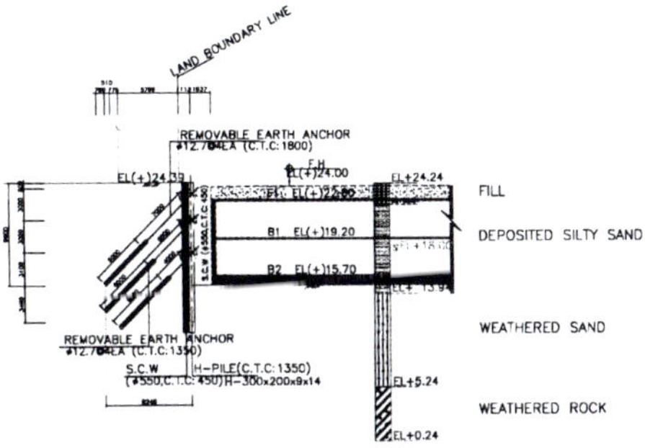

Fig. 3. Representative section and layers

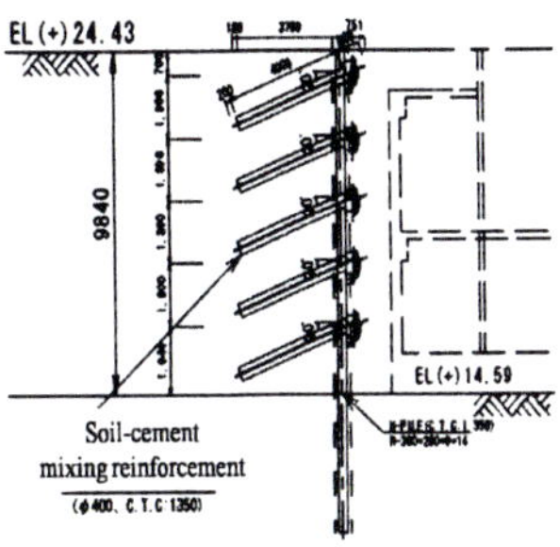

Fig. 4. Final section

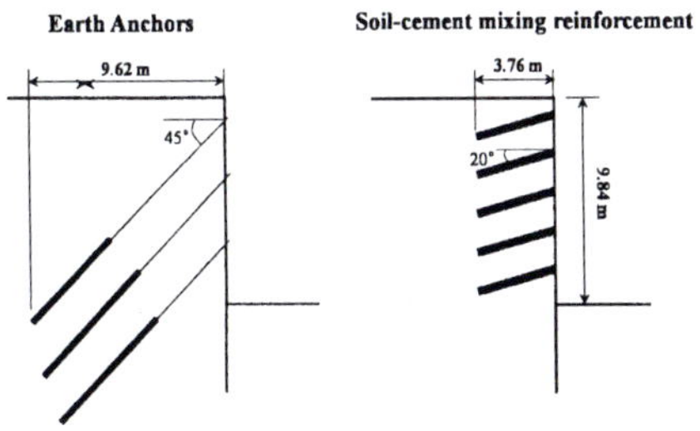

Fig. 5. Comparison with existing earth anchor section

Field Measurements

In order to evaluate the stability and behavior of the retaining wall with soil-cement mixing reinforcement during excavation, field measurements were carried out using 1 inclinometer, 3 load cells, and 1 piezometer. Fig. 6 shows the horizontal displacements of the retaining wall reinforced with the large diameter soil-cement mixture. As excavation stages progressed, the wall movement on the front side of the reinforcement increased. The displacement of the retaining wall monitored up to 9.0 m of excavation (up to 4th bench) was 15 mm, which is about H/560 (H: excavation depth). This displacement is less than the general requirement of horizontal displacement (H/500 ~ H/300). However, the maximum displacement at 9.8 m depth was increased up to 45 mm. The excessively increased displacement was attributed to the excessive excavation that was carried out to lower the water level at the 5th bench. Soil was washed out with water through the weep hole after an extra wall was built to accommodate the weep hole or the well point method used to lower water level. As a result, the ground at the 3rd and 4th bench was loosened and the cohesion of the soil-cement mixture and friction resistance were reduced, and the horizontal deformation of the wall was increased. After the final excavation, additional deformation of 5 mm occurred, due to the grouting, which was carried out to isolate the water from the wall.

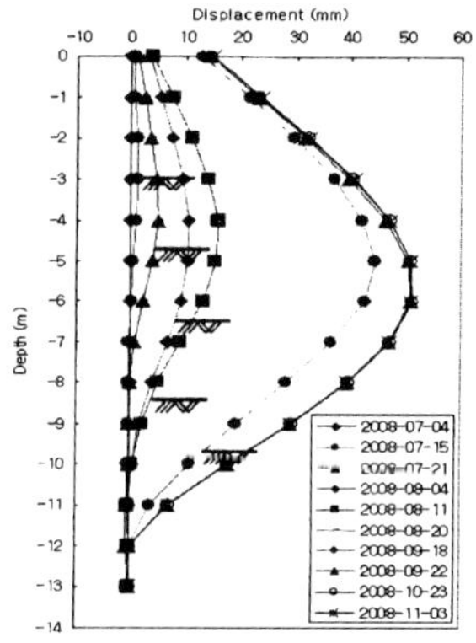

Fig. 6. Horizontal displacements of wall by inclinometer

The excavation depth, water level, and maximum horizontal displacements of the wall are shown together in Fig. 7. The horizontal displacement of the wall increased up to 50 mm as the excavation progressed and the maximum horizontal displacement occurred during excavation at the 5th bench (Aug 10 ~ 20, due to groundwater). As described, when applying the soil-cement mixture below the groundwater level, some soil was washed out by the groundwater, so additional reinforcement was required (after Aug 11, the water level was lowered by the weep hole or the well point method). The field measurement results indicated that the soil-cement mixing retaining wall can be applied to bracing work in urban areas, since the displacement by excavation was about 25 mm. However, excessive deformation was observed when the soil was washed out by the groundwater through the holes. To improve the applicability, therefore, the method should be modified so that it can be applied below the water level as well.

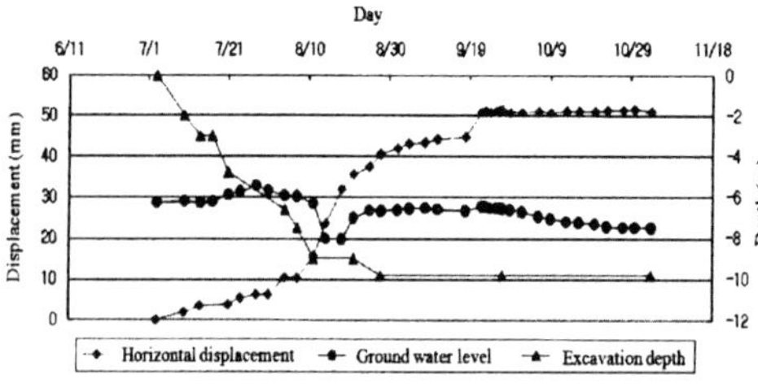

Fig. 7. Measured behavior of retaining wall during excavation

CONCLUSIONS

A novel retaining wall system with soil-cement mixing reinforcement was developed. In this paper, the applicability of the soil-cement mixing reinforcement to a retaining wall was evaluated from the field measurements at a 10 m deep excavation site in a Korean urban area. In order to evaluate the applicability, some of the earth anchors were changed to the soil-cement mixtures of 400 mm diameter, 4 m penetration depth, and 20° of angle for a part of the retaining wall. The conclusions from the case study of retaining wall with soil-cement mixing reinforcements are as follows.

1) The novel soil-cement mixing retaining wall system had an advantage of reducing the length and amount of reinforcements. For the trial excavation site, the required length of soil-cement mixing reinforcement was 3.76 m whereas the original required length of earth anchor was 9.62 m.
2) Field measurement results indicated that the method can be applied as reinforcement to a retaining wall unless groundwater exists behind the wall. The displacement of the retaining wall monitored up to 9.0 m of excavation was 15 mm (H/560), which is less than the general requirement of horizontal displacement (H/500 ~ H/300).
3) When the groundwater exists behind the wall, however, some soil was washed out with water through the weep hole after an extra wall was built to accommodate the weep hole or the well point method used to lower water level. As a result, the maximum displacement at 9.8 m excavation was increased up to 45 mm. Therefore, further investigations are necessary to apply the method below the groundwater level.

ACKNOWLEDGMENTS

This research was supported by Samsung Corporation and grant (07-UR-B04) from High-tech Urban Development Program funded by Ministry of Land, Transport and Maritime Affairs.

REFERENCES

Japan Advanced Construction Technology Center (1997), "Development of a Radish Anchor method", *Report of ACTC*.

Kim, Y.S., Cho, Y.S., Kang, I.C., and Kim, I.S. (2009), "Application of earth retaining structure using soil cement-mixing method", *Proc. Korean Geotechical Society Conference 2009, KGS*: 883-887.

Miura, N., Shen, S.L., Koga, K., and Nakamura, R. (1998), "Strength change of the clay in the vicinity of soil cement column", *J. Geotechnical Engrg., JSCE*, Vol. 596 (43): 209-221.

Tateyama, M and Taniguchi, Y. (1993), "Development of a large diameter reinforced anchor by cement-mixing method", *Report of Japan Railway Technical Research Institute*, Vol. 7 (4): 1-8.

A Preliminary Study on the Behavior of Axially Loaded Single Pile Subjected to Lateral Soil Movement behind a Retaining Wall

Yu Feng [1], Chu Feng [2, 3], Liang Fa-Yun [4*]

[1]Graduate Student, Department of Geotechnical Engineering, Tongji University, Shanghai 200092; colincalling@163.com
[2]Ph.D Student, Department of Geotechnical Engineering, Tongji University, Shanghai 200092; chufeng@ljz.com.cn
[3]Project Manager, Shanghai Lujiazui Finance & Trade Zone Development Co., Ltd., Shanghai, 200127; chufeng@ljz.com.cn
[4*]Corresponding Author, Associate Professor, Department of Geotechnical Engineering, Tongji University, Shanghai 200092; fyliang@tongji.edu.cn

ABSTRACT: Based on the flexural differential equations of elastic piles and Winkler's spring model, the equation of axially loaded piles subjected to lateral soil movement is solved first. In the solution, the flexural differential equations are established using two-stage method and solved by the finite difference method, which makes the equation solving more easily when dealing with the pile foundation in layered soils. Numerical method using commercial software "LPile" was conducted to verify the effectiveness of the presented solution. The results of parametric study show that the situation of retaining wall and the distribution pattern of soil elastic modulus along the pile depth can affect the behavior of axially loaded piles subjected to lateral soil movement significantly. The analyzing conclusion in this paper should be useful for the design in practice.

INTRODUCTION

With rapid development of cities, transportation systems, and wharfs, existing axially loaded piles are unavoidably subjected to lateral soil movement in practice. Lateral movement and loads mostly have a negative influence on the behavior of axially loaded piles. Leussink and Wenz (1969) reported a structural failure of existing piles due to clay deformations caused by surcharge loading. Another failure of existing piles caused by surcharge loading in an industrial factory was also reported by Yang et. al. (2002). Situations might be even worse in big cities where more projects are implementing. A collapse accident of a 13-storey building in Shanghai was reported recently which was caused by the lateral soil movement. API (2000) pointed out that during the offshore foundation design, the effect of soil movement, which is often ignored in common engineering design, should be taken into account. Subsequently, it has become a hot topic in geotechnical engineering to consider and analyze the effect of lateral soil movement coupled with existing axial loading. To investigate the effects of relative soil-pile displacement on pile response, large-scale load tests were performed on relatively slender piles elements by White et.al. (2008). Poulos (2007) investigated the nonlinear behavior of axially loaded piles undergoing ground movement. Guo and Ghee (2006) carried out model tests on axially loaded pile groups subjected to lateral soil movement. The Finite Element Method (FEM) also has been widely used to simulate this kind of problem using a certain constitutive model. However, FEM requires the discretization of soils and piles in the analysis, which are limited in the real three-dimensional size of the piled foundation.

Poulos and Chen (1997) presented solutions to piles subjected to soil movement by two-stage method. Fan (1986) investigated piles with vertical and horizontal loads concurrently applied on by solving flexural differential equations. This paper combined these two approaches mentioned above together to obtain solutions of axially loaded piles in laterally moving soil.

ANALYSIS METHOD

Two-stage method is introduced to analyze the axially loaded piles subjected lateral soil movement. The first stage is to obtain the soil movement when there are no piles installed, which is called the free soil movement hereinafter. Then in the second stage, the soil movement is treated as a known and taken into the horizontal flexural differential equation to calculate the final response of piles.

To analyze this problem, the following three assumptions should be made according to Winkler's foundation model and two-stage method:

(a) the pile is considered as an elastic foundation beam;

(b) the surrounding soil is treated as continuum media; and

(c) deformations of piles and soil mass are considered to be compatible when soil movements and axial loads are applied.

Free movement of soil mass is restricted by existing piles. Combining the free soil movement $h_s(z)$ and the final lateral displacement of the existing pile $U(z)$, the relative displacement of soil and the pile, or the restricted displacement can be obtained as $\Delta = h_s(z) - U(z)$, so that the lateral pressure imposed on the pile according to Winkler's foundation model is:

$$F = k_z \Delta \tag{1}$$

where k_z—modulus of reactive force of foundation at a certain depth (kN/m^2), in this paper, $k_z = \frac{0.65E_s}{1-\nu^2}\sqrt[12]{\frac{d^4 E_s}{E_P I_P}}$ as recommended by Vesic (1961), in which E_s is the elastic

Modulus of soil (kPa);

E_P—elastic modulus of pile (kN/m^2).

$E_P I_P$—bending stiffness of pile (kN·m^2) ;

ν—Poisson's ratio of soil ;

d—diameter of pile (m);

F—pressure imposed on unit length of pile at a certain depth (kN/m).

Horizontal flexural differential equation of pile can be achieved if the pile is regarded as an elastic foundation beam:

$$E_P I_P \frac{\mathrm{d}^4 U(z)}{\mathrm{d}z^4} + N_z \frac{\mathrm{d}^2 U(z)}{\mathrm{d}z^2} = F \tag{2}$$

where N_z—axial force existing at a certain depth of z (kN).

Analytical solution can be easily worked out by combining equation (1) and (2) to evaluate piles in uniform foundations. Considering the characteristics of the layered foundation, finite difference method (FDM) can be applied to solve the equations. Internal forces and lateral displacement of pile can be calculated at a given relation between them.

COMPARISON OF NUMERICAL ANALYSIS

Numerical analysis using a software named "LPile" is conducted to verify the effectiveness of the proposed solution. LPile software, developed by Reese et.al. (2006) has been one of most widely used horizontal pile analyzing programs in geotechnical engineering.

A free-headed, axially loaded pile with a constant cross-section and length-diameter ratio L/d =25, installed in the uniform foundation is subjected to the lateral soil movement as shown in Fig. 1. The ratio of elastic modulus and cohesive force (i.e., E_s / c_u) of soil around the pile is equal to 143, while the Poission's ratio of the soil is 0.5. The ratio of the compressive strength of the pile material (i.e., f_c) to the elastic modulus of pile is 1/1000 in this case. The elastic modulus of the pile assumed to be constant and equal to 5,000 times of that of the soil.

Fig. 2 is the comparison of bending moment of the pile along the depth calculated by the proposed method with the numerical method using LPile. It can be concluded that the proposed method is effective to analyze the existing pile subjected to the lateral soil movement.

In the following figures, the axial load, displacement and bending moments of pile have been written in dimensionless forms as below:

$$K_N = \frac{N}{f_c A} \times 100\% , \quad K_U = \frac{U(z)}{d} , \quad K_M = \frac{ML}{EI} \tag{3}$$

where N —the axial load applied on the pile head(kN);

A —area of the cross section (m^2);

M —bending moment at a certain depth(kN·m);

L —length of the pile(m).

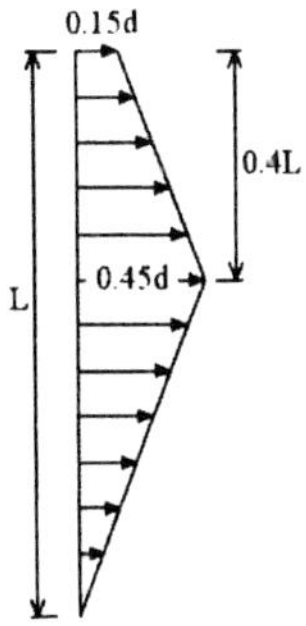

Fig. 1 Displacement mode of lateral soil movement

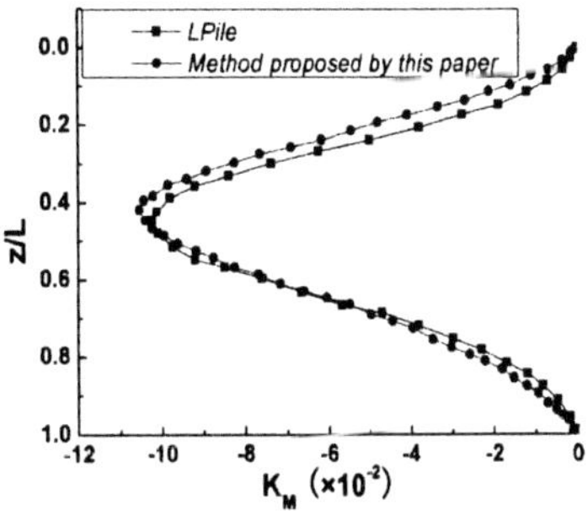

Fig. 2 Comparison of dimensionless bending moments of pile

PARAMETRIC ANALYSIS

Effect of Retaining Walls

A retaining wall is quite essential to prevent built structures from over-sized deflections and internal forces by reducing the displacement of surrounding soil (Fig.4). On the contrary, it poses a severe threat to pile foundations when retaining structures break down or are not properly constructed. The existence of the retaining wall can cause several or even tens of times increase of lateral soil movement when the retaining wall collapses (Leung, et. al., 2000).

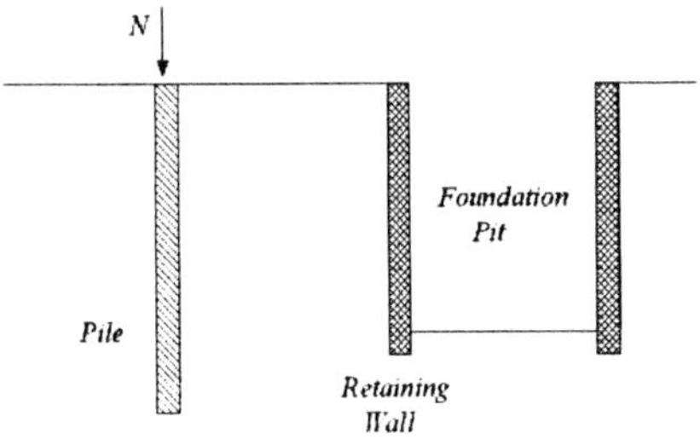

Fig.4 Sketch of the pile and the retaining wall

This conclusion is verified in the studies of Leung et. al. (2000), which shows that when the retaining wall collapses, there can be a great development of lateral soil displacement around the ground surface, while much smaller increase in deeper depths in the soil. One possible lateral soil displacement mode is illustrated in Fig. 5, where soil displacements are referred to as K_h and written in dimensionless form by being divided by the diameter of pile.

A free-headed, axially loaded pile with a constant cross-section and length-diameter ratio L/d=60, installed in the soil mass with Poisson's ratio of 0.5, is subjected to the lateral soil movement as shown in Fig. 5. And the compressive strength (f_c)of the pile material is 1/1000 of the elastic modulus of pile in this case. The ratio of elastic modulus of soil and that of the pile varies linearly with the depth, following the rule in equation (4).

$$\frac{E_s}{E_p} = 8\times10^{-4}\frac{z}{L} \tag{4}$$

where z —depth under the surface of the ground (m).

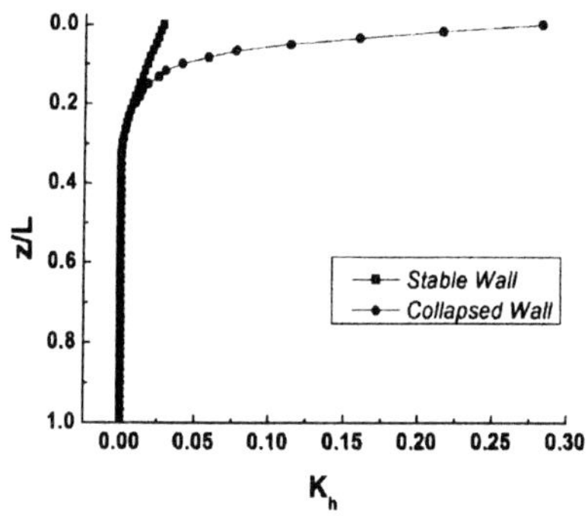

Fig. 5 Comparison of lateral soil movement

Comparisons of displacements and bending moments are shown in Fig. 6 and Fig. 7 under the different situations of retaining walls and applied axial loads. When the pile without an axial load is installed behind a stable wall, the maximum dimensionless displacement and bending moment caused by soil movement are 0.0242 and 0.0062, respectively. And the two results K_U and K_M increase significantly by 228% and 315%, to 0.079 and 0.0257 respectively after the retaining wall collapses.

On the other hand, larger additional horizontal displacements will be induced by axial loads resulting from P-Δ effect. As it can be seen in Fig. 6, when the pile is behind a stable wall and subjected to dimensionless axial load K_N =30%, the maximum dimensionless horizontal displacement of the pile head amounts to 0.0294, which is augmented by 21.5% compared to the case without axial loading. And this increase will be even aggravated up to 25.5% as the retaining wall breaks down.

As shown in Fig. 7, with the axial load K_N =30% applied on the pile, maximum dimensionless bending moments are raised from 0.0062 to 0.0074 (increasing by 19.4%) behind a stable wall, and from 0.0257 to 0.0364, (increasing by 41.6%) behind a collapsed wall, respectively.

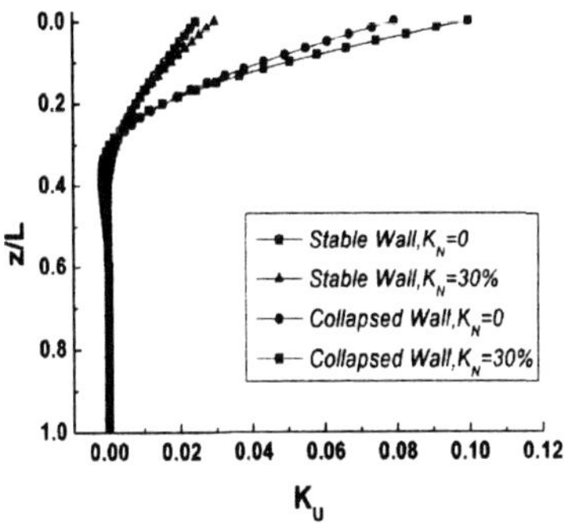

Fig. 6 Comparison of dimensionless lateral displacements of pile

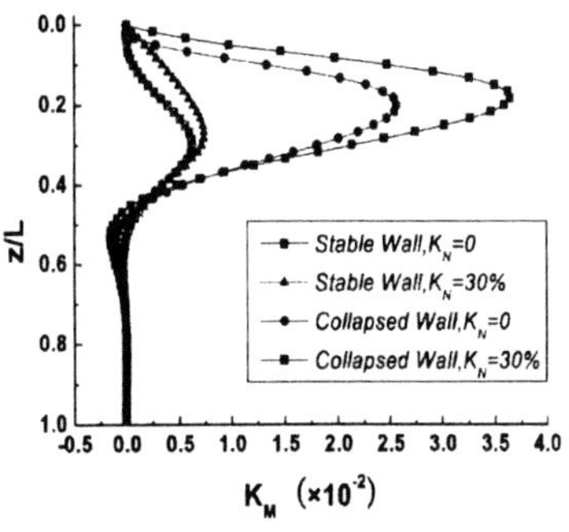

Fig. 7 Comparison of dimensionless bending moments of pile

Effect of Layered Soils

In the following example, the types of foundation are layered and uniform. The modulus ratio of layered soil and the pile can be determined using equation (4), while in uniform foundation this ratio equals 4×10^{-4}. Fig. 8 and Fig. 9 show the effect of layered soils on the displacements and bending moments of piles.

As indicated in Fig. 8 and Fig. 9, for piles without axial loading, maximum displacements and maximum bending moments of piles in uniform soils are greater than those in layered soils which provide weaker resistance at the ground surface. However, when dimensionless axial loading K_N =30% and lateral soil movement both applied to the pile in a layered foundation, the maximum displacement and bending moment of pile increase by 21.5% and 19.4%, respectively, which is much more than the corresponding increments 5.8% and 2.6% of the pile in uniform soils.

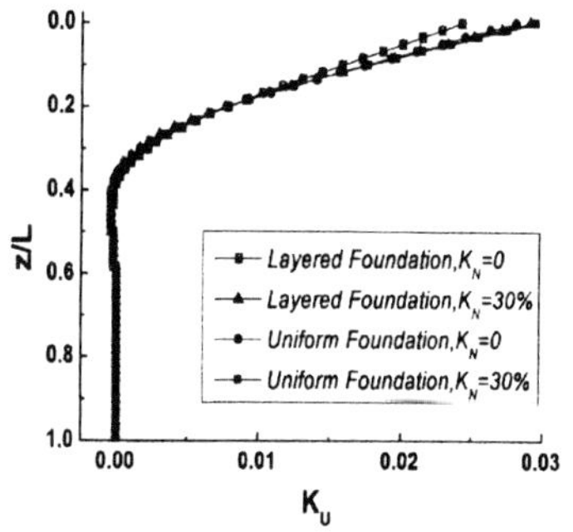

Fig. 8 Comparison of dimensionless lateral displacements of pile

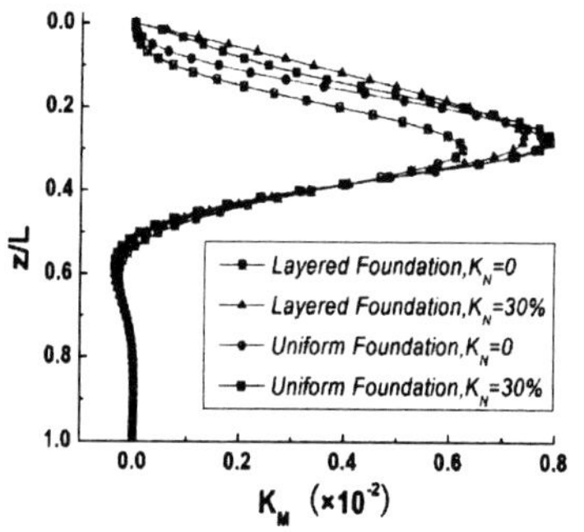

Fig. 9 Comparison of dimensionless bending moments of pile

CONCLUSIONS

The behavior of piles subjected to lateral soil movement coupled with axial loading is investigated in this paper. The equation of axially loaded piles subjected to lateral soil movement is solved first. Then, numerical method is used to verify the effectiveness of the proposed solution. Based on the calculation results, the following conclusions can be obtained:

(1) The existence of the vertical loading can influence the horizontal displacement and bending moment of the piles which is subjected to the lateral soil movement. The influence significance of the existing vertical loading depends on the situation of retaining walls and geological conditions of the foundation.

(2) In uniform soil foundation, single piles behind a retaining wall would suffer much bigger soil displacement when the retaining wall collapses compared with the case of a stable wall, especially at the place near the ground surface.

(3) When the failure of the retaining wall happens, some portion of the additional displacements and bending moments is caused by the existence of the vertical loading at the pile.

(4) The behavior (i.e., displacements and bending moments) changing of piles in layered soil foundation under lateral soil movement is more significant than it described

above in the uniform soil foundation.

ACKNOWLEDGMENTS

The work reported herein was supported by the National Natural Science Foundation of China (Grant No. 50708078), Innovation Program of Shanghai Municipal Education Commission (10ZZ24), and Key Technology Research and Development Program supported by Science and Technology Commission of Shanghai (Grant No.09231200900). The authors wish to express their gratitude for the above financial supports.

REFERENCES

American Petroleum Institute (2000). *Recommended practice for planning, designing and constructing fixed offshore platforms-working stress design.* API RP 2A-WSD.

Fan, W.T. (1986). "Analysis of flexible pile under axial and horizontal loads", *Journal of Southwest Jiaotong University*, Vol. 21 (1): 39-44 (In Chinese).

Guo, W.D., and Ghee, E.H. (2006). "Behavior of axially loaded pile groups subjected to lateral soil movement", *Proceedings of Sessions of GeoShanghai* (GSP 153), ASCE, Shanghai: 174-181.

Leung, C.F., Chow, Y.K., and Shen, R,F. (2000). "Behavior of Pile Subject to Excavation-induced Soil Movement", *Journal of Geotechnical and Geoenvironmental Engineering*, Vol. 126 (11): 947-954.

Leussink, H. and Wenz, K. P. (1969). "Storage Yard Foundations on Soft Cohesive Soil", the *7th International Conference on Soil Mechanics and Foundation Engineering*, Mexico City, Vol. 2: 149-155.

Poulos, H.G. (2007). "Ground Movement-A hidden Source of Loading on deep foundations", *DIF Journal*, Vol. 1 (1): 39-42.

Poulos, H.G., and Chen L.T. (1997). "Pile response due to excavation-induced lateral soil movement", *Journal of Geotechnical and Geoenvironmental Engineering*, Vol. 123 (2): 94-99.

Reese, L., Wang, S.T., and Isenhowe, W.M. (2006). "LPILE+5.0 Manual - A program for the analysis of piles and drilled shafts under lateral loads". ENSOFT, Inc., Houston, USA, 2006.

Vesic, A.S. (1961). "Bending of beams resting on isotropic elastic solids", *Journal of Soil Mechanics and Foundation Engineering*, Vol. 87 (2): 35-53.

White, D.J., Thompson, M.J., Suleiman, M.T., and Schaefer, V.R. (2008). "Behavior of slender piles subject to free-field lateral soil movement". *Journal of Geotechnical and Geoenvironmental Engineering*, Vol. 134 (4): 428-436.

Yang, M., Zhu B.T., Chen, F.Q. (2002). "Pilot study on collapse of an industrial building due to adjacent surcharge loads". *Chinese Journal of Geotechnical Engineering*, Vol. 24 (4): 446-450.

A Combined Retaining Structure and Its Application in Deep Excavation

Yuwen Yang

Senior Geotechnical Engineer, Wuhan Geotechnical Engineering and Surveying Institute, Wan Song Yuan Road 209#, Hankou, Wuhan, 430022, China; wayneywyang@hotmail.com

ABSTRACT: In a combined structure, soil-nailing wall retains the upper and shallow soil while propped piles at the toe of the wall augments the overall stability of ground in deep excavation. Although such a structure developed from soil nailing wall plus internally-braced pile has widely been employed, it is regarded as not an integral structure, but a separate one to analyze and calculate. This treatment is unreasonable. By the numerical experiments, the combined structure is explored about its space earth-retaining mechanism as the integrated system. Analytical results indicate that the actual earth pressure exerted to the structure is different with that determined by Rankine's theory. The distribution of Rankine's pressure is adjusted to accommodate the calculation results. A case study proves the validity of the adjustment.

INTRODUCTION

The combined structure in this paper refers to the earth-retaining wall developed from soil nailing wall plus internally-strutted piles, where soil nailing wall is used to retain the upper soils while the internally propped piles provide overall stability to ground in deep excavation. Such a structure is usually applied to the geological conditions: (1) top soil is the high strength fill, underlying which, there is soft soil and (2) underground facilities or nearby buildings are in close to the excavation and necessary to protect.

For stability and deformation analysis of soil nailing wall, many scholars have developed a variety of methods (Bridle, 1989; Yang et al, 1998; Yuan et al 2003), among which, the limit equilibrium method is popular. Pile wall is a traditional earth-retaining type, and the elastic spring resistant approach is widely used to predict pile deformation and earth pressure. However, to the combined structure which will be discussed, its retaining mechanism is different with either pile wall or soil nailing wall. Due to the upper soil restrained by soil nailing wall, the pile can be designed to be shorter. Correspondingly, pile wall adjacent to the toe of soil nailing wall improves the overall stability of ground. It is difficult to consider the influence of stability of soil nailing wall on the lower pile wall or effect of pile resistance at the wall toe to soil nailing wall above. So the mechanism of the combined structure has still remained unknown. In this paper, use FLAC for numerical simulation to explore its features such as deformation, soil stress, and pressure during excavation.

NUMERICAL SIMULATION

Assumptions

In the combined structure, soil nails, facing, pile, binding beam, internal props constitute a spatial structure as a whole to bear earth pressure and restrain excessive deformation. In order to reduce the volume of calculations and highlight main factors, the reasonable simplification for the issue is necessary. The assumptions are made in the following.

(1) it is plane strain problem in homogeneous soil;
(2) pile top is directly connected with prop neglecting existence of binding beam;
(3)except soil, all the structural elements are assumed to be elastic. Soil adheres to Mohr - Coulomb yield criterion;
(4) neglect overload on the ground, and
(5) neglect groundwater.

The binding beam directly linked with the tops of the adjacent piles transmits axial force of prop to pile tops and almost does not play a role in retaining. In addition, in the following numerical simulation, there is only one pile to involve. The effect of neglecting the binding beam on the calculation is limited.

Element types

Table 1 illustrates the types of structural elements to represent the combined structure. Soil nails are slender, flexible, and can be simulated by cable. The 100mm-thick facing may be separate with soil in active zone if large displacement occurs in soil nailing wall. There is the coupling effect, and so liner structural element is applied to represents the facing layer. Pile wall is directly in use of pile element. Pile and the surround soil can be separate or close and there is also the coupling effect. Use beam element to represent the prop. Soil body is discretized into a series of cubes of hexahedral elements to improve the calculation accuracy. These structural elements are of basic features: beam element, the two-node straight-line element, can withstand axial force, shear, bending moment; cable element, the two-node straight-line unit, only to bear axial force, is mobilized with the surrounding soil through the interaction of frictional resistance; pile element, the two-node straight-line unit, similar to beam element, can also interact with the surrounding soil in separation or closure; liner element, three-node plane element, interacts with the surrounding soil in normal and tangent directions to separate or slide according to Mohr-Coulomb yield criterion. The combined structure consists of soil, a pile, 4 rows of soil nails, facing and one prop.

Table 1. Element Types

structures	Soil nail	facing	pile	strut	soil
Structural element	cable	liner	pile	beam	hexahedral entity

In numerical simulation, construction steps can be taken account. For soil nailing wall, while excavating, soil nails are installed one row by one row simultaneously forming the shotcrete facing. When excavation proceeds to a given depth, install a pile at the toe of the wall and then set a prop to link the top of the pile. At the sixth step of excavation, install the pile; at the seventh step, set up the internal prop.

Simulation extent

As shown in Figure 1, the combined structure of the soil nailing wall plus pile wall is employed in a 50m-wide rectangular excavation in homogeneous soil with the final excavation depth of 10m. There are total steps of 10 to excavate, 1m deep each step. The structure is assumed to be thickness of 2.0m as a plane strain problem. The area of simulation is 50m × 30m. The simulation boundary is extended to 20.0m under the base of pit at last step (see Figure 1).

The values of parameters

Figure 1 shows the geometry of the structure and some of parameters used in simulation. The values of the parameters are in line with the actual conditions. The remaining parameters are given as below.

(1) Soil nailing wall. In the extent of simulation, there are four 5m-long soil nails in spatial arrangement $S_V \times S_H = 2.0 \times 1.5m^2$, made from Φ22 steel bars with elastic modulus $E = 2 \times 10^8$kPa and yield strength $fy = 1.86 \times 10^7$kPa. Soil nails are installed in inclination of 15 ° to the horizontal. Pre-drilled holes with diameter D = 0.12m to insert soil nails are filled with cement mortar in C20. Soil nails are mobilized with surrounding soil up to the ulltimate frictional force τ= 15kPa. 100mm-thick facing layer makes up of reinforced concrete in C20 with elastic modulus $E = 2.55 \times 10^7$kPa and Poisson's ratio $\mu = 0.25$. The normal and tangential stiffness between facing and the surrounding soil are assumed to be 100kPa / m , respectively, and their cohesive strength 2kPa.

(2) Pile. There exists a single pile with diameter of 1000mm and the central spacing of 2.0m, made from reinforced concrete in C30. Pile interacts with surrounding soil in the tangential and normal stiffness of 100kPa / m, 10000kPa / m, cohesive strength 10kPa, friction angle 20 °, respectively.

(3) Prop. Only one prop is assumed to support the extent of 20 piles. The prop is rectangular in cross-section of 500 × 700mm^2, made from reinforced concrete in elastic modulus of 2.55×10^7kPa and Poisson's ratio of 0.25.

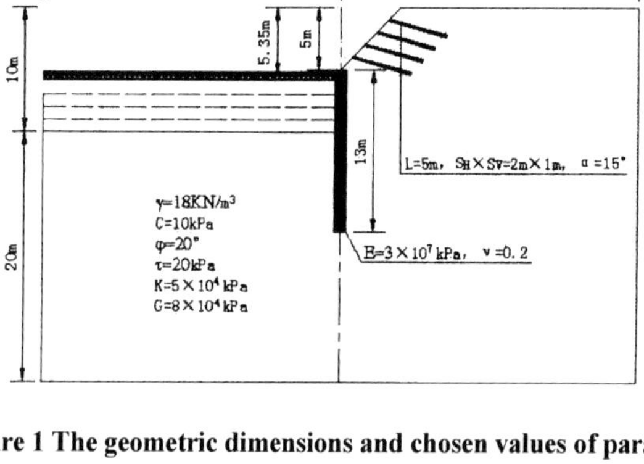

Figure 1 The geometric dimensions and chosen values of parameter

Computational results

Figure 2 illustrates settlement δ_V and horizontal displacement δ_H at depth H = 5m, 10m, respectively. At H = 5m, the maximum horizontal displacement is below the base of pit, δ_{Hmax} = 25mm, and the largest settlement is at the boundary, δ_{vmax} = 15mm. At H = 10m, δ_{Hmax} is near the bottom (δ_{Hmax} = 70mm), and the largest settlement δ_{vmax} = 35mm also locates in the boundary. With increase in the excavation depth, deformation of soil nailing wall increases, but the distribution shape is kept to be similar. Location in maximum horizontal displacement is not near ground surface, but close to the base of pit. This is due to soft soil to cause bottom uplift. At the different depths of excavation, if deformation in pile wall increases, displacement for soil nailing wall also increases.

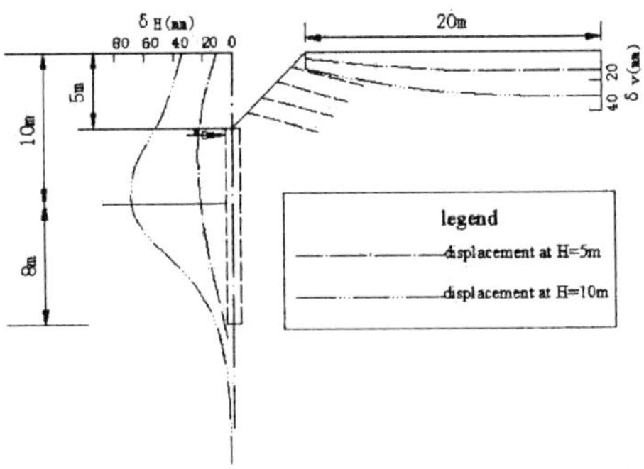

Figure 2 Deformation at depths of 5m and 10m

Figure 3 illustrates coefficient distribution of the earth-pressure at depth H = 5m, 10m, respectively, where K_0 is coefficient of earth pressure at-rest, K_0 = 0.7; K_a indicates coefficient of active earth pressure , Ka = 0.49; K_p is coefficient of passive earth pressure, K_P = 2.04; σ_{H5} and σ_{H10} represent horizontal stresses exerted to the retaining structure at depths of H = 5m, 10m, respectively. σ_{v0}^{B} and σ_{v0}^{F} represent the self-weight stresses back and in front of the combined structure. σ_{v0}^{B} is calculated from ground surface and σ_{v0}^{F} from the base of pit.

Figure 3 shows that at depth of H = 5m, the earth pressure exerted to the facing of soil nailing wall is far less than the values of Rankine's active earth pressure. At H = 10m, the distribution of σ_{H10} is complicated and there are two turning points, locating around the prop and near the middle of passive zone, respectively. Under the base of pit, the value of earth pressure from simulation is greater than that from the Rankine's theory and close to the earth pressure at-rest. Close to pile base, σ_{H10} does not increase and essentially remains unchanged. Accordingly, within the extent of 3m in the passive zone, σ_{H10} is larger than that from Rankine's theory. Comparing to the passive earth pressure in pile wall (Hashash et al, 2002), the value of passive earth pressure in pile wall in the combined structure increases by about 20%. This is due to the existence of soil nailing wall above the pile. The curves in Figure 3 also illustrate that the values of earth pressure applied to the facing of soil nailing wall decreases with increase in excavation depth, but the maximum value decreases by almost 50%, compared with that from Rankine's theory. This proves that the pile and the soil nailing wall interact to keep stability of ground during excavating.

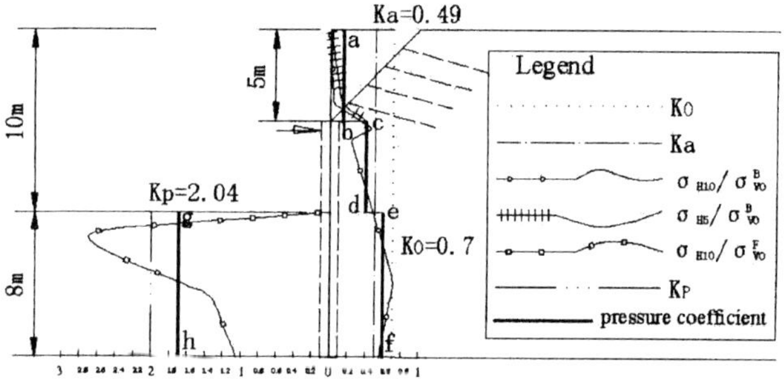

Figure 3 Pressure distributions at excavation depth of 5m and 10m

To braced excavation, Terzaghi and Peck proposed a semi-empirical formula to predict the load on retaining structures (Craig, 2002). This paper involves the combined structures, whose characters are more complicated than only pile in braced excavation. Due to the prop to restrain deformation of ground, the soil in the vicinity of the prop may not have entered the plastic state and so Rankine's theory is not applicable to predict the earth pressure. Soil body near the base of pit is affected less by the prop and likely to enter the plastic state and the Rankine's theory is appropriate. Near the pile base, because of the embedding effect of soil in active zone, soil body is inadequate to move and it seems to not apply Rankine's theory. Numerical simulation results show that (see Figure 3), at H = 5m, the facing in soil nailing wall is subjected to the earth pressure less than that determined by Rankine's theory. With increase in excavation depth, due to the pile near its toe to limit displacement of soil around, the pressure acting on the facing decreases, as shown in Figure 3. at H = 10m, around the base of pit, the earth pressure is close to that conducted by the Rankine earth pressure theory, while near the prop and pile base, the earth pressure is larger. Therefore, as shown in Figure 3, the pile along its depth in the active zone can be divided into three sections: ab, cd, ef, and in passive zone gh to amend coefficients of the earth pressure determined by Rankine theory as follows:

$$\begin{cases} K_{ab} = 0.1K_a \\ K_{cd} = 0.8K_a \\ K_{ef} = 1.2K_a \\ K_{gh} = 0.85K_p \end{cases} \tag{1}$$

In equation (1), the signs k_a, k_p represent the average coefficient of active earth pressure and passive earth pressure coefficient from Rankine's theory along excavation depth, respectively.

CASE STUDY

Overview

The Jianghua residential complex building was surrounded by Sanyang Road to the north, Zhongshan Road to the east, a tall building near Jinghan Road to the west. The project consisted of one 13-storey, one 15-storey, one 2-storey office buildings, as well as three 18-storey residential buildings, with two-storey basements to extend the total area. Layout of these six buildings was similar to the "courtyard", China's traditional house type. The basement in perimeter of 516m is rectangular in shape occupying an area of about 14400m^2 with depth of 9.4 m to10.6m. To the north and to the east, there existed underground pipes or cables very close to excavation boundaries. These nearby underground facilities are necessarily protected to avoid damage during excavation.

Take the section fa, close to Sanyang Road, as an example to illustrate the geology on site. Strata related to excavation from ground surface included (1) fill; (1-2) fill; (2-1) silt; (2-2) silty clay; (3) silt and fine sand; (4-1) fine sand; (4-2) fine sand . Groundwater involves the upper perched water and confined water, the former in fill and the latter in the fourth layer of fine sand. Deep well is designed to lower water head to 2m below the base of pit. Along boundaries of excavation, cement-soil mixing piles in 6m long were arranged to shutoff the perched water in fill layers. The internally-propped pile wall was selected in deep excavation. In order to reduce the bending moment within piles, the 4m-deep upper fill was removed to form a 4m-wide platform, as shown in Figure 4. The cut slope was retained using soil nailing wall (Figure 4). The profile of excavation included two types of retaining structures: pile wall and soil nailing wall, namely, the combined structure. According to the geology and environment along excavation boundaries, earth-retaining plans are different slightly. Design in the section fa is shown in Figure 4.

Some of monitoring results were illustrated in Figure 4 in the section fa on September 23, 2007 at final depth of 10.6m: the maximum bending moment of 390kN • m, maximum displacement of 26mm, both locating near the base of pit, the horizontal displacement of 39mm, 26mm at the top and toe of soil nailing wall, respectively, and settlement 18.1mm at the crest.

Numerical simulation

The profile in the section fa is selected to carry out numerical simulation. The assumptions are similar to that previously introduced in this paper except heterogeneous soil instead of homogeneous soil. Neglect the resistance of cement-soil mixing piles.

Excavation is proximately rectangular in shape with 185.8m long and 96.9m wide. The section fa is located in the middle of the long side. Take it as a plane strain problem with simulation thickness of 1.2m. Choose the simulation extent of 100m × 35m. The section fa is the middle of 100m and 50m in and outside away from the excavation side, respectively. From the final depth of excavation, the simulation boundary is extended 24.4m. Set 11-step excavation, 1m deep each step for former 10 steps and 0.6m deep for the last one. The values of the soil parameters are shown in Table 2. In Figure 4, layer thickness, excavation geometry and retaining structure are depicted. The remaining parameters for structural elements are similar to that previously-stated in the case in homogenous soil.

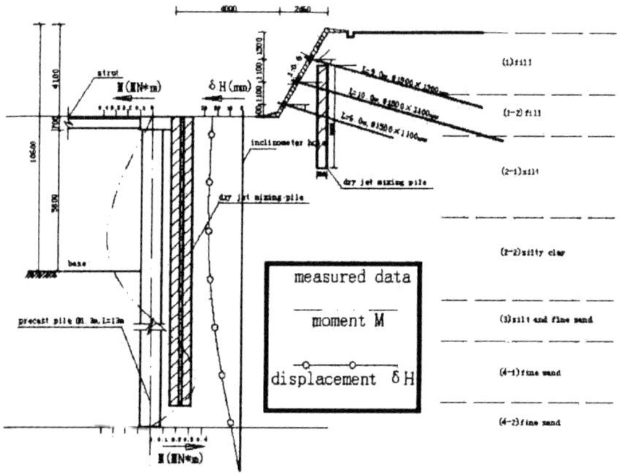

Figure 4 Profile and monitoring data in the section fa

Table 2 Parameter values of soil

Soil	K(kPa)	G(kPa)	C (kPa)	φ(°)	τ(kPa)
(1) fill	5×10^4	2×10^4	5	22	30
(1-2) fill	5×10^4	2×10^4	5	6	20
(2-1) silt	6×10^4	3×10^4	12	20	
(2-2) silty clay	7×10^4	4×10^4	16	9.5	
(3) silt and fine sand	7×10^4	2×10^4	9	16	
(4 - 1) fine sand	7×10^4	4×10^4	0	20	
(4 - 2) fine sand	10×10^4	6×10^4	0	35	

Part of computational results is shown in Figure 5. At depth of H = 4.1m, maximum horizontal displacement in soil nailing wall reaches to 19.7mm. At H = 10.6m, horizontal displacements in crest and the toe are 32mm, 22mm, respectively. In crest, settlement is equal to 22mm.The maximum horizontal displacement is 32mm near the base of pit. It can be found out that the deformation computed in the combined structure has very close to the values of monitoring results on September 23, 2007. Maximum bending moment in pile is calculated to be equal to 409.6kN.m, very close to values of the measured data, too.

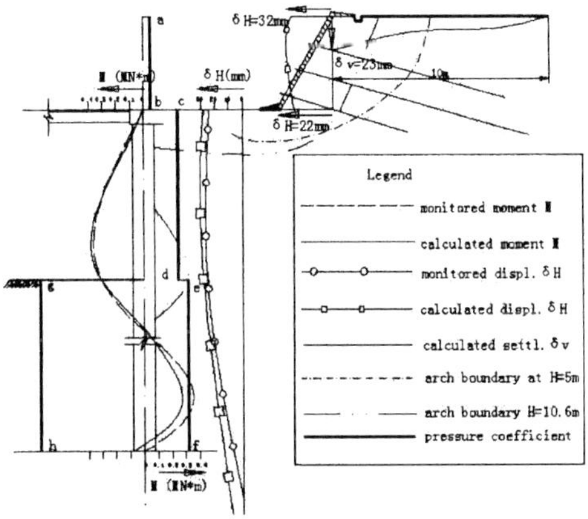

Figure 5 Numerical simulation results

CONCLUSIONS

The earth-retaining structure is subjected to the earth pressure, which is related to many factors such as soil elasto-plastic states, strength, stress-related path as well as structure own stiffness. Deformation and earth pressure in the combined structure are discussed in this paper to regard it as a whole. Within the combined structure, soil nailing wall and pile interact to bear earth pressure. For example, due to effect of the pile, the facing in soil nailing wall is subjected to the earth pressure much reduced. On the contrary, due to the existence of soil nailing wall, the passive earth pressure to pile becomes larger. In the combined structure, the distribution of earth pressure is inconsistent with that from Rankine theory. Based on numerical simulation results, the depth of excavation in the active zone is found to divide into three sections. Coefficients of active earth pressure are adjusted to meet the results of simulation each section.

REFERENCES

Bridle, R. J.(1989) "Soil nailing—analysis and design", *Ground Engineering*, Vol. 22(6): 52–56

Craig R.F. (2002) *Soil mechanics* (6th edition), Spon press

Hashash, Y.M.A., and Whittle, A.J.(2002). "Mechanisms of load transfer and arching for braced excavations in clay", *J. Geotech. Eng.*, Vol. 128(3): 187-197

Yang, Y., Yuan, J. (1998) "Limit equilibrium method of soil nailing walls in deep excavation", *Geotechnical Investigation and Surveying*, (6): 9-11

Yuan, J., Yang Y., Tham,L.G., Lee P.K.K. and Tsui Y. (2003), "A new approach of limit equilibrium and reliability analysis of soil nailed walls", *International Journal of Geomechanics*, ASCE, Vol.3(4): 145 -151

Interaction effect of retaining wall and existing foundations in braced excavation

H.M. Shahin[1], A.M. ASCE, T. Nakai[2], M. Kikumoto[3], Y. Uetani[4] and F. Zhang[5].

[1]Associate Professor, Nagoya Institute of Technology, Gokiso-cho, Showa-Ku, Nagoya 466-8555; shahin@nitech.ac.jp
[2]Professor, ditto; nakai.teruo@nitech.ac.jp
[3]Assistant Professor, ditto; kikumoto@nitech.ac.jp
[4]Graduate Student, ditto; @nitech.ac.jp
[2]Professor, ditto; cho.ho@nitech.ac.jp

ABSTRACT: Two-dimensional model tests are conducted to investigate the deformation mechanism of the ground and the earth pressure of retaining wall. Numerical simulations with finite element method using FEMtij-2D are also carried out for the same scale of the model tests. Subloading t_{ij} model is used in the analyses to model the ground material. Several patterns of the model tests are performed varying the length of the retaining wall and changing the distance between the foundation and wall. It is revealed in this research that maximum surface settlement does not always occur just behind the wall, but mostly at the position of the existing building. The rotation of the foundations depends on the distance between the foundation and wall. The numerical analyses can well simulate the observed earth pressures, surface settlements and deformation mechanism of the ground.

INTRODUCTION

In urban area an open excavation causes problem to the deformation of surrounding ground and adjacent structure together with the stability of the retaining wall and the ground. At present practice, earth pressure of retaining wall and stability are predicted with rigid plasticity theory such as Rankine's earth pressure theory, the deflection of wall is predicted using beam spring model, and the deformation of the ground is estimated by elastic finite element method. These methods ignore the interaction of retaining wall and nearby existing structures, which is an important factor to be considered in open excavation problem. In this research, two-dimensional model tests are conducted to investigate the deformation mechanism of the ground and the earth pressure of retaining wall as a continuous work of the previous researches (Nakai et al., 2007 and Iwata et al., 2008). Numerical simulations with finite element method using FEMtij-2D are also carried out for the same scale of the model tests. Subloading t_{ij}

model (Nakai and Hinokio, 2004) is used in the analyses to model the ground material. This model can describe typical stress deformation and strength characteristics of soils such as the influence of intermediate principal stress, the influence of stress path dependency of plastic flow and the influence of density and/or confining pressure.

OUTLINE OF MODEL TESTS AND NUMRICAL SIMULATIONS

Figure 1 shows the schematic diagram of the two-dimensional apparatus. Figure 2 represents a photograph of the ground during excavation. The size of the model ground is 68cm in width and 45cm in height. Aluminum rods of 5cm in length, having diameters of 1.6mm and 3.0mm and mixed with the ratio of 3:2 in weight, are used as the model ground (unit weight of the mass is 20.4kN/m^3). The retaining wall is 30cm in depth, 6cm in width and 0.5mm in thickness, which is a plate of aluminum material (EI = 0.88 N*m^2/cm, EA = 4.22 kN/cm). It is set in the ground before excavation and located at 20cm from the right boundary of the ground. In the experiment the model ground was excavated with a layer of 1.5cm in thickness up to 18cm depth of the ground. Two struts, located at the levels of 1.5cm and 7.5cm from the ground surface, are set in the excavated ground when the excavation level reached at the depths of 3cm and 9cm, respectively. Both struts are made of aluminum consisting springs in between two aluminum rods. The stiffness of the upper strut is 3.64 kN/m/cm and for the lower strut it is 3.04kN/m/cm. The dimension and stiffness are chosen considering the real ground condition assuming aspect ratio of 1:100 between the model and prototype structures. The right side of the model ground is supported with 15 sliding blocks where each block contains a load cell which measures axial force of the strut and horizontal earth pressure of the ground. A laser type displacement transducer is used to measure surface settlement of the ground. By taking photographs with a digital camera and using PIV technique the ground movements and consequently the strain of the ground are measured.

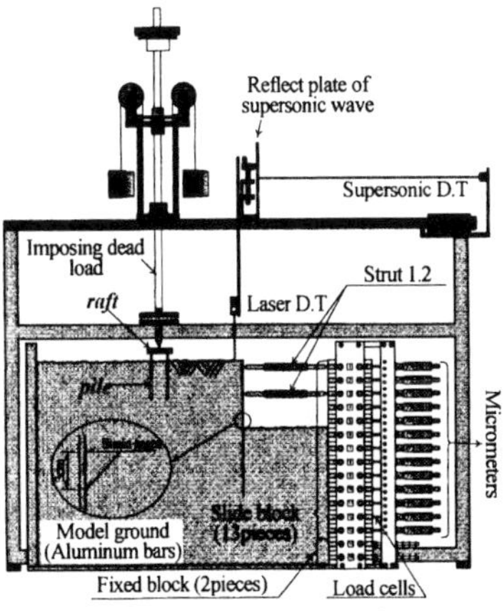

FIG.1. Schematic diagram of the apparatus

FIG.2. Photo of a model test

Table 1 shows the experiment patterns considered in this research. Case 1 represents the experiment pattern where no building load is considered, while in Case 2 to 4 building loads are considered. Two types of foundations, strip foundation and pile foundation, are employed. The model tests are carried out varying the lateral distance between the retaining wall and the foundation (L_w) for the values of 4cm, 8cm and 12cm. In the case of pile foundation three sizes of pile length (10cm, 15cm and 20cm) are considered to investigate the effect of the pile dimension. The dead load applied in this research is about 1/3 of the maximum bearing capacity ground.

Figure 3 shows a typical mesh used in the finite element analyses. Isoparametric 4-noded elements are used in the mesh. Both vertical sides of the mesh are free in the vertical direction, and the bottom face is kept fixed. Analyses are carried out with the same scale of the model tests considering plane strain drained conditions. The strip foundation is modeled as elastic element and the pile is modeled as hybrid element consisting elastic solid and beam elements. Elastoplastic joint element (Nakai 1985) is used as an interface element between the ground and the foundations using the friction angle $\delta=18^0$. The constitutive law used in this numerical analysis is the subloading t_{ij} model (Nakai & Hinokio 2004). The program code FEMtij-2D was developed in our laboratory which is used for all finite element analyses. Model parameters for the aluminum rod mass are shown in Table 2. The parameters are fundamentally the same as those of the Cam clay model except the parameter a, which is responsible for the influence of density and confining pressure. The parameter β represents the shape of yield surface. The parameters can easily be obtained from traditional laboratory tests. Figure 4 shows the results of the biaxial tests for the mass of aluminum rods used in the model tests. The figure shows the positive and negative dilatancy of aluminum rods mass; and it is clear that the strength and deformation behavior is very close to those of dense sand. The dotted lines represent the numerical results for a confining pressure of 1/100 times the confining pressure of experiments. From the stress-strain behavior of the element tests simulated with subloading t_{ij} model, it is noticed that this model can express the dependency of stiffness, strength and dilatancy on the density as well as on the confining pressure. The initial stresses, correspondent to the geostatic (self-weight) condition, are assigned to the ground in all numerical analyses. After that the existing

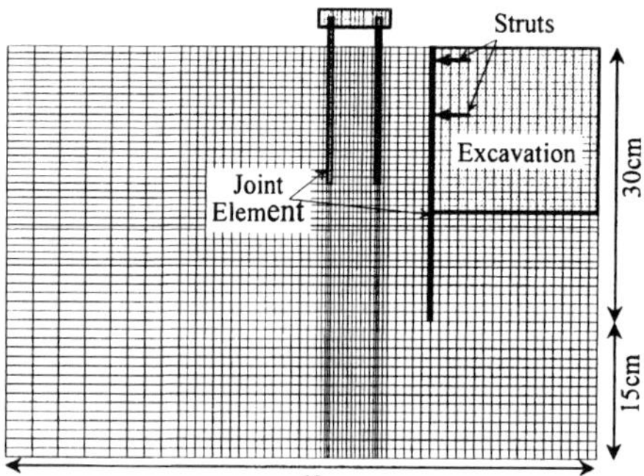

FIG.3. Mesh for finite element analysis

Table 1. Experiment patterns

		Building loads		
		Case2	Case3	Case4
Case1 Greenfield	Strip foundation	L_w=4cm	L_w=8cm	L_w=12cm
	Pile foundation	L_p=10cm	L_p=10cm	L_p=10cm
		L_p=15cm	L_p=15cm	L_p=15cm
		L_p=20cm	L_p=20cm	L_p=20cm

load is applied to the foundation. This state of ground is assumed as the initial condition of the excavation. In this paper, the results of Case1 and Case2, in the pile foundation the length of the pile L_p=10cm, will be described.

Table 2. Material parameters for aluminum rods

λ	0.008	Same parameters as Cam-clay model
κ	0.004	
N (e_{NC} at p=98kPa & q=0kPa)	0.3	
R_{CS}=$(\sigma_1/\sigma_3)_{CS(comp.)}$	1.8	
ν_e	0.2	
β	1.2	Shape of yield surface (same as Cam-clay at β=1)
a	1300	Influence of density and confining pressure)

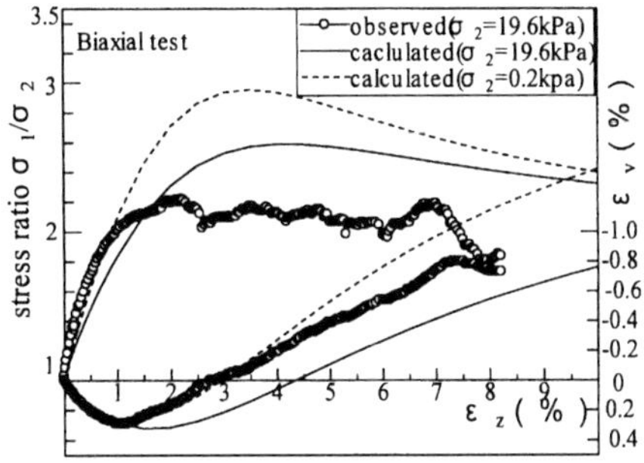

FIG. 4. Stress-strain-dilatancy relation

RESULTS AND DISCUSSIONS

Figure 5 shows the observed (top) and computed (bottom) horizontal displacement of the retaining wall for Case1 and Case2. Here, the length of the pile L_p=10cm. The abscissas represent horizontal displacement and the vertical axes show the ground depth from the surface. The legend represents excavation depth (d). The arrows in the figures indicate the position of the upper and lower struts. It is seen in the figures that for Case2 where nearby structures exist the wall displaces more compare to Case1 where building load is not considered. Comparing the results of the foundation types it is found that for the strip foundation displacement is larger than that for the pile foundation. With the advancement of the excavation the occurrence of maximum horizontal displacement goes deeper position. In the case of pile foundation, if the excavation depth is shallower than the pile length displacement of the retaining wall is restrained. However, when the excavation depth passes the pile tip, larger displacement of the wall occurs at the level below the tip of the foundation. The results of the numerical analyses show the same tendency of the model tests not only in shape but also in quantity

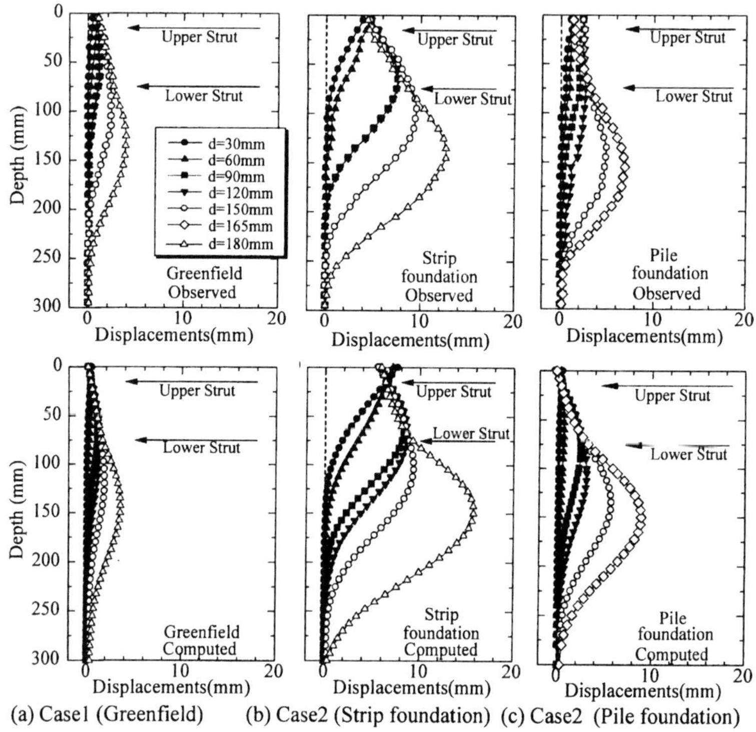

FIG.5. Distributions of wall horizontal displacements

Figure 6 shows the observed (top) and computed (bottom) surface settlement troughs for Case1 and Case2. The abscissas represent distance from the retaining wall while the vertical axes show surface settlement of the ground. The legend represents excavation depth (d). It is seen in the figures that for Case2 the settlement of the ground is larger than that for Case1, and the maximum surface settlement occurs at the position of the foundation regardless the foundation types. Comparing the results of the foundation types it is found that for the strip foundation surface settlement at the position of the foundation is larger than that for the pile foundation. In contrast, for the pile foundation surface settlement adjacent to the wall is larger than that for the strip foundation. At the initial stage of the excavation, i, e. when the excavation depth is shallower, the foundation tills toward the excavation, but, when the excavation depth is getting deeper the foundation inclines to the opposite direction of the excavation regardless the foundation types. The numerical simulation slightly over-predicts the surface settlement at d=180mm as some elements in the ground undergo into failure state. In the case of pile foundation when the excavation depth passes the pile tip, the foundation rotates to the opposite direction of the excavation side. The rotation depends on the distance of the foundation from the retaining wall. The finite element analyses can capture well the displacement and the rotation of the foundation.

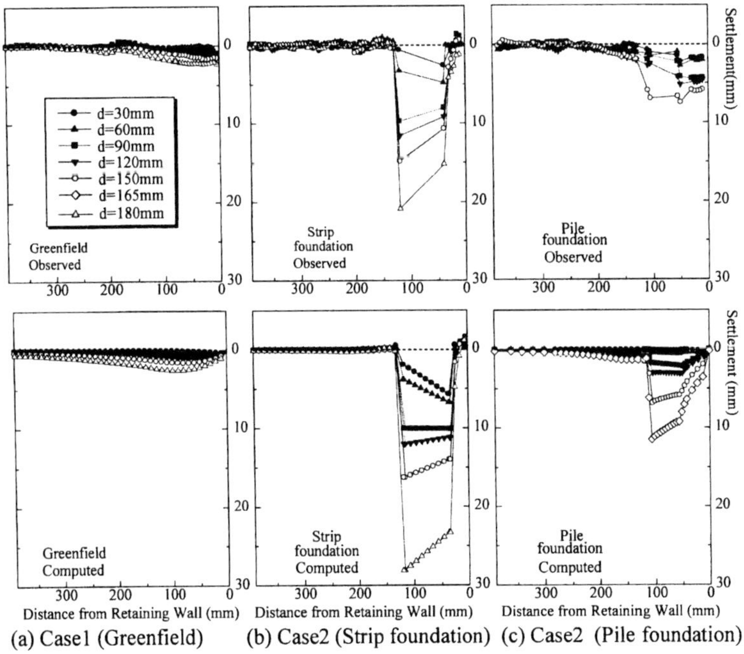

FIG.6. Profiles of surface settlement troughs

Figure 7 illustrates the change of axial force of the upper and lower struts for Case1 and Case2. Here, the length of the pile L_p=10cm. The abscissas represent excavation depth and the vertical axes show axial force in Newton. The legend represents excavation depth (d). It is seen in the figures that for Case2 where the existing load is considered axial forces for both lower and upper struts are larger than those for Case1 where building load is not considered. Comparing the results of the foundation types it is found that for the pile foundation the maximum value of axial force of the lower strut and proportion of axial force with the upper strut are significantly larger than those for the strip foundation. As the upper strut is located at 1.5cm from the ground surface and the pile restricts the wall displacement at this location the peak value of the axial force for this strut is comparatively smaller. On the other hand, as the wall displacement becomes larger at 7.5cm where the lower strut is located for the excavation depth passes the pile tip the axial force of the lower strut increases. The numerical analyses perfectly capture the observed axial force of the struts.

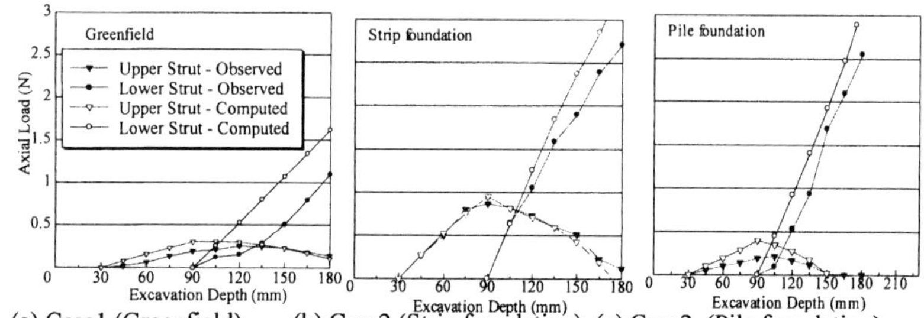

(a) Case1 (Greenfield) (b) Case2 (Strip foundation) (c) Case2 (Pile foundation)

FIG.7. Axial forces of the upper and lower struts

Figure 8 illustrates the distribution of shear strain of the model tests. The distributions of shear strain of the model tests are obtained using Particle Image Velocimetry (PIV) technique. In this paper, two images are divided into a finite area; the average movement rate of the mass of aluminum rods of each area is extracted as nodal displacements. The strain for one grid is calculated from these displacements by using the shape functions and the B matrix (strain- displacement matrix) that is usually used in finite element method to relate displacements and strains. Figure 9 shows the distribution of shear strains obtained from the finite element analyses. It is seen in the figures that for the building loads (Case 2) shear band develops from the excavation part and reaches both side of the foundations. The concentration of shear strain for the nearby structures is higher than that for the Greenfield condition. In Case 2, a significant ground deformation is observed in the edges of the strip foundation. In the pile foundation, shear band develops toward the pile tip with the advancement of the excavation. For the excavation depth deeper than the pile tip, concentration of shear strain increases towards the pile tip which is farther from the retaining wall. Therefore, differential settlement occurs in the foundation during excavation. The shear strain of the numerical analyses shows very good agreement with the results of the model tests.

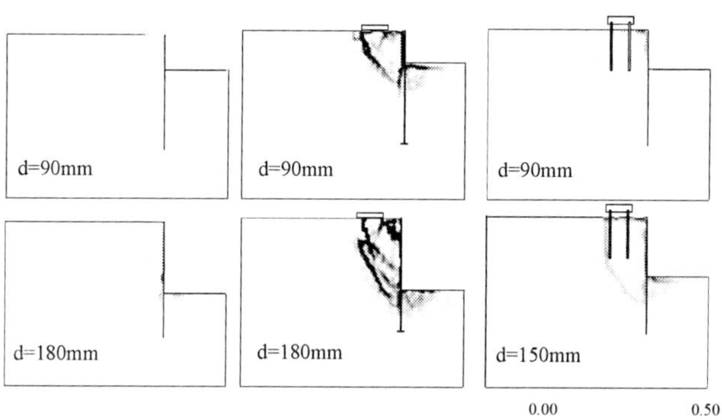

(a) Case1 (Greenfield) (b) Case2: Strip foundation (c) Case2: Pile foundation

FIG.8. Distribution of shear strain – model tests

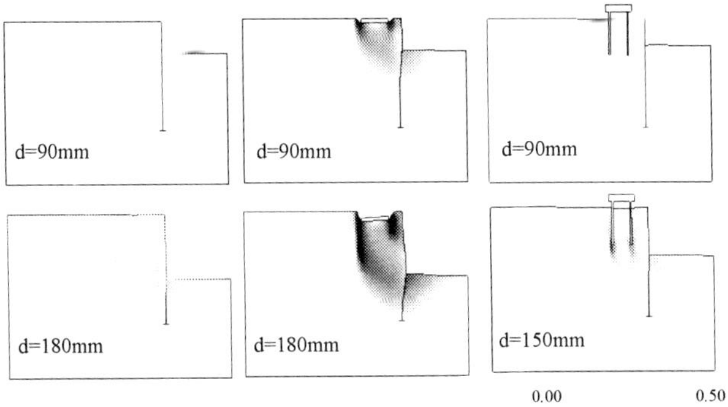

(a) Case1 (b) Case2: Strip foundation (c) Case2: Pile foundation

FIG.9. Distribution of shear strain – FE analyses

CONCLUSIONS

To investigate interaction effect of nearby structures on the retaining wall during excavation model tests and two-dimensional finite element analyses are carried out. It is revealed in this research that maximum surface settlement does not always occur just behind the wall, but mostly at the position of the existing building. The rotation of the foundations is dependent on the distance between the foundation and wall. The numerical analyses can well simulate the observed wall displacement, shape of the displacement mode, surface settlements and deformation mechanism of the ground. It can also be concluded that the computed results in which typical stress-strain behavior of soils is appropriately taken into account agree well with the experimental results qualitatively and quantitatively and can be a powerful tool to predict ground deformation, displacement of wall during the interaction effect of the existing structure and the retaining wall.

REFERENCES

N. Iwata, H.M. Shahin, F. Zhang, T. Nakai, M. Niinomi & Y.D.S. Geraldni (2008): Excavation with stepped-twin retaining wall: model tests and numerical simulations, Proc. of the 6th International Symposium on Geotechnical Aspects of Underground Construction in Soft Ground, Shanghai, China, April, pp.655-661

T. Nakai (1985): Finite element computations for active and passive earth pressure problems of retaining wall. *Soils and Foundations*, 25(3), pp. 98-112

T. Nakai and M. Hinokio (2004): A simple elastoplastic model for normally and over consolidated soils with unified material parameters, Vol.44, No.2, 53-70.

T. Nakai, H. M. Shahin, N. Iwata, M. Niinomi and H. Takei (2007): Influence of deflection process and mode of wall and existing building load on retaining wall, 13th Asian Reg. Conf. on Soil Mech. and Geotechnical Engineering, Kolkata, India, December, pp. 481-484.

Deformation Prediction of Deep Excavation Based on Unequal interval Grey Model GM(1,1)

E-chuan YAN , Ying LI, Yu-lei LI , Ting-ting ZHANG

Faculty of Engineering, China University of Geosciences, Wuhan, 430074, China;yecyec6970@163.com

ABSTRACT: The Grey Model GM (1, 1) is extensively used in deformation prediction of deep excavation, which is very important for design and construction of foundation excavation. The Grey Model GM (1, 1) is based on the equal interval data. So it is difficult to predict when the data is unequal. For the shortcoming, the Grey Model GM (1, 1) is improved. The difference which is formed when unequal interval is transformed to interval is adjusted by unit time coefficient. So the unequal interval Grey Model GM (1, 1) is obtained after reversing the equal interval sequence. The deformation prediction is made by a deep excavation in Wuhan, Hubei, China. Through comparison and analysis on the prediction and actual data, it shows that unequal interval Grey Model GM (1,1) is more closed to the reality.

INTRODUCTION

The impact of the deformation of excavation on urban environment has become prominent with the increasing excavation depth. Accurate prediction and effective control of deformation are important for excavation.

Deformation prediction of deep excavation include: deformation of building envelope, surface subsidence, base heave, changes around the pipeline, and so on. Since 1982, through the establishment of the excavation system as a Grey system, Grey theory are widely used and achieved good results on deformation prediction of excavation. But there're also many deficiencies. Grey model GM (1, 1) is based on the equal interval original data. So it is difficult to predict when the data is unequal interval. In this paper, GM(1,1) is improved by the establishment of unequal interval Grey Model GM(1,1) .And it is applied to deformation prediction of a deep excavation .It shows that unequal interval Grey Model GM(1,1) can meet prediction accuracy of engineering requirements.

ESTABLISHMENT OF UNEQUAL INTERVAL GREY MODEL GM (1, 1)

The difference which is formed when unequal interval is transformed to interval is adjusted by unit time coefficient. The difference part and the initial part are accumulated respectively. Then the interval model is established. The unequal interval Grey Model GM (1, 1) is obtained after reversing the equal interval sequence.

Transform unequal interval sequence into equal

Let there are unequal interval original sequence

$X^{(0)}=\left(x^{(0)}(k_1),x^{(0)}(k_2),\cdots,x^{(0)}(k_n)\right)$, Which $k_i \quad k_{i-1} \neq$ constant.

(1)Seek average time interval Δk

$$\Delta k=\frac{1}{n-1}(k_n-k_1) \tag{1}$$

(2) Seek unit time coefficient of every interval $\mu(k_i)$

$$\mu(k_i)=\frac{k_i-(i-1)\Delta k}{\Delta k} \qquad i\in(1,2,\cdots,n) \tag{2}$$

(3) Seek total difference of every unequal interval

$$\Delta x^{(0)}(k_i)=\mu(k_i)\left[x^{(0)}(k_{i+1})-x^{(0)}(k_i)\right] \tag{3}$$

Which $x^{(0)}(k_i)$ is original observed data of k_i

(4)calculate grey value of every equal interval

$$x_1^{(0)}(k_i)=x^{(0)}(k_i)-\Delta x^{(0)}(k_i) \qquad i\in(1,2,\cdots,n) \tag{4}$$

Then we can get equal interval sequence

$$X_1^{(0)}(t)=\left\{x_1^{(0)}(1),x_1^{(0)}(2),\cdots,x_1^{(0)}(n)\right\} \quad (t=1,2,\cdots,n) \tag{5}$$

(5)Get time response sequence and prediction equation of equal interval $x_1^{(0)}(k_i)$

$X^{(1)}$ is double accumulative of $X_1^{(0)}$

$$X^{(1)}=\{x_1^{(1)}{}_{(1)},x_1^{(1)}{}_{(2)},\cdots x_1^{(1)}{}_{(n)}\} \tag{6}$$

Which $x_1^{(1)}{}_{(k)}=\sum_{i=1}^{k}x_1^{(0)}{}_{(i)}=x_1^{(0)}{}_{(1)}+x_1^{(0)}{}_{(2)}+\cdots+x_1^{(0)}{}_{(k)} \quad (k=1,2\cdots n)$ (7)

$Z^{(1)}$ is consecutive neighbors sequence of $X^{(1)}$

$Z^{(1)}=\left\{z^{(1)}{}_{(1)},z^{(1)}{}_{(2)}\cdots z^{(1)}{}_{(n)}\right\}$

Which $z^{(1)}{}_{(k)}=\frac{1}{2}x_1^{(1)}{}_{(k)}+\frac{1}{2}x_1^{(1)}{}_{(k-1)} \quad (k=2,3\cdots n)$ (8)

Let $\vec{a}=(a,b)^T$ and define

$$Y=\begin{bmatrix}x_1^{(0)}{}_{(2)}\\x_1^{(0)}{}_{(3)}\\\vdots\\x_1^{(0)}{}_{(n)}\end{bmatrix} \qquad B=\begin{bmatrix}-z^{(1)}{}_{(2)} & 1\\-z^{(1)}{}_{(3)} & 1\\\vdots & \vdots\\-z^{(1)}{}_{(n)} & 1\end{bmatrix} \tag{9}$$

So the least square estimation parameters sequence of grey differential equation $x^{(0)}{}_{(k)}+az^{(1)}{}_{(k)}=b$ meet

$$\vec{a}=(B^TB)^{-1}B^TY \tag{10}$$

$$(a,b)^T=(B^TB)^{-1}B^TY \tag{11}$$

Time response series of grey differential equation $x^{(0)}{}_{(k)}+az^{(1)}{}_{(k)}=b$ is

$$\hat{x}^{(1)}{}_{(k)}=\left[x^{(1)}{}_{(0)}-\frac{b}{a}\right]\times e^{-a(k-1)}+\frac{b}{a} \quad (k=1,2\cdots n) \tag{12}$$

The prediction equation is

$$\hat{x}^{(0)}(k)=\hat{x}^{(1)}(k)-\hat{x}^{(1)}(k-1)=\left(1-e^{a}\right)\left[x^{(0)}(1)-\frac{b}{a}\right]\times e^{-a(k-1)}\quad (k=1,2\cdots n) \tag{13}$$

(7)The prediction equation is reduced. And the unequal interval Grey Model GM(1,1) is established.

$$\hat{x}_1^{(1)}(k_i)=\left(x_1^{(0)}(1)-\frac{b}{a}\right)e^{-a\times\frac{k_i-k_1}{\Delta k}}+\frac{b}{a} \tag{14}$$

$$\hat{x}^{(1)}(k_i)=\begin{cases}\hat{x}^{(1)}(k_1) & i=1\\ \hat{x}^{(1)}(k_i)-\hat{x}^{(1)}(k_i-\Delta k) & i=2,3,\cdots,n\end{cases} \tag{15}$$

Formula (15) is used to reduce.

The improvement of method of parameter calculation

Grey system prediction, which is based on interaction and mutual influences of system factors, establishes grey model group between main behavior and correlation factor of system. And the prediction is carried out by solving equation. Generally speaking, the more data samples we have, the more precise prediction we get. However, if all the elements are used to get the averaging, there is no meaning of the matrix, as we can't get the value of a and b.

In this paper, the reverse of the average of three continuous elements is the first row elements compared with two in (18).So the more prediction we get.

Keep the matrix Y invariant .Construct

$$z^{(1)}(k)=\frac{1}{3}x^{(1)}(k-2)+\frac{1}{3}x^{(1)}(k-1)+\frac{1}{3}x^{(1)}(k)\quad (k=3,4,\cdots,n) \tag{16}$$

Get the value of a and b by (11) .

Precision inspection of unequal interval Grey Model GM (1, 1)

(1)Residual test

Absolute error of prediction:

$$\varepsilon^{(0)}(k)=X^{(0)}(k)-\hat{X}^{(0)}(k)\quad (k=1,2\cdots n) \tag{17}$$

Relative error of prediction:

$$q=\left|\varepsilon^{(0)}(k)\Big/X^{(0)}(k)\right|\times 100\% \tag{18}$$

(2)Test of post-difference

$$S_1^2=\frac{1}{n}\sum_{k=1}^{n}\left[x^{(0)}(k)-\bar{x}\right]^2 \text{ Where } \bar{x}=\frac{1}{n}\sum_{k=1}^{n}x^{(0)}(k) \tag{19}$$

$$S_2^2=\frac{1}{n}\sum_{k=1}^{n}\left[\varepsilon^{(0)}(k)-\bar{\varepsilon}\right]^2 \text{ 其中 } \bar{\varepsilon}=\frac{1}{n}\sum_{k=1}^{n}\varepsilon^{(0)}(k) \tag{20}$$

Posterior error ratio $C=S_2\big/S_1$ (21)

Small error probability

$$P=P\left\{\left|\varepsilon_k-\bar{\varepsilon}\right|\langle 0.6745\, S_1\right\}=1 \tag{22}$$

Prediction accuracy is divided into four kinds by the value of P and C. It is found that if the accuracy grade of this model is excellent, the forecast results are reliable (Table 1).

Table 1. Reference list of precision grade of posterior- variance-test

Grade of model precision	C	P	Result
Primary	C≤ 0.35	P≥ 0.95	Excellent
Secondary	0.35<C≤ 0.50	0.80<P≤ 0.95	Good
Tertiary	0.50<C≤ 0.65	0.70<P≤ 0.80	Qualified
Band Four	C>0.65	P<0.70	Unqualified

ENGINEERING EXAMPLE

Take the excavation of a building in Wuhan China for example. The building is with 26 floor and 2 layers underground. Excavation depth is 9 ~ 10m.And it's a deep excavation. The main Soil layer constructions are miscellaneous fill and silty clay. It is stratified excavation and pile-anchor joint supporting.

Excavation monitoring is throughout the whole process of excavating and constructing Observation frequency is encrypted and reduced according to the deformation. Take significant horizontal deformation of M15 for instance. The displacements from 7/11/2005 ~ 25/11/2005 are Predicted by interval Grey Model GM (1, 1) and unequal interval Grey Model GM (1, 1) (Table 2 and Table 3).

Grey Time-series established by equal interval Grey Model GM (1, 1) is

$$\hat{x}_1^{(1)}(k_i)=\left(x_1^{(0)}(1)-\frac{b}{a}\right)e^{-a(k-1)}+\frac{b}{a}=-18.04e^{-0.1444(k-1)}+23.24 \quad (k=1,2\cdots n) \tag{23}$$

$$\hat{x}_1^{(1)}(k_i)=\left(x_1^{(0)}(1)-\frac{b}{a}\right)e^{-a\times\frac{k_i-k_1}{\Delta k}}+\frac{b}{a}=-24.48e^{-0.092\times\frac{k_i}{3}}+29.68 \quad (i=1,2\cdots n) \tag{24}$$

The unequal interval Grey Model GM (1, 1) is compared with the unequal interval Grey Model GM (1, 1) by using the posterior-variance-test. The results are as following (table 4):

Table 2. Horizontal displacement prediction of M15 by equal interval Grey Model GM (1, 1)

Time/day-month	time interval /d	measured displacement/mm	Predicted displacement /mm
7-11	0	5.20	5.20
10-11	3	7.61	7.62
13-11	3	9.70	9.73
16-11	3	11.20	11.54
19-11	3	13.01	13.20
22-11	3	14.44	14.50

25-11	3	15.70	15.50

Table 3. Horizontal Displacement prediction of M15 by unequal interval Grey Model GM (1, 1)

Time/day-month	time interval /d	measured displacement/mm	Predicted displacement /mm
7-11	0	5.20	5.20
11-11	4	8.20	8.03
14-11	3	9.80	9.93
16-11	2	11.20	11.11
20-11	4	13.20	13.25
23-11	3	14.50	14.70
25-11	2	15.70	15.80

Table 4. Precision of two method

method	C	P
unequal interval Grey Model GM(1,1)	0.0562	1
Traditional Grey Model GM(1,1)	0.0653	1

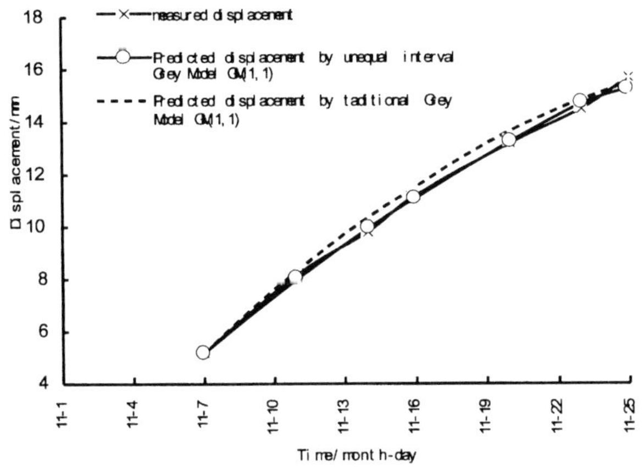

FIG. 1. Displacement prediction of M15 by unequal and equal interval Grey Model GM (1, 1)

As a comprehensive index, C value of unequal interval is smaller, which suggests its error swing range changes less significantly when P is the same. Seen above in the table 4, we can get that unequal interval Grey Model GM (1, 1) has smaller C value. So, unequal interval Grey Model GM (1, 1) has higher precision. The Predicted displacement is more closed to the actual value (Fig 1).

Seen from the predict trend of last two time intervals of unequal interval Grey Model GM (1, 1), prediction curve deviate from practical curve more pronounced. So the error of long-term deformation prediction of excavation greater than short-term for the unequal interval Grey Model GM (1, 1).Therefore, it's better to use it for short-term prediction and guiding construction activities.

CONCLUSIONS

(1)The Grey Model GM (1, 1) is based on the equal interval original data. So it is difficult to predict when the data is unequal interval. Therefore, it's necessary to establish unequal interval Grey Model GM (1, 1).

(2)The difference which is formed when unequal interval is transformed to interval is adjusted by unit time coefficient. In this paper, the reverse of the average of three continuous elements is the first row elements compared with two in the traditional Grey Model GM (1,1) .The unequal interval Grey Model GM(1,1) is established from the above way. The validity of the unequal interval Grey Model GM (1, 1) is verified by the excavation of a building in Wuhan China. It shows that unequal interval Grey Model GM (1, 1) has higher precision. And the Predicted displacement is more closed to the actual value.

REFERENCES

Zhao, C.G. and Wang, J.L. (2008). " Evaluation on Deformation Prediction of Foundation Pit Based on Grey Model." *Journal of Water Resources and Architectural Engineering* (in Chinese). Vol. 12: 118-120.

Liu, Z.B. and Shi, X. (2008). " Prediction of Rock Slope Based on Modified GM (1, 1) Model." *Journal of China Three Gorges University. (Natural Sciences)* (in Chinese)., Vol. 30(5): 33-35.

Deng, J.L. (2002). "Grey Theory " *Huazhong University of Science and Technology Press* (in Chinese). Page: 20-35.

Ruan, Y.F. and Ye, L.Y. (2002). "Determination of Deep Foundation Pit Supporting Scheme by Using the Grey System Theory and Method " *Chinese Journal of Rock Mechanics and Engineering* (in Chinese). Vol. 22 (7): 1203-1206.

Zhao, H.P. and Ye, L. (2005). "Holographic Forecast of Deformation Field of Rock under Blasting by Grey Theory " *Chinese Journal of Rock Mechanics and Engineering* (in Chinese). Vol. 24 (1): 39-43.

Zhou, Z.G. and Yao, W.S. (2009). "Genetic analysis and grey theoretical prediction of land subsidence in Fushun power plant " *The Chinese Journal of Geological Hazard and Control* (in Chinese). Vol. 20 (1): 83-87.

Three Dimensional Performance Observed in an Irregular Deep Excavation in Shanghai Soft Clay

Hua Yuan[1, 2] and Qinghe Zhang[1, 2]

[1]Department of Geotechnical Engineering, Tongji University, Shanghai, 200092, P.R. China. Tel: 0086-21-65985249, E-mail: zzl6108089@163.com
[2]Key Laboratory of Geotechnical and Underground Engineering of Ministry of Education, Tongji University, Shanghai, 200092, P.R. China.

ABSTRACT: Pudong first excavation region of Shanghai South Xizang Road River-crossing Tunnel, which is the transportation hub of 2010 Shanghai World Expo, is irregular shaped and located in soft clay. Diaphragm wall deflection, ground surface settlement and strut axial force during excavation have been monitored. The results show that except general recognized factors, such as excavation depth and width, wall stiffness, strut spacing, and soil characteristics, the joint angel of two adjacent diaphragm walls also has a significant impact on diaphragm wall deflection and ground surface settlement of irregular shaped deep excavation; Three dimensional effect and corner effect in the researched irregular excavation can't be neglected.

INTRODUCTION

Many researchers over the world have studied the observed performance in deep excavation in the past years (peck 1969; Mana and Clough 1981; Clough and O'Rourke 1990;). Clough and O'Rourke (1990) advanced a semi empirical chart to estimate δ_{hm} for soft to medium clays and this chart is widely used in engineering practice. Finno and Roboski (2005) approached an empirical relation for ground surface settlement parallel to an excavation wall. However, most of these researches mentioned above focus on regular shaped excavations. In order to reveal the difference of diaphragm wall and soil deformation behavior between regular and irregular shaped excavation, the performance of a 21.6m deep internally braced irregular shaped excavation in Shanghai soft clay is researched in this paper.

SITE DESCRIPTIONS

Shanghai South Xizang Road River-crossing Tunnel is the transportation hub of 2010 Shanghai World Expo. The excavation project reported in this paper is Pudong first excavation region of this tunnel, which is located in the east coast of Huangpu River, southeast of Shanghai city. This excavation is composed of Pudong shield working shaft

(PD0) and tunnel buried segment (PD1~PD3). PD0 is rectangular in shape whereas the tunnel buried segment is irregularly shaped. PD1 ~ PD3 are retained by 0.8 m thick, 32 m, 30 m, and 28 m deep reinforced concrete diaphragm wall, respectively. The final excavation level inclines at an angle. Maximum excavation depth is 17.3 m, and the minimum depth is 14.3 m. The maximum width of Pudong first excavation region is suited in PD3 and reaches 45 m (Fig.1). The site is underlain by relatively soft to medium silty clay (Table 1), the undrained shear strength (c_u) varies from 27 to 52 kPa.

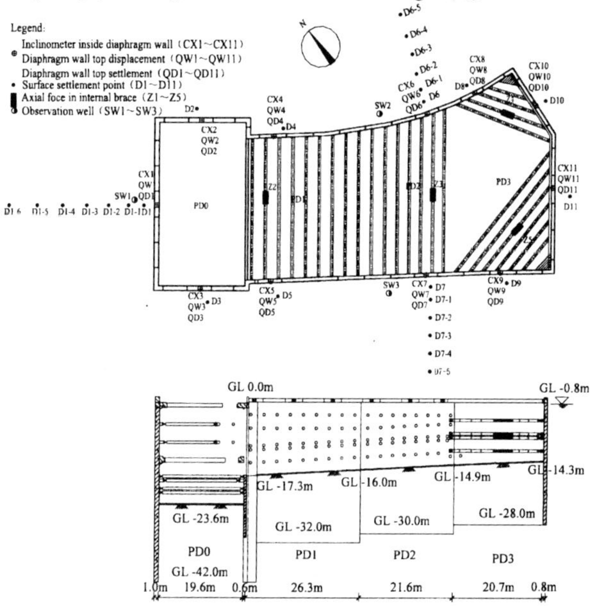

FIG. 1. Monitoring plan and cross section of the first excavation region

Table 1. Soil profile and mechanical parameters

Soil layer	Depth(m)	γ(kN/m^3)	ω(%)	φ'(degree)	c_u(kPa)	Q_u(kPa)	E_s(MPa)
Back fill	0~-1.4	16.9	——	——	——	——	——
Silty clay	-1.4~-3.2	18.8	32.6	22.3	28	42	4.10
Mud-silty clay	-3.2~-7.4	17.4	42.7	34.3	27	40	2.82
Mud-silty clay	-7.4~-14	16.8	49.8	28.3	26	51	2.25
Gray clay	-14~-17	17.4	40.9	24.0	31	54	3.09
Sand silty	-17~-25.2	18.1	32.7	25.2	38	58	7.78
Silty sand	-25.2~-37	19.5	24.1	18.9	41	66	6.70
Stiff silty clay	-37~-45	19.0	25.9	20.6	52	——	15.23

The construction was carried out using cut and cover method; details of the construction sequence are listed in Table 2. Passive soil of foundation base was strengthened by high pressure rotary jet grouting. Each prestressed steel strut was installed 0.3 m below the corresponding excavation level and reinforced concrete support placed 0.5 m below the excavation level. Prior to each excavation stage, groundwater level in pit must be lowered at least 2 m blow the excavation level. When dig to final level, grade C20 early strength concrete cushion layer in 0.2 or 0.3 m thickness was cast immediately. The preload applied to each steel strut was adjusted based on the total vertical stress. The duration time of each operation condition such as soil excavation and support installation can't be more than 16 hours. Upright column pile and connection beam were erected between steel struts in order to strengthen the system stiffness and avoid probable strut buckling caused by large strut width.

Table 2. Excavation Stage

Stage	Construction Activity	Period(day/month/year)			
		PD0	PD1	PD2	PD3
1a	Excavate to level 1	1/11			
1b	Cast 1st level prop				
2a	Excavate to level 2	11/11~12/11	08/11~09/11	10/11	19/11~20/11
2b	Install 2nd level prop	11/11~12/11	08/11~12/11	10/11	21/11~22/11
3a	Excavate to level 3	23/11~24/11	12/11~14/11	15/11~16/11	30/11
3b	Install 3rd level prop	24/11~25/11	13/11~15/11	15/11~16/11	04/12~05/12
4a	Excavate to level 4	26/11~28/11	26/11~29/11	30/11~01/12	13/12
4b	Install 4th level prop	30/11~06/12	28/11~30/11	30/11~02/12	14/12~15/12
5a	Excavate to level 5	17/12~19/12	05/12~07/12	08/12~11/12	——
5b	Install 5th level prop	19/12~20/12	06/12~08/12	08/12~12/12	——
6a	Excavate to level 6	21/12~22/12	——	——	——
6b	Install 6th level prop	23/12~24/12	——	——	——
7a	Excavate to final level	24/12~26/12	10/12~12/12	16/12~17/12	23/12~24/12
7b	Cast base slab	26/12~03/01	11/12~15/12	18/12~21/12	24/12~27/12

As shown in Fig. 1, some instruments including inclinometers inside the diaphragm wall and soils, settlement markers on the ground surface and the top of diaphragm wall, vibration steel wire gauges and electrical piezometer on the struts were used to monitor soil movement and the deflection of diaphragm wall.

DEFLECTION OF DIAPHRAGM WALL

Six inclination points (CX1, CX2, CX6, CX8, CX10 and CX11) are chosen to study the three-dimensional effect and corner effect on the wall deflection of irregular shaped deep excavation. Fig. 2 displays the normalized wall deflection at the six researched inclination points for all excavation stages. A positive ratio of observed diaphragm wall deflection (δ_h) to final excavation depth (H_d) represents inward movement toward the excavation. The deflection curves shown in Fig. 2 exhibits an arched profile and consistent with the typical diaphragm wall deformation pattern. As the construction proceeded, maximum wall deflection (δ_{hm}) increases and position of maximum deflection (H_m) occurs gradually descends. Maximum wall deflection at each excavation stage generally occurs within a few meters near the corresponding excavation level. When the final excavation depth is reached, H_m in the six researched casings are all smaller than H_d. Moreover, the smaller δ_h/H_d, the shallower maximum deflection occurs. Maximum toe deflection of the six points occurs in CX1 and is only 0.022%H_d, which could be attributed to passive soil reinforcement by high pressure rotary jet grouting. As the angle between two panel diaphragm walls CX10 and CX11 are placed on is nearly 135 degree, far more than 90 degree, the mutual support and constrain effect between these two panel is highly weak, thereby the overall maximum

δ_h/H_d (0.15%) of the deep excavation takes place in CX11, which exactly reflects the corner effect of irregular shaped deep excavation. Because of the relatively smaller length of diaphragm wall CX2 is installed on, inclination monitoring depth in CX2 is limited to 26m. The discrepancy of δ_h/H_d between CX1 and CX2 illustrates the significant influence of diaphragm wall stiffness on diaphragm deflection. Therefore, besides conventional factors such as the excavation depth and width, wall stiffness, strut spacing, active and passive soil characteristics, diaphragm wall deformation behavior of an internally braced deep excavation is also affected by angel between two adjacent diaphragm walls.

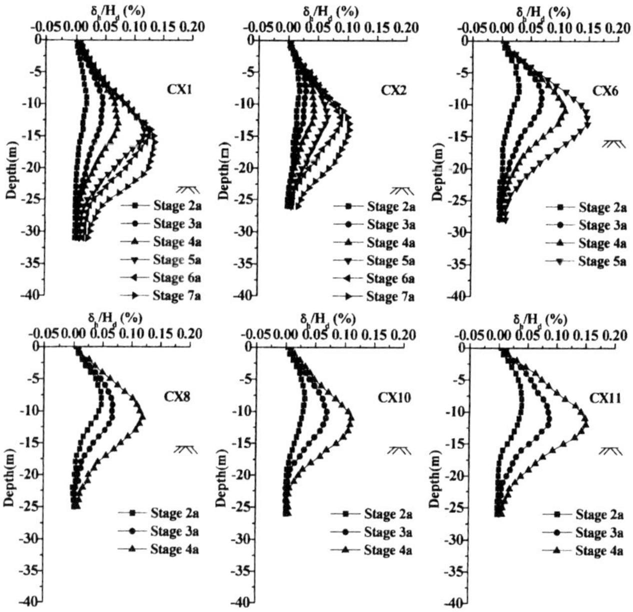

FIG. 2. Normalized wall lateral movement during various excavation stages

SOIL RESPONSES TO EXCAVATION

The relationship between normalized observed surface ground settlement (δ_v/H_d) and normalized distance from the diaphragm wall (d/H_d) at D1、D6 and D7 sections is shown in Fig. 3. Negative δ_v means ground subsidence. As expected, ground settlement increases gradually with the excavation stage, just as the variation of diaphragm wall deflection. Moreover, ground subsidence curve develops into a spoon shape from the original inclined line step by step. The position maximum surface settlement occurs ranges from $0.6H_d$ to $0.8H_d$ away from the wall. Significant increase in d/H_d between some stages may be due to soil creep, groundwater seepage and deformation of support system. Since the diaphragm wall D6 section measure is outer convex, not usual straight line, surface settlement at D6 section appears obviously higher than that at D1 and D7 sections. Maximum δ_v/H_d reaches up to 0.138%. Peck (1969) approached an empirical chart for excavation-induced settlement estimation, based on field observed data, and classified the settlement chart into three zones. Evidently, all measured data in the discussed deep excavation fall within zone Ⅰ (sand and soft to hard clay), which is consistent with result Wang and Ng (2005) reported. Hsieh and Ou (1998) suggested a trilinear relationship between δ_v/δ_{vm} and d/H_d based on monitoring data in soft soils in Taipei, and divided the observation into primary and secondary influence zones. From Fig. 3, it can be discovered that the ground subsidence curve is closer to trilinear proposed by Hsieh and Ou (1998).

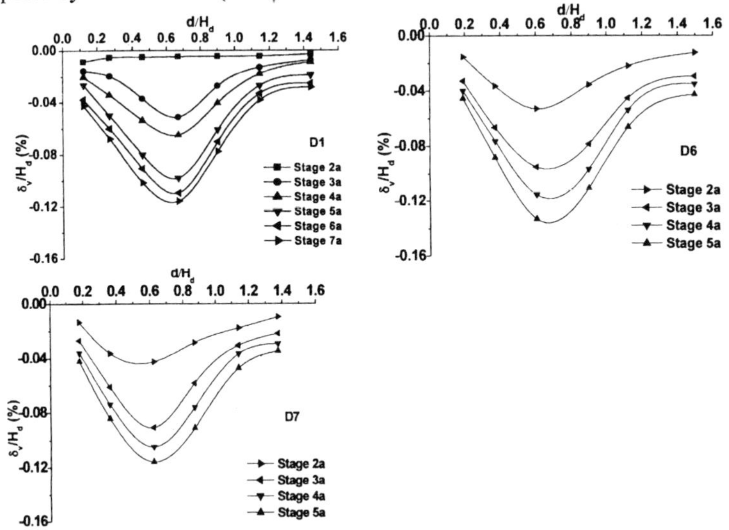

FIG. 3. Normalized ground surface settlement at various excavation stages

RELATIONSHIP BETWEEN DIAPHRAM WALL DEFLECTION AND GROUND SURFACE SETTLEMENT

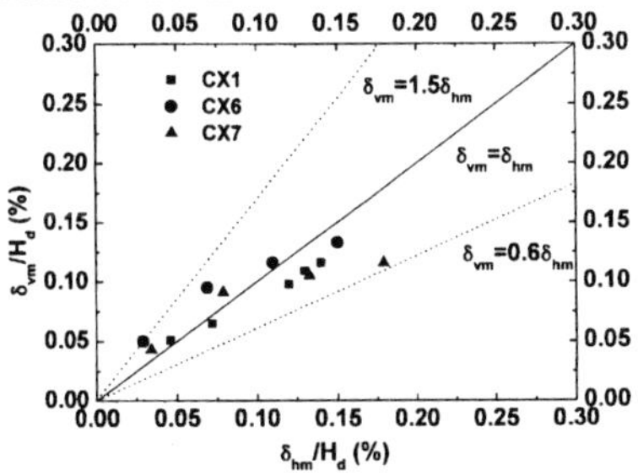

FIG. 4. Relationship between normalized maximum wall deflection and normalized maximum ground surface settlement

The relationship between maximum wall lateral deformation and maximum ground surface settlement (δ_{vm}) at CX1, CX6 and CX7 section is shown in Figure 4. It can be noted that there is a comparatively high scatter in the ratio of δ_{vm} to δ_{hm}. The values of δ_{vm}/δ_{hm} all fall in the range of 0.6 to 1.5, further wider than 0.5 to 1.0 suggested by Hsieh and Ou (1998). It is partially owing to the great final excavation depth, the irregular shape of this excavation, and the special excavation sequence. Finno and Roboski (2005) discovered that δ_{vm}/δ_{hm} decreased with the excavation proceeded and the maximum value was more than 1, which may be attributed to the volume changes in the surficial granular soils adjacent to the cut as a result of the excavation-induced stresses. The ratios of δ_{vm} to δ_{hm} in CX6 are all greater than that in CX1 and CX7. The more significant of corner effect, the smaller the value of δ_{vm}/δ_{hm}.

INTERNAL BRACING SYSTEM

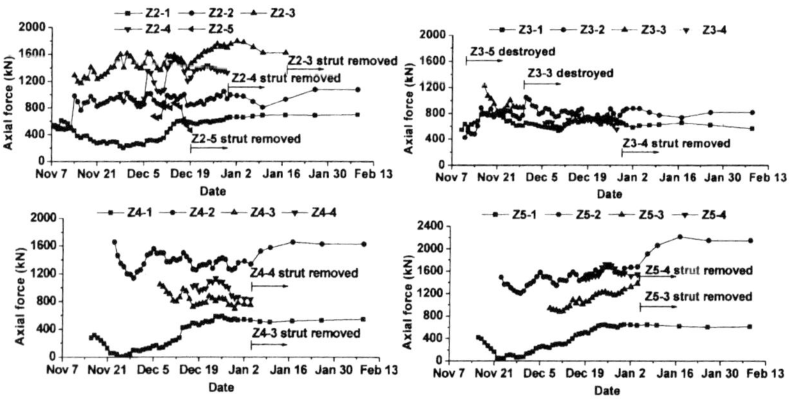

FIG. 5. Observed strut loading during excavation

Fig. 5 shows the observed axial loads in struts during excavation. Unfortunately, some survey gauges can't survive the entire excavation process. As Blackbum and Finno (2007) approached, the variation curves of measured axial forces in diagonal struts (Z4 and Z5) are silimar in each level of strut. When the lower strut is removed, axial forces in upper strut begin to increase more or less as expected. Considering the outer convex shape of diaphram CX6 section is suited in and its possible instability, supports in this region are arranged more densely and with larger cross section during construction. It's could be noted that axial forces of each support in Z3 are nearly equal, and the fluctuation amplitude is about 400kN, far less than that in other struts (Z2, Z4 and Z5).

CONCLUSIONS

Based on the field observations of diaphragm wall deflection, ground surface settlement and axial force at an irregular deep excavation of Shanghai South Xizang Road River-crossing Tunnel, the following conclusions are drawn:

(1) Maximum toe diaphragm wall deflection occurs in CX1 and is only 0.022%H_d. The relatively small toe deflection can be attributed to passive soil reinforcement by high pressure rotary jet grouting. The angle between two panel diaphragm walls CX10

and CX11 are installed on is nearly 135 degree, so the maximum δ_h/H_d of the researched excavation takes place in CX11, which reflects the corner effect of irregular shaped deep excavation. Except excavation depth and width, wall stiffness, strut spacing, and soil characteristics, diaphragm wall deflection of an irregular shaped deep excavation is also affected by the joint angel between two adjacent diaphragm walls.

(2) The distance between diaphragm wall and position maximum surface settlement occurs ranges from $0.6H_d$ to $0.8H_d$. Observed data all fall in zone I of the empirical chart Peck (1969) proposed, which just demonstrate the excavation occurs in soft clay. The ground subsidence curve is close to trilinear proposed by Hsieh and Ou (1998). The more significant of corner effect, the smaller the value of δ_{vm}/δ_{hm}.

(3) The ratios of δ_{vm} to δ_{hm} at CX1, CX6 and CX7 are all in the range of 0.6 to 1.5, more than the value reported by Hsieh and Ou (1998). It's probably related to the irregular shape of the excavation and its special excavation sequence.

(4) Axial forces in upper strut begin to increase, after the lower strut is removed. During the period of strut removal and backfill, the observed groundwater level rises gradually. Deformation behavior of support system and soil behind diaphragm wall in irregular shaped deep excavation becomes more complicate. As for the discussed irregular deep excavation, the coordination of soil excavation and support installation, especially in PD0 and PD3 must be paid more attention to.

ACKNOWLEDGMENTS

The authors appreciate the support of the Shanghai South Xizang Road River-crossing Tunnel Project Management Department.

REFERENCES

Peck, R.B. (1969). "Deep excavation and tunneling in soft ground. State-of-the-art-report." *Proc., 7th International Conf. Soil Mechanics and Foundation Engineering*, 225-281.

Mana, A.I. and Clough, G.W. (1981). "prediction of movements for braced cuts in clay." *J. Geotechnical & Geoenv. Engrg., Div., Am. Soc. Civ. Eng.*, 107 (6): 759-777.

Clough, G.W., and O'Rourke, T.D. (1990). "Construction induced movements of in situ walls." *Proc., Design and Performance of Earth Retaining Structure*, Geotechnical Special Publication No. 25, ASCE, New York: 439-470.

Finno, R.J., and Roboski, J.F. (2005). "Three-dimensional responses of a tied-back excavation through clay." *J. Geotechnical & Geoenv. Engrg.*, Vol. 131 (3): 273-282.

Wang, Z.W., Ng, C.W.W. and Liu G.B. (2005). "Characteristics of wall deflections and ground surface settlements in Shanghai." *Canadian Geotechnical Journal*, Vol. 42 (9): 1243-1254.

Heish, P.G., and Ou, C.Y. (1998). "Shape of ground surface settlement profile caused by excavation." *Canadian Geotechnical Journal*, Vol. 35 (6): 1004-1017.

Blackbum, J.T., and Finno R.J. (2007). "Three-dimensional responses observed in an internally braced excavation in soft clay." *J. Geotechnical & Geoenv. Engrg.*, Vol. 133 (11): 1364-1373.

Evaluation of Stability of Tailings Dam Based on Evolutionary Artificial Neural Network

Zai-hong Li[1], Rui Chen[2] and Wei-dong Lei[3]

[1]Postgraduate student, Department of Urban and Civil Engineering, Shenzhen Graduate School, Harbin Institute of Technology, Shenzhen University Town, Shenzhen, China; zaihonglee@live.cn
[2]Assistant professor, Department of Urban and Civil Engineering, Shenzhen Graduate School, Harbin Institute of Technology, Shenzhen University Town, Shenzhen, China; chenrui@hitsz.edu.cn
[3]Associate professor, Department of Urban and Civil Engineering, Shenzhen Graduate School, Harbin Institute of Technology; Also at Key Laboratory of Geotechnical Mechanics and Engineering of the Ministry of Water Resources, China; wdlei@hitsz.edu.cn

ABSTRACT: The stability of tailings dam is a multifactor-influenced and highly nonlinear problem. In this paper, an Evolutionary Artificial Neural Network (EANN) model was proposed to deal with this problem. In the proposed model, 9 influence factors are adopted as input variables, which are divided into three main categories, i.e., topography, geology and external influence factors. The Genetic Algorithm is used to optimize initial weight distribution of a Back Propagation network. The factor of safety predicted by the EANN model has an error less than 3% as compared with that value calculated by Bishop's method. This justifies that the EANN is an effective method for assessing the stability of tailings dam. From saliency analysis, it is found that the gradient of tailings dam, friction angle of tailings and the surface flow are the predominant influence factors to the factor of safety.

INTRODUCTION

Tailings dams are distributed worldwide. Their failures can induce significant economic loss and casualties and then their stability is concerned by owners and the public. Due to daily mineral processing operation, tailings dams are highly stratified and heterogeneous. In addition, tailings dams are characterized by uncompleted consolidation as the height rises continuously. These features mean that the stability of tailings dam is a multifactor-influenced and highly nonlinear problem. It is necessary to find out an effective and efficient approach to evaluate stability of tailings dam.

For ordinary slopes, Artificial Neural Network (ANN) methods have been used for slope stability analysis during the past decades. Feng *et al.* (1995) proposed a slope evaluation model based on Back Propagation (BP) neural network and this type of model was utilized to evaluate Yudonghe landslide by Wang *et al.* (2005). Some other neural networks also were employed, such as Self-Organizing Feature Map (Xue *et al.*,

2008); Evolutionary Artificial Neural Network (EANN) based on Genetic Algorithm (GA) (Shangguan *et al.*, 2006). EANN has been proven to be of sufficient ability to map nonlinear slope problem (Shangguan *et al.*, 2006) and GA in the model can establish a stable network with first convergence performance.

Considering the complexity of stability problem of tailings dam, ANN method may be a promising approach. In this paper, an EANN model developed from BP network was proposed to evaluate the stability of tailings dam. Furthermore, saliency analysis was carried out to investigate the relative importance of input factors for the stability of tailings dam.

EANN MODEL FOR STABILITY ANALYSIS OF TAILINGS DAM

Input variables

In most existing ANN models for ordinary slopes, the mostly used input variables are unit weight, cohesion, friction angle, slope angle, height, pore pressure ratio (Lu and Rosenbaum, 2003). A schematic diagram of cross-section of a typical tailings dam is shown in FIG. 1. A tailings dam is generally stratified and properties of filling materials in different layers are different. Therefore, shear strength parameters, i.e. cohesion and friction angle are varied in a range. Due to the continuous accumulation of tailings materials, the phreatic line in a tailings dam is generally high accompanied with uncompleted consolidation. Sometime, there is surface flow. This greatly reduces the stability of tailings dam. Therefore, the emergence of surface flow is used as an input variable in the proposed EANN model.

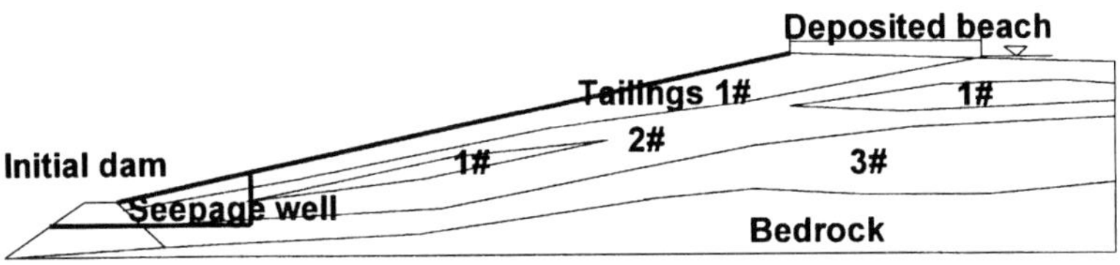

FIG. 1. A typical tailings dam

Based on previous analysis, in the proposed EANN model, 9 input variables are used. They can be classified into three main categories, i.e. geology, topography and external influence factors. The category of geology includes (1) the average value and (2) standard deviation of cohesions; (3) the average value and (4) standard deviation of friction angles, also (5) the average unit weight of saturated soils. The category of topography expresses the shape of tailings dam and it contains (6) the height of tailings dam; (7) the gradient. The variables, (8) the surface flow and (9) the average depth of phreatic line below slope surface are used as external influence. These two variables express the influence of pore water pressure on slope stability.

Evolutionary artificial neural network

ANN is a mathematic method to simulate characteristics of human brain. For the complex nonlinear mapping problem with 9 input variables, EANN was selected to evaluate the stability of tailings dam. EANN was developed to optimize the initial weights of BP network by using GA (Maniczzo, 1994). This method can result in excellent network performances.

In 60's, 20th century, GA was defined by Holland (Yuan *et al.*, 2002). GA firstly generates a set of random values as "population"; then begin to search optimized solution. After several repeated operations, GA converges at best "chromosome".

For neural network structure design, one hidden layer is proved to be a universal function approximator by Hornik (1993). The number of neurons in hidden layer directly affects mapping function of neural network in complicated problems. Generally, the number is selected as $2N+1$ (Hecht-Nielson, 1987).

Before training, usually the data cannot be applied directly to train network and some pretreatment is required. Firstly, a correlation coefficients analysis is needed for input vectors. If the correlation coefficients of couples of vectors are large, an uncorrelated handling is needed. Then, input variable vectors need normalized processing to have a value between 0 and 1, because the input variables belong to different dimensions. After simulating the network, results can be non-normalized. Moreover, in practical training, some samples are reserved to test network avoiding overtraining, and training parameters are determined through comparing results.

APPLICATION OF EANN MODEL

Data preparation for training

As an example for developing such an empirical model to evaluate the stability of tailings dam, 34 cases collected from the literatures (e.g. Zhang *et al.*, 2003; Li *et al.*, 2008) were utilized to calculate the factors of safety (FS) based on Bishop's method. Then above database was employed to train the proposed EANN model.

The correlation coefficients of input variables are shown in Table 1. Compared with the correlation coefficients of input variables used in ordinary slopes (Lin *et al.* 2008), the correlation coefficients of tailings dam model are relatively large, e.g. variables (1) and (2). This is related to geometry complexity and heterogeneity of tailings dams. Therefore, network input matrix must be preprocessed by applying a principal component analysis and input vectors were uncorrelated to reduce complexity (Jolliffe, 1986).

Table 1. Correlation Coefficients of Input Vectors

Issue	(1)	(2)	(3)	(4)	(5)	(6)	(7)	(8)	(9)
(1)	1	0.895	-0.46	-0.07	-0.13	0.041	0.011	0.121	-0.04
(2)	0.895	1	-0.41	0.014	-0.19	-0.04	0.129	0.136	-0.09
(3)	-0.46	-0.41	1	-0.2	0.181	0.274	-0.14	-0.09	0.25
(4)	-0.07	0.014	-0.2	1	-0.02	-0.26	-0.05	-0.17	-0.15
(5)	-0.13	-0.19	0.181	-0.02	1	0.231	-0.01	-0.03	0.205
(6)	0.041	-0.04	0.274	-0.26	0.231	1	-0.1	-0.02	0.53
(7)	0.011	0.129	-0.14	-0.05	-0.01	-0.1	1	0.155	-0.19
(8)	0.121	0.136	-0.09	-0.17	-0.03	-0.02	0.155	1	-0.54
(9)	-0.04	-0.09	0.25	-0.15	0.205	0.53	-0.19	-0.54	1

Prediction results

Based on the proposed model, a single hidden layer structure with 9 input variables, 19 hidden layer neurons and 1 output is adopted. Powerful nonlinear mapping function of the network lies in trigonometric function as the transfer function of hidden layer and logarithmic function as the transfer function of output layer. Training function is Levenberg-Marquardt, fast but needing more computer memory. The Mean Square Error (MSE) as train goal is 0.00001, "population" adopts the value of 100 and "generation" is 150 in optimization search by GA.

To demonstrate the performance of proposed EANN model, 4 samples were reserved to compute the Mean Relative Error (MRE). MRE is used to express the performance value of test samples in a random training and it is defined as follows.

$$\sum_k \frac{|T_k - O_k|}{T_k \times k} \times 100\% \tag{1}$$

Where T_k is target value and O_k is evaluation result.

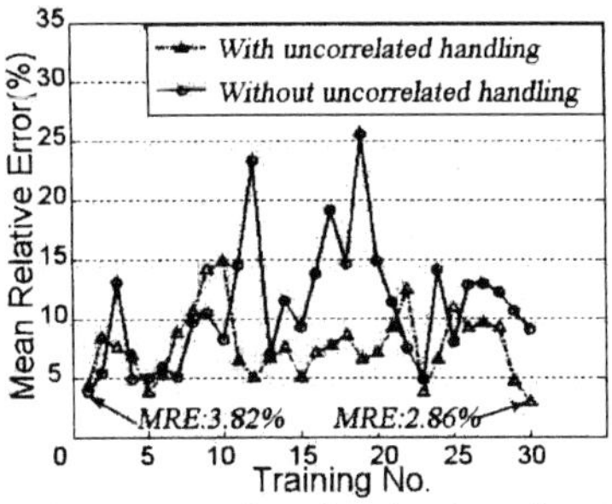

FIG. 2. Performances of evaluation based on EANN model.

As shown in FIG. 2, two curves respectively illustrate 30 times of random continuous trainings of EANN model through uncorrelated handling or not. From the figure, through a principal component analysis the performance values of repeated tests concentrate below 10%. As compare, MRE of the network without uncorrelated handling varies greatly, so the network possesses a more stable performance through a principal component analysis. Furthermore, FIG. 2 reveals that the best performance values of two sets of tests (MRE: 2.86% and 3.82%) are approximate, that attribute to the same network structure and train function.

Table 2 shows the detailed data of test samples in a test, the test samples' MRE is 2.86%. Previous 9 columns are input variables described in former section, FS is target value and EV is evaluation result. The largest error of FS is 0.08. Due to the tiny errors, the network possesses high accuracy.

Table 2. Evaluation Results of Test Samples based on EANN

(1)	(2)	(3)	(4)	(5)	(6)	(7)	(8)	(9)	FS	EV
4.1	4.5	29.5	3.1	20	94	1/2.2	1	2.5	0.72	0.74
9.6	10.2	31.6	3.6	20.9	154	1/2.5	1	2.4	0.86	0.84
8.6	8.4	28.8	6.7	19.5	60	1/4.2	1	2	1.38	1.34
26.5	19.1	31.3	3.2	19.8	59	1/4.2	0	6	2.57	2.65

Saliency analysis

The saliency analysis of EANN model was employed to assess the relative importance of inputs for the stability of tailings dam. According to training of neural network, nonlinear mapping function has been stored in the model. Then the saliency analysis is operated when network trains in the cases of incomplete input variables. Lee *et al.* (2009) applied this technique to estimate the relative importance of each input variable for slope stability based on BP network. Large relative importance depends on large difference with MRE of the network of complete input variables. FIG. 3 compares the performances of EANN when an input variable is absent. Just for factor of safety, the gradient, the friction angle of tailings and surface flow are the variables contributing to primary influence. The phreatic line related with incomplete consolidation, height of tailings dam and unit weight of saturated soils act as secondary influence. FIG. 3 also illustrates complex soil layers reducing FS of tailings dam from the variety of MRE when standard deviation of tailings strength is absent.

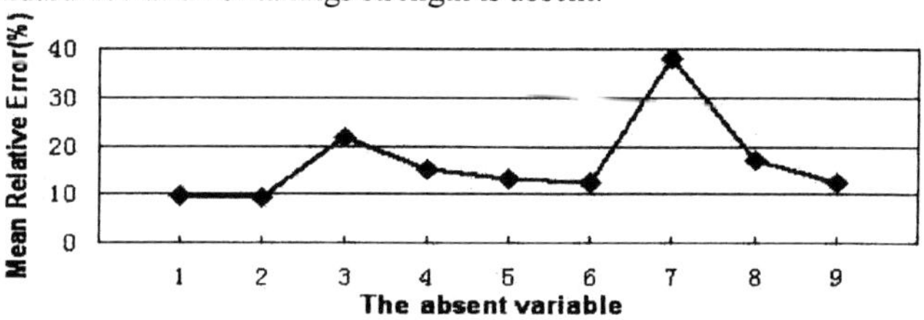

FIG. 3. Performances of incomplete input variables.

CONCLUSIONS

An EANN model was developed to assess the stability of tailings dam. In the model, 9 input variables characterizing the features of tailings dam were proposed. Due to the large correlation coefficients of input variables, a principal component analysis is necessary and it leads to faster convergence and more stable performance in the training stage. The factor of safety predicted by the EANN model has an error less than 3% as compared with that value calculated by Bishop's method, indicating an accurate evaluation for stability of tailings dam. This suggests that the proposed EANN model is an effective and efficient approach for assessing the stability of tailings dam. The saliency analysis demonstrates that the gradient of tailings dam, friction angle of tailings and the surface flow are the predominant influence factors to the factor of safety.

ACKNOWLEDGMENTS

This research was supported by the Natural Science Foundation of Guangdong Province, China (Grant No. 9451805707003007).

REFERENCES

Feng, X.T., Wang, Y.J. and Lu, S.Z. (1995). "Neural network evaluation of slope stability." *Journal of Engineering Geology*, Vol. 3 (4): 54-61.

Hecht-Nielsen, R. (1987). "Counterpropagation networks." *Applied Optics*, Vol. 26 (23): 4979-4984

Hornik, K. (1993). "Some new results on neural network approximation." *Neural Networks*, Vol. 6 (9): 1069-1072.

Jolliffe, I.T. (1986). "Pincipal component analysis." *New York: Springer*, ISBN: 0-387-96269-7.

Lee, T.L., Lin, H.M. and Lu, Y.P. (2009). "Assessment of highway slope failure using neural networks." *Zhejiang Univ SciA*, Vol.10 (1): 101-108.

Lin, H.M., Chang, S.K., Lee, D.H. and Wu, J.H. (2008). "Evaluating slope failures in Ali-shan, Taiwan by artificial neural networks." *GeoCongress 2008*, 862-869.

Lin, H.M., Chang, S.K., Wu, J.H. and Juang, C.H. (2009). "Neural network-based model for assessing failure potential of highway slopes in the Alishan, Taiwan Area: Pre- and post-earthquake investigation." *Engineering Geology*, Vol.104: 280-289.

Li, H.R., Hu, Z.Q., Chen, C.L. and Dang, F.N. (2008). "Numerical simulation and slope stability in Jingduicheng tailings dam to be designed to increase the dam height." *Rock and Soil Mechanics*, Vol. 29 (4): 1138-1142.

Lu, P. and Rosenbaum, M.S. (2003). "Artificial neural networks and grey systems for the prediction of slope stability." *Natural Hazards*, Vol.30: 383-398.

Maniezzo, V. (1994). "Genetic evolution of the topology and weight distribution of neural networks." *IEEE Trans Neural Netw*, Vol. 5 (1): 39-53.

Rico, M., Benito, G., Salgueiro, A.R., Diez-Herrero, A. and Pereira, H.G.(2008). "Reported tailings dam failures A review of the European incidents in the worldwide context." *Journal of Hazardous Materials*, Vol. 152: 846-852.

Samui, P. (2008). "Slope stability analysis: a support vector machine approach." *Environ Geol*, Vol 56: 255-267

Shangguan, Z.C., Li, S.J. and Zhao, J.Z. (2006). "Evolutionary artificial neural network for forecasting slope stability." *Journal of Harbin Engineering University*, Suppl: 92-96.

Wang, W.L., Yang, C.H. and Yan, J.A. (1992). "Investigation and stability analysis of a tailings dam." *CJRME*, Vol.11 (4):332-344.

Wang, H.B., Xu, W.Y. and Xu, R.C. (2005). "Slope stability evaluation using Back Propagation Neural Networks." *Engineering geology*, Vol. 80: 302-315.

Xia, Y.Y and Xiong, H.F. (2004). "Sensibility analysis of slope stability based on artificial neural network." *CJRME*, Vol.23 (16):2703-2707.

Xue, X.H., Zhang, W.H. and Liu, H.J. (2008). "Evaluation of slope stability based on SOFM neural network." *Rock and Soil Mechanics*, Vol. 29 (8): 2236-2240.

Yuan, X.M., Li, H.Y., Liu, S.S. and Cui G.T. (2002). "Application of neural network and genetic algorithm in the field of hydroscience." *China WaterPower Press*, ISBN: 7-5084-1085-8.

Zhang, D.J., Lei, X.J. and Zhang, X.T. (2003). "Stability analysis and treatment research of Longxingzhai impounding dam." *Rock and Soil Mechanics*, Vol. 24 (4): 670-672.

Zhang, C., Yang, C.H. and Xu, W.Y. (2004). "Reliability analysis of tailings dam stability." *Rock and Soil Mechanics*, Vol. 25 (11): 1706-1711.

Zhao, H.B. and Feng, X.T. (2003). "Application of support vector machines function fitting in slope stability evaluation." *CJRME*, Vol. 22 (2): 241-245

Identification of landslide susceptible slopes and risk assessment using a coupled GIS-FEA-module

Franz-Xaver Trauner[1], Conrad Boley[1] and Eva Nuhn[2]

[1]Institute for Soil Mechanics and Geotechnical Engineering, University of the Bundeswehr Munich, Werner-Heisenberg-Weg 39, 85577 Neubiberg, Germany; franz-xaver.trauner@unibw.de, conrad.boley@unibw.de
[2]Geoinformatics Research Group, University of the Bundeswehr Munich, Werner-Heisenberg-Weg 39, 85577 Neubiberg, Germany; eva.nuhn@unibw.de

ABSTRACT: Periodically, natural disasters like landslides cause many casualties and great damage. Therefore warning systems have been developed to protect people and goods. But to be able to set-up and to apply these systems, landslide susceptible areas need to be identified first and appropriate methods for hazard and risk assessment must be available.

In the research project a coupled module, consisting of a geographical information system (GIS) and a Finite-Element-Analysis-tool (FEA) for the investigation of slopes has been developed. Therewith, potentially unstable slopes can be analyzed from a geotechnical point of view. Physical processes, which may lead to slope failures and may cause the development of landslides can be investigated in detail. Moreover, the possibility to define criteria indicating slope failure and to determine critical magnitudes of events is supported by the module.

The set-up of geotechnical models for FE-Analyses is supported by the GIS as well as the evaluation and visualization of the computation results. Advanced visualization and analysis methods, regarding the hazard and risk assessment of slopes are merged in a decision support system, which is integrated in the GIS to support decision-makers in time.

INTRODUCTION

An increasing number of catastrophic landslides caused great damage worldwide during the last decades. Thus, more and more warning systems are installed for some exceptionally endangered areas, mostly based on deformation sensor measurements to alert prior to an event. However these systems are not available for the bulk of endangered areas and don't provide a deep insight into the physical processes causing slope failures. Thus, reliable and mechanically well-founded prognoses of slope deformations in future or estimations of the point in time, when slopes may fail and landslides are initiated, is hardly possible.

As a consequence, the GEOTECHNOLOGIEN-initiative has been launched by the German Ministry of Education and Research to enhance information systems for the purpose of early warning to catastrophic events.

Our research work is carried out jointly by the Institute for Soil Mechanics and Geotechnical Engineering and the Geoinformatics Research Group at the University of the Bundeswehr Munich. It is part of the multi-disciplinary project "Development of suitable information systems for early warning systems", which combines approaches from the fields of geology, geotechnics and geoinformatics for the identification and investigation of landslide susceptible areas and the early warning of disastrous events. Therein, our work is enriched by statistical and linguistic analyses (Gallus et al., 2008) and a 3D/4D geo-database for the storage and management of spatial and time-related data (Breunig et al., 2009). For detailed information on the joint project Breunig et al. (2007) refers.

GIS-FEA-MODULE

For the investigation of slopes the method of Finite-Element-Analysis (FEA) is applied. This analysis tool is combined with a geographical information system (GIS) to enable a user-friendly and assisted investigation of slopes regarding their stability and susceptibility to the development of landslides. Influences of various action effects (e.g. loads, accelerations, geometrical changes, etc.) on the stability and deformation of slopes are examined in numerical simulations, while the set-up of the geotechnical models and the processing and assessment of computation results is supported by the GIS component (Fig. 1).

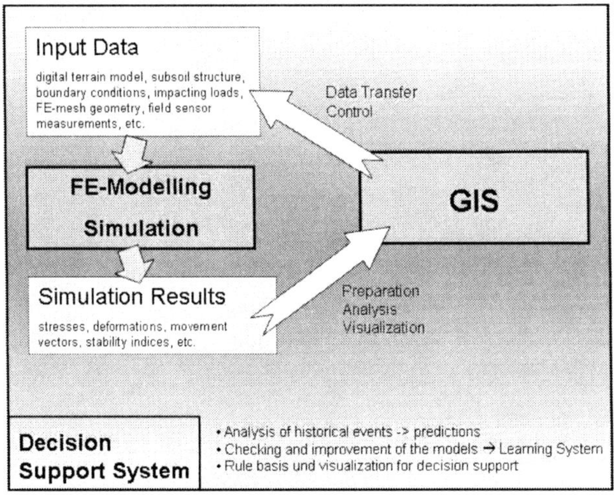

FIG. 1. Architecture of the coupled GIS-FEA-module

When an area on local scale shall be examined in detail, first the respective area of interest is selected within the GIS on basis of topographical maps, orthophotos or land development plans (Fig. 2). But of course, any other source may be chosen, which may, for instance, already suggest the distribution of increased landslide susceptibility determined by statistical analyses.

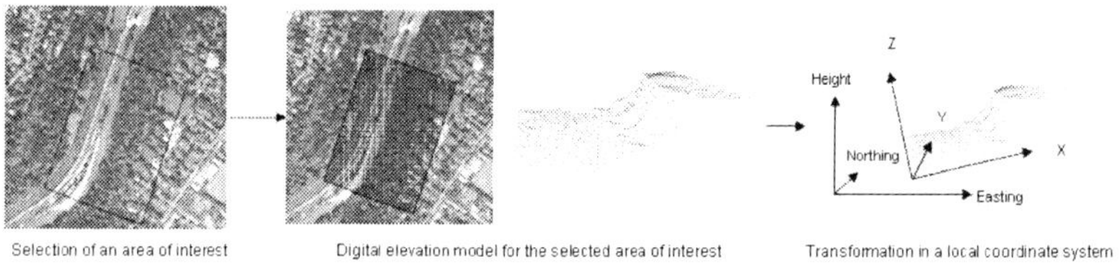

FIG. 2. Selection of the area of interest within the GI system

Thereafter the topography of the slope is extracted from a digital elevation model and supplemented with information about the subsoil structure, e.g. stratigraphic or lithologic units, ground water levels or civil works, either by query to a data base or by direct entry of the user. The data base stores all information about the subsurface structure like data from site investigations (e.g. profiles from bore holes, ground water level measurements, geodetic surveys, etc.) and field or laboratory tests (e.g. densities, grain size distributions, consistency and plasticity, etc.).

Finally, the geometrical description is supplemented by boundary conditions, action effects and other details to compile the geotechnical model of the slope. A simplified three-dimensional model for a segment of the slopes in the Isar valley south of Munich is shown in Fig. 3. Alternatively, two-dimensional profiles can be compiled in the same manner as described above.

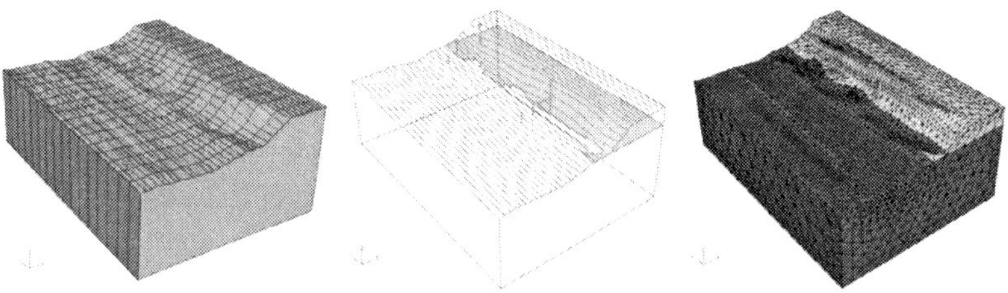

FIG. 3. Generation of the geotechnical model for FE-analysis

The stability and deformation of slopes due to action effects are investigated within the simulation component by Finite-Element-Analyses. Therefore geotechnical models, which represent the slope in a realistic manner to a large extent, have to be generated.

For modeling of slopes in a realistic manner, an appropriate characterization of the material involved is essential. Since different materials favor different failure mechanisms, the geotechnical models have to reflect this association. Therefore the decision between time-invariant or time-variant models has to be made considering both, the material's characteristics (change of material properties with time, e.g. degrading materials or dependence of strength on moisture content) and the probable failure mechanism (e.g. spontaneous failure or creep deformations).

The generation of models for prognosis purposes includes the validation and verification of these models. Prior to the deployment of these models it must be shown that their reaction to implemented action effects is similar to the ones observed in nature. This is assured during the calibration process, when events observed in the past, e.g. prior slope failures or slope deformations are simulated with the geotechnical model and then the computed results are compared to recorded deformations. To adjust the behavior of the models and to bring them into line with the natural slope, modifications (e.g. variation of material parameters, change of constitutive equations or degree of geometry's simplification) may be necessary.

Once the geotechnical model has been calibrated successfully and shows satisfying consistency in its behavior compared to the real slope, it can be applied for prognosis purposes. The outcomes and consequences of various scenarios (characteristics of deformations, change in material strength utilization, reduction of safety factor, etc.) for the slope can be determined in a simulation. Moreover the simulations can be used to determine critical magnitudes or thresholds for loads, accelerations, ground water levels or any other action effects. The knowledge of these thresholds is essential for early warning and purposes and hazard management, since they characterize the slopes' ultimate limit states.

Subsequently to the simulations, relevant computation results regarding the assessment of the endangering by possible slope failures, are compiled in an output file and transferred to the linked GIS for further processing and visualization of the data. Typically, the significant output variables are strains or deformations, deformation rates in the case of time-variant simulations, degrees of material strength utilization or local and global safety factors. But also data indicating failure mechanisms or particularly weak and highly stressed zones provide information about the system characteristics and probable type of failure. To support the assessment of the results, a rule-base containing, for example, thresholds for critical deformation rates is established within the GIS and assists users in similar situations in the future.

A main focus within the project is put on the data preparation and visualization in such a manner that the results are understandable and useable for decision makers.

DATA PREPARATION FOR DECISION SUPPORT

The outputs of the simulations are extensive and may confuse, if no assistance is provided how to obtain the required information to support a decision. A simple but plausible example is the determination of deformation vectors, representing the movements of the nodes of the meshed geotechnical model (Fig. 4a).

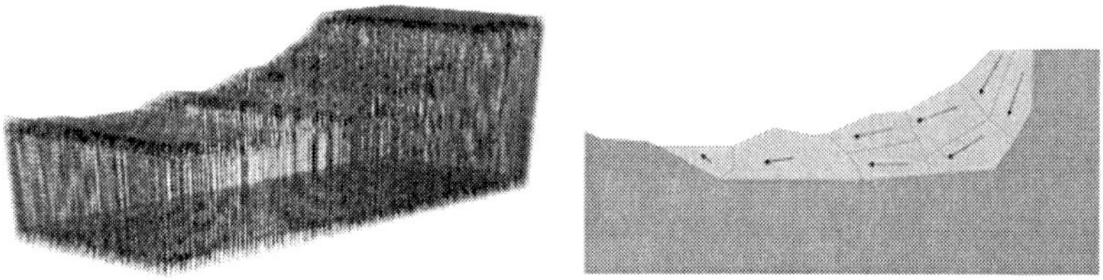

a) b)

FIG. 4. Deformation vectors on nodes of meshed geotechnical model

Usually it is not of great importance to have deformation vectors covering the whole model, but to distinguish between regions of different movements, both regarding the magnitude and direction.

Therefore methodologies have been developed to organize deformation vectors in classes for cluster identification (Ortlieb et al., 2009). A cluster defines a region characterized by points of comparable movements, i.e. similar direction and magnitude of deformation, and which are adjacent to each other. As a result, one deformation vector for each cluster is obtained and can sufficiently show the dominant deformation characteristics of a slope. A simplified two-dimensional example is given in Fig. 4b. In addition, this methods allow the determination of the boundary of the sliding bodies, to separate the surrounding area with no or minor movements or to calculate the depth of a sliding plane or total volume of unstable masses.

Moreover, the processed computation results can be enriched by additional data from any other resources in the GI system to support the decision-making process. This information can be used in conjunction with the cluster identification to identify buildings or structures on and close to the sliding bodies (Fig. 5).

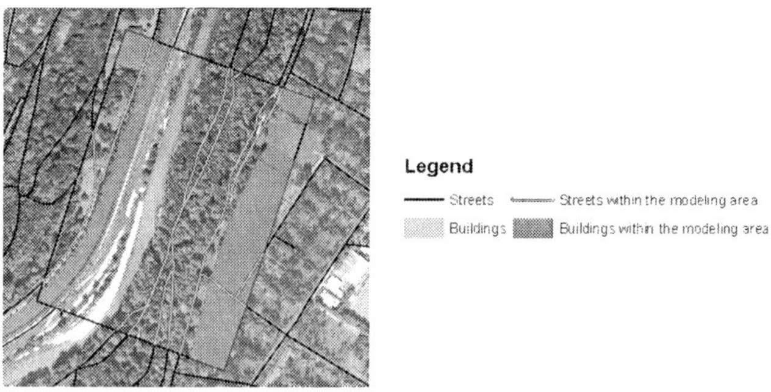

FIG. 5. Modeling area with overlain topographical data

CONCLUSIONS

The application of the GIS-FEA-module allows for the geotechnical-based determination of stability indices and thresholds for slope failures and therewith facilitates the identification of landslide susceptible slopes. The user-friendly set-up of the system including a GIS environment allows for a quick set-up of the geotechnical models compared to conventional approaches and assisted analysis of computation results to support decision-makers in the hazard management and risk assessment.

The conception of the module with its possibility for connection to other modules based on various approaches (statistics, analytical computation methods, etc.) via GIS and the integrated assistance functions for decision-makers provide a user-friendly and reliable medium for warning purposes and contribute to the reduction of threats posed by landslides.

ACKNOWLEDGMENTS

The authors appreciate the funding by the German Ministry of Education and Research (BMBF) by grant no. 03G0645A within the framework of the GEOTECHNOLOGIEN-initiative (www.geotechnologien.de). We also thank the Bavarian Environment Agency (www.lfu.bayern.de) for providing data for the application area. This is publication no. GEOTECH-1297 of the R&D-Programme GEOTECHNOLOGIEN.

REFERENCES

Breunig, M., Reinhardt, W., Ortlieb, E., Mäs, S., Boley, C., Trauner, F.-X., Wiesel, J., Richter, D. (2007). "Development of suitable information systems for early warning systems (EGIFF)" *Geotechnologien Science Report (Early Warning Systems in Earth Management)*, Vol. 10, 113-123.

Breunig, M., Broscheit, B., Thomsen, A., Butwilowski, E., Jahn, M., Kuper, P.V. (2009). "Towards a 3D/4D Geo-Database supporting the Analysis and Early Warning of Landslides" *Proceedings of Cartography and Geoinformatics for Early Warning and Emergency Management: Towards Better Solutions*, Prague/Czech Republic, January 19-22, 100-110.

Gallus, D., Abecker, A., Richter, D. (2008). "Classification of landslide hazard/ susceptibility in the development of early warning systems" *Proceedings of the 13th International Symposium on Spatial Data Handling*, Montpellier/France, June 23-25, 55-75.

Ortlieb, E., Reinhardt, W., Trauner, F.-X. (2009). "Development of a coupled Geoinformation and Simulation System for Early Warning Systems" *Proceedings of Cartography and Geoinformatics for Early Warning and Emergency Management: Towards Better Solutions*, Prague/Czech Republic, January 19-22, 483-492.

Deformation of anchor-sheet pile wall retaining system at deep excavations in soft soils overlying bedrock

Jianqin Ma[1], Bo S. Berggren[2], Håkan Stille[3] and Staffan Hintze[3]

[1]College of Highway, Chang'an University, Xi'an, China, 710064; majq@gl.chd.edu.cn
[2]Swedish Geotechnical Institute, Linköping, Sweden, 58193
[3]Royal Institute of Technology, Stockholm, Sweden, 10044

ABSTRACT: This contribution gives a case study on the deformations of anchor back-tied steel sheet pile walls and retained soils at deep excavations in soft soils overlying bedrock in Stockholm. Instrumentation results show the deformation of the retained soils is larger than that from similar case histories. The monitored anchor stresses increase quickly in the early stages of excavation and decease gradually to a constant value during later stages, while the lateral displacements of the sheet pile walls increase gradually. The results of both monitoring and numerical analysis show that this dynamic feature becomes weak as the thickness of the underlying soils is small. The decrease of the anchor stresses is mainly due to the flectional behavior of the sheet pile wall and the stress rotation in the retained soils.

INTRODUCTION

The idea of the efficient use of underground space has been accepted since 1970s. Deep excavations are being carried out to meet the surging need for infrastructure, especially in urban areas. During construction, a retaining wall system is often used to control the movements of adjacent ground. Of a successful deep excavation, the behaviors of the wall and the adjacent ground must be considered during design. The behaviors are related to a number of factors for a deep excavation in soft soils (Peck, 1969; Mana and Clough, 1981). The estimation of the behaviors is generally a combination of analytical and empirical methods (Peck, 1969; Mana and Clough, 1981; Clough and O'Rourke, 1990 and Ou et al., 1993), combined with judgment and experience. Recently, Hashash and Whittle (2002), Finno and Calvello (2005) gave examples for numerical analysis on deep excavation. Shao and Macari (2008) reviewed the application of instrumentation and numerical feedback analysis. The deformation feature of a retaining system depends not only on the properties of the excavated soils, but also on the properties of the underlying layers. Case histories in multilayered soils overlying rock (Wong et al., 1997; Yoo, 2001 and Long, 2001) showed relatively small deformation. However, there are significant differences in deformation magnitude between the cases histories, which are of regional characteristic. It is of interest to know that whether the benefit of underlying bedrock is applicable to other cases.

This study aims at understanding the behavior of anchor-sheet pile walls constructed in multilayered soils overlying hard rocks. To meet this goal, measured data from the South Link 10 (SL10) in Stockholm are analyzed. A 2D finite element analysis with Plaxis code on a test section is used to show the performance of the retaining system.

PROJECT DESCRIPTION

This case is from a cut and cover concrete tunnel and its ramp. The geotechnical conditions can be simply expressed as soft soils overlying bedrock. The soils consist of a fill layer overlying a sequence of glacial deposits, with about 1 m thick dry crust in top layer. At the bottom of the soils, there is granular soil or dense moraine about 1 m thick. The bedrock is mainly unweathered gneiss and its surface elevation varies within working area. The maximum thickness of the soils is about 25 m. Groundwater level is around 2 m below surface. The ground surface level is around +15 m. The undrained shear strength of the clay is around 16 kPa, with an increase of 2 kPa/m below the level of +10 m (Hintze et al., 2000). The overconsolidation ratio (OCR) of the clay decreases from approximately 1.4 at surface to 1.0 at a level of +6 m and below.

Deep excavations, with depth varing from 3 to 16, are supported by sheet pile walls with back-tied anchors. Anchors are keyed into bedrock and pile toe is grouted to prevent groundwater leakage (Hintze et al., 2000). The number and intervals of anchors vary with the length of the wall and site conditions, respectively.

The construction activities can be presented as wall installation, excavation and support. Excavations inside the sheet pile walls are proceeded in stages, which are alternated by the installation and preloading of anchors.

The site instrumentations include 11 inclinometers and 11 settlement points. 21 load cells are used to monitor anchor stresses at 10 test sections, respectively. The inclinometers are mostly 2 m from the sheet pile wall and the settlement points are 1 m from the wall. Inclinometers are vertically installed and keyed into bedrock. The interval of the sensors in an inclinometer is 2 m and its toe is considered totally fixed in displacement calculation.

DEFORMATIONS OF THE WALLS AND RETAINED SOILS

The deformations of the maximum lateral displacement ($\delta_{h,m}$) and settlement ($\delta_{v,m}$) are generally shown as a value in relation to excavation depth (H) (Peck, 1969). The monitored results from SL10 show $\delta_{h,m}$ is between 0.4%H and 1.7%H, and is in tendency of around 1%H. The $\delta_{v,m}$ is between 0.6%H and 2.7%H, and is in tendency of around 1.5%H. The magnitude of $\delta_{v,m}$ is larger than that of $\delta_{h,m}$ in a section. This indicates that the retaining system at SL10 is effective to control the movement of the retained soils.

The studies on deep excavations in soft clay overlying stiff soil layer or bedrock show that the underlying stiff layers could have a dominating influence on the displacement of a retaining wall system (Mana and Clough, 1981; Wong et al., 1997; Yoo, 2001 and Long, 2001). Yoo (2001) made a conclusion regarding deep excavation in soft soils overlying bedrock in Korea. Figure 1 shows that all of the data from Korea (Yoo, 2001) are below the ratio 0.5 of $\delta_{h,m}$ /(10H), while the data from SL10 are mainly above it. This implies that the displacement feature at SL10 is different from that of Korea. In Figure 1, the magnitude of $\delta_{h,m}$ from SL10 is similar to the data of the propped walls with h>H, as shown Long (2001), where h is the thickness of soft soil layers. This shows that the relative thickness of the soft soils may have a key influence on the deformation of the sheet pile wall-anchor system.

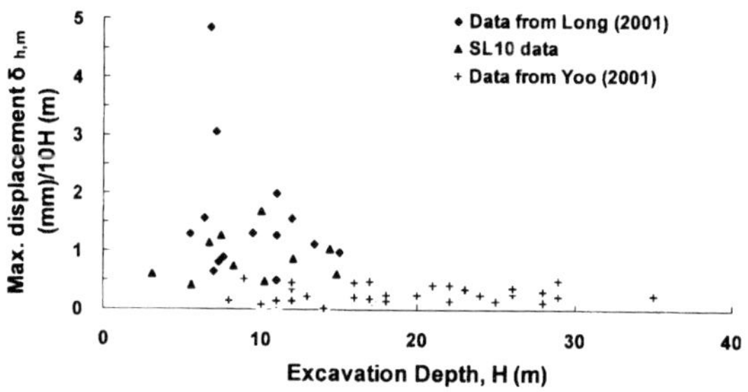

FIG. 1. Variation of $\delta_{h,m}/H$ with Excavation Depth.

The effect of system stiffness for a propped retaining wall system on the maximum lateral displacement can be significant for deep excavation in soft soils. For a sheet pile wall-anchor system, the definition of system stiffness as $k_s = EI/(\gamma_w h^4_{avg})$ is well accepted, where γ_w is the unit weight of water and h_{avg} is the average vertical spacing of the anchors, E is the Young's modulus of the wall and I is the structural internal moment of inertia per unit length of the wall. Figure 2 shows that most of the data from Korea are below a boundary line as $\delta_{h,m}/H$ (%)= 0.5exp(-0.007k_s) (Yoo, 2001) and that relative lateral deformation, $\delta_{h,m}/H$, decreases with system stiffness k_s increasing.

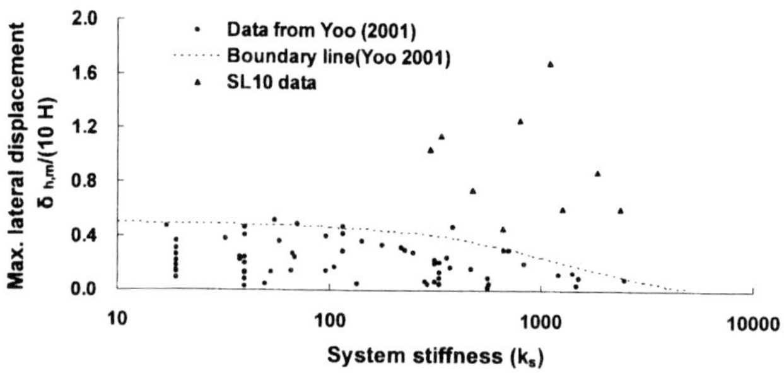

FIG. 2. Variation of $\delta_{h,m}$/H with stiffness of support system k_s.

In Figure 2, the data from SL10 is far above the boundary line and there is no clear relationship between $\delta_{h,m}/H$ and k_s. The scatter is greater for the data from SL10 than that from Korea. This implies that the lateral wall movements at SL10 are more prone to site conditions. Once sufficient stiffness is available, movement is determined by the magnitude of base heave, preloading (Ma et al., 2006) and also if the soils have significant capacity to "arch" (Long, 2001).

Figures 1 and 2 indicate that the influence of the bedrock may be overshadowed as the depth of the excavation bottom to bedrock (D) is large enough. In order to check this deduction, the lateral deformation of retaining wall with excavation stage in test section 1/840N is shown in terms of the results of both instrumentation and numerical analysis with Plaxis code (Ma et al., 2006), as shown in Figure 3. In Figure 3, the stages of even number and odd number correspond to excavation and the installation and preloading of anchors, respectively. Figure 3 shows that most of the lateral displacement increment from preceding stage takes place in the excavations of stages 4, 6 and 8. At stage 8, the increment occurs mainly in the lower part of the test section. At stages 10 and 12, the increment is small or even negative.

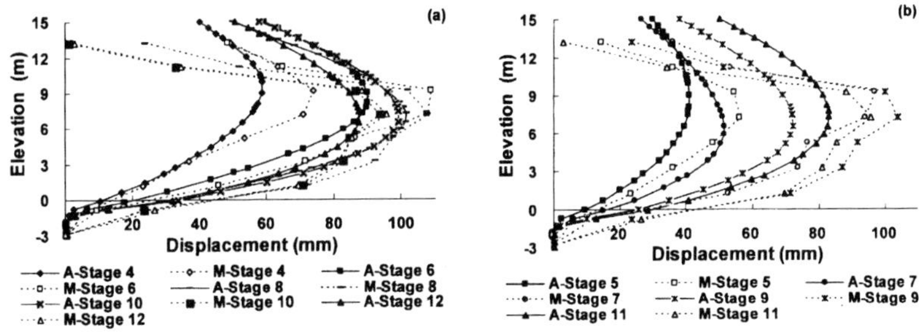

FIG. 3. Comparison of numerical analysis (A-) and monitored (M-) displacement.

The influence of the bedrock on the lateral displacement increases with the excavation bottom getting nearer to bedrock. The data from SL10 indicate that the maximum lateral displacement increases with the formation level more above the bedrock. In section 1/840N, the normalized maximum displacement decreases with the ratio of ratio of H/h increasing, as shown in Figure 4 (a). The stage increment of lateral displacement is small provided the ratio of D/H is less than 1.0 as shown in Figure 4 (b).

The apparent earth pressure on the retained wall is related to the deformation feature of the wall, as well as to the resistance against wall movements provided by bedrock. The apparent earth pressures of the upper part of the test sections from SL10 are larger than the value of the Peck (1969) envelope value of $0.89\gamma H$, while data from the lower level is less than the envelop value, where γ is average unit weight of the soils.

During the stage excavations, the monitored anchor stresses show dynamic feature with the excavation proceeding. At the beginning of a stage excavation, the stresses of the anchors increase quickly. But the stresses decrease gradually to a constant value with excavation proceeding while the lateral deformations of the sheet pile walls increase gradually. This dynamic feature become weak as the ratio of D/H approaching to unit, as indicated by results of both monitoring and numerical simulation in Figure 3.

The relatively large apparent earth pressure of the upper part of the walls at SL10 is relevant to load transfer and arching in retained soils. Numerical analysis result of the test section 1/840N indicates that there occurs arching effect and significant stress redistribution in the upper part of the retained soils, though the wall toe is fixed into the bedrock. Dry crust in the upper part of the test section and the anisotropic yield strength of the retained soils play a key role in stress transferring through arching mechanism.

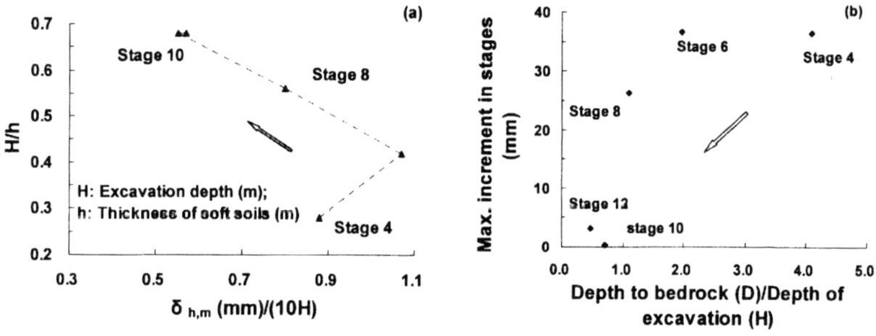

FIG. 4. Maximum lateral displacement and its increment in excavation stages.

The decrease of the anchor stresses is mainly due to the flectional behavior of the sheet pile wall and the shear deformation of the retained soils, where there occurs stress rotation. The numerical and case histories (Hashash and Whittle, 2002) indicated that arching mechanism may also occur in lightly overconsolidated soil layers. Similarly, the largest displacement point occurs in the middle part of a test section (Fig. 3) at SL10, with the upper part of the sheet pile wall well propped by top anchor (Ma et al., 2006). The magnitude of the displacements of the sheet pile wall and its toe section can meet the requirement of the toe not being fixed.

In Figure 3, there occurs significantly negative increment of lateral displacement from preceding stage at the later excavation stages in the upper part of the test section. Part of the load from top anchor should be transferred through the sheet pile wall flexural deformation as shown in Figure 3. The retained soils should be in passive earth pressure state behind the top anchor. As the load of top anchor is transferred by sheet pile walls to the retained soils through wall backward flexure, the rotations of the principal stress directions and reversal in shear direction are relatively more significant for the upper part of the wall in the early excavation stages. This is similar to the results of Hashash and Whittle (2002). However, at SL10, the apparent earth pressures in the middle and lower parts of the test walls are less than the stresses at rest in the soils. This is mainly due to the low strength of the retained soils and arching effect is even not. The back deflection in the upper part and the larger lateral deformation in the middle part of the wall (Fig. 3) are harmonious with this deduction. The results from SL10 imply that caution should be paid to interpreting the stress paths for soil elements adjacent to the steel sheet pile wall, where arch mechanism occurs depending on anchor stresses and the lightly overconsolidated feature of the soft soils.

CONCLUSIONS

This study on the deformations of retaining wall and retained soils at deep excavations in soft soils overlying bedrock at SL10 in Stockholm gives following conclusions.

The deformations of the steel sheet pile walls and retained soils at SL10 are larger than the magnitude from similar case histories. The monitored anchor stresses show dynamic feature with excavation proceeding and the feature become weak as the excavation is near to bedrock. The decrease of the anchor stresses is mainly due to the flectional behavior of the sheet pile wall and the stress rotation in the retained soils.

The low strength of the retained soils and the thickness of the soft soil layers h are the most prominent factors of controlling the deformations. The beneficial influence of the bedrock may be overshadowed by the low strength of the soils if h is large enough.

REFERENCES

Clough, G.W. and O'Rourke, T.D.(1990). "Construction induced movements of in situ walls." Proc. on Conf. on Design and Performance of Earth Retaining Structures, ASCE, Geotechnical Special Publication No. 25: 439-470.

Finno, R.J. and Calvello, M. (2005). "Supported Excavations: the Observational Method and Inverse Modeling." *J. Geotech. and Geoenvir. Engrg.*, Vol.131(7): 826-836.

Hashash, Y.M.A. and Whittle, A.J.(2002). "Mechanisms of load transfer and arching for braced excavations in clay. *J. Geotech. and Geoenvir. Engrg.*, 128(3):187-197.

Hintze, S., Ekenberg, M. and Holmberg, G.(2000). "Southern Link Road Construction: Foundation and Temporary Constructions." Proc. 16th IABSE. Switzerland.

Hsieh, P.G. and Ou, C.Y. (1998). "Shape of ground surface settlement profiles caused by excavation." *Can. Geotech. J.*, Vol.35(6): 1004–1017.

Long, M. (2001). "Database for Retaining Wall and Ground Movements due to Deep Excavations." *J. Geotech. and Geoenvir. Engrg.*, Vol.127(2):203-224.

Ma, J.Q., Berggren, B.S., Bengtsson, P.E., Stille, H. and Hintze, S.(2006). "Back analysis on a deep excavation in Stockholm with finite element method." Proc. 6th European Conference on Numerical Methods in Geotechnical Engineering, 423-429.

Mana, A.I. and Clough, G.W.(1981). "Prediction of movements for braced cuts in clay." *J. Geotech. Engrg. Div.*, Vol.107(6): 759–777.

Ou, C.Y., Hsieh, P.O. and Chiou, D.C.(1993). Characteristics of ground surface settlement during excavation." *Can. Geotech. J.*, Vol.30(5):758–767.

Peck, R. B.(1969). "Deep excavations and tunneling in soft ground". Proc., 7th Int. Conf. on Soil Mech. and Found. Engrg., State-of-the-Art Rep., Vol., 225–290.

Shao, Y. and Macari, E.J. (2008). "Information Feedback Analysis in Deep Excavations." *International Journal of Geomechanics*, Vol. 8(1): 91-103.

Wong, I.H., Poh, T.Y.and Chuah, H.L.(1997). "Performance of Excavations for Depressed Expressway in Singapore." *J. Geotech. and Geoenvir. Engrg.*, Vol.123(7): 617-625.

Yoo C.S. 2001. "Behavior of braced and anchored walls in soils overlying rock." *J. Geotech. and Geoenvir. Engrg.*, Vol. 127(3): 225-233.

Centrifuge Model Study of Impact on Existing Undercrossing Induced by Deep Excavation

X.Y. Xie[1], Z.W. Ning[1], X.R. Liu[1] ,F.Z. Liu[1]

[1] Key Laboratory of Geotechnical Engineering, Department of Geotechnical Engineering, Tongji University, No.1239 Siping Road Shanghai, 200092; xiexiongyao@tongji.edu.cn

Abstract: A 33-meter deep excavation with multi-level struts for a ventilation shaft of Shanghai metro line 9 was simulated by a series of centrifuge model tests at Tongji Centrifuge Laboratory in 100g acceleration filed to study its impact on an adjacent undercrossing and the effect of barrier piles-wall installed between them. The draining-fluid method was used to simulate the process of excavation while a new hydraulic-control device to simulate the installation of multi-level inner struts in-flight was developed. The observed settlements of undercrossing in tests were compared with those obtained in field monitoring in the first place. Then the behaviours of the undercrossing in tests differing in parameters of barrier piles-wall were analyzed. Finally, the impact range of the excavation and the recommended geometric parameters of barrier piles-wall were discussed.

Keywords: centrifuge model test, deep excavation, in-flight, inner struts, undercrossing, barrier piles-wall.

1. INTRODUCTION

As the massive construction of metro system goes on in Shanghai, deep excavation in urban area becomes a great threaten to the safety of surrounding environment. A 25m×20m deep excavation for Yuanshen Road ventilation shaft of Shanghai metro line 9 was constructed adjacent to an existing undercrossing structure. The excavation is 33m deep and the minimum distance between the excavation and the undercrossing is only 3m. To make sure the safety of the excavation, 59m deep diaphragm wall with a thickness of 1.2m was chosen as the retaining structure and 9 levels of inner struts were installed. In order to protect the undercrossing, a barrier piles-wall made up of φ 850mm mixing piles was constructed between the excavation and the undercrossing. The depth of the piles-wall is 40m and extends 10m laterally from the boundary of the excavation on each side.

Centrifuge modelling is a useful tool to study the geotechnical problems because of its ability to reproduce the same stress level in a small-scale model. However, centrifuge modelling of deep excavation issues is strongly limited by proper approaches to simulate the process of excavation and struts installation. Different techniques have been used over the years. The first and simplest method is to remove the soil and construct the struts by hand when stopping the centrifuge and then restart it to a required acceleration. This process is repeated until all the construction phases are completed. It however does not reproduce the real stress path of soil correctly. Another widely used method simulates the excavation by replacing soil to be excavated by certain liquid with a unit weight close to the soil and then draining the liquid with accurate control. The short coming of this method is that the horizontal and vertical stress of soil can not be correctly simulated at the same time since ko is fixed in fluid as units. The best but complicate way is to design special devices or on-board robots to perform the desired operation, which are commonly used in many centrifuge laboratories around the world nowadays.

In present study, a new device to simulate the installation of multi-level inner struts in-flight was developed and the draining-fluid method was combiningly used to study the impact of this deep excavation on the adjacent undercrossing. All the tests stated here were carried out at Tongji Centrifuge Laboratory in 100g acceleration filed.

2. Centrifuge modeling

2.1 Centrifuge tests facility

The centrifuge tests presented in this study were carried out on the 150g-tons beam centrifuge with a effective radius of 3m at Tongji Centrifuge Laboratory. A hydraulic-control in-flight excavation and struts installation device provides the key technology in this research. All the operations of the device can be achieved through a uniform control platform in the control room.

The replacing liquid was contained in a chamber enclosed by diaphragm wall model and a steel rigid plate. A hydraulic-control valve was connected to the outlet of the chamber with a load-cell installed to monitor the level of the liquid surface. Thus, the liquid can be drained step by step to any level according to the real excavation procedures. The diaphragm wall was supported by struts at several levels with two baring points at each level. Hydraulic-control lock was installed at the end of each strut to lock or free the strut. All the struts contacted the diaphragm wall with hydraulic-locks kept open in the beginning. Thus the struts can move freely in horizontal direction without applying any force to the diaphragm wall. When the surface of liquid fell beneath struts at a certain level, the position of those struts would be fixed by activating the locks on them. This process repeated till the liquid was completely drained off. Figure 1 shows the schematic drawing of this device.

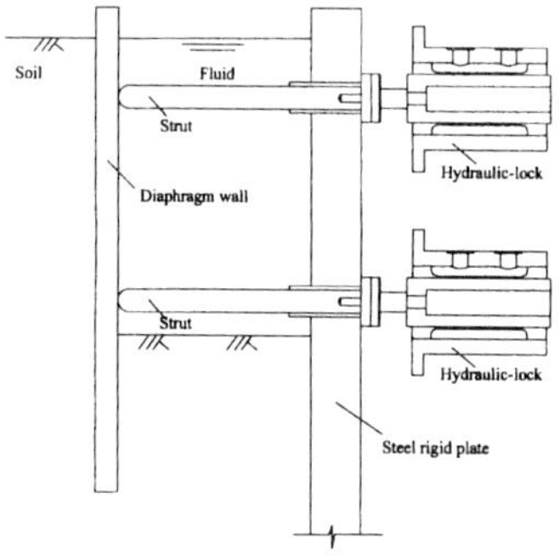

FIG. 1. Schematic drawing of excavation and struts installation simulation device

2.2 Tests set-up

In present study, four 1:100 scale centrifuge model tests were performed in a strongbox with internal dimension of 900mm×700mm×700mm. Test1, Test2 and Test3 differed in parameters of barrier piles-wall while Test 4 was an extra test with no excavation. Table 1 gives brief specifications of each test.

Because of the symmetric layout of the real project, only half part of the prototype was modelled. Such an arrangement helped in constructing a model of reasonable size to minimize the boundary effect. A rigid steel plate which located in the position of the central axis of the excavation in prototype was set up to divide the strongbox into two sections. The bigger one was for the arrangement of soil, liquid and structures while the smaller one was for the accommodation of hydraulic-lock, valve and accessorial oil pipes.

Table 1. Specifications of Tests

Test No.	Soil excavation	Barrier Piles-wall	Depth of Piles-wall	Width of Piles-wall
Test 1	Yes	Yes	40m	40m
Test 2	Yes	Yes	20m	30m
Test 3	Yes	No	-	-
Test 4	No	No	-	-

Test 1 referred to the equivalent specification of real project.

Due to the limit free space in the strongbox, the struts were reduced into 5 rows by two columns. To make sure an equivalent lateral rigidity of the whole retaining structure to the prototype, compensation were made both in the thickness of the diaphragm wall and the radius of the struts.

The instrumentation included 6 LVDTs to measure the settlement of the undercrossing, 10 load-cells to measure the struts' stress and 5 eddy current sensors to measure the displacements of the struts at 5 different levels(-2m, -9m,

-16m, -23m ,-30m). The following discussion mainly focuses on the observed settlement of the undercrossing.

The plan view and cross-section 1-1 of the model and instrumentation arrangement are shown in Figure 2 and Figure 3.

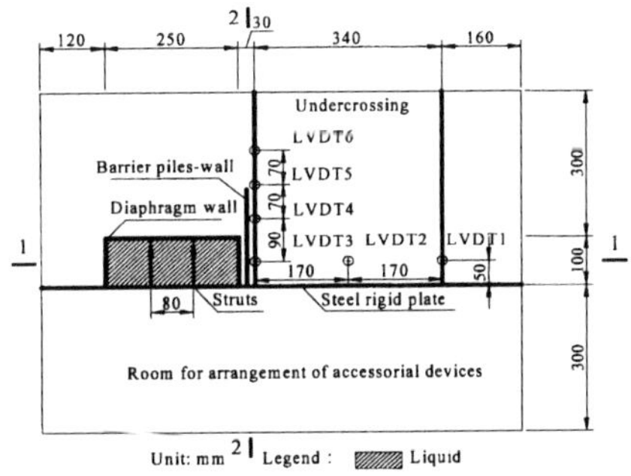

FIG. 2. Plan view of model and instrumentation arrangement

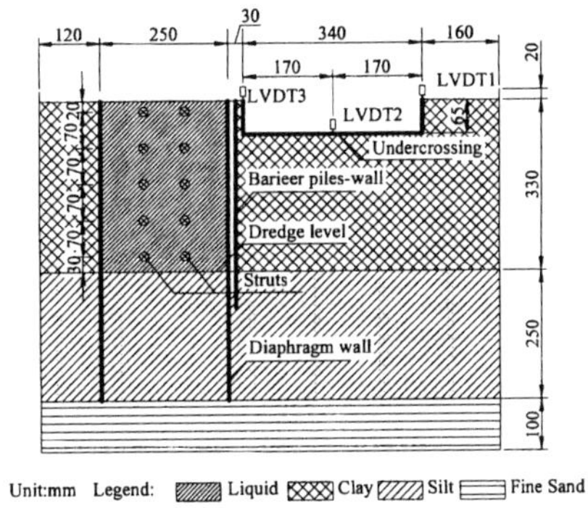

FIG. 3. Cross-section 1-1 of model and instrumentation arrangement

2.3 Model preparation and test procedures

The ground model consisted of three layers in this study: a 330mm thick layer of clay above the dredge level, a 250mm thick layer of silt below the dredge level and a 100mm thick layer of find sand at the bottom, which represented a simplified ground profile of the real site. The clay and silt were saturated at first and then were consolidated in the centrifuge under 100g for 2-3 hours until at least 90% degree of consolidation was achieved. The $Na_2O.3SiO_2$ solution with a density of 1.36g/ml was employed in these tests providing a lateral pressure close to the soil. Aluminum plates with various thicknesses were used to model the diaphragm wall, undercrossing structure and the barrier piles-wall according to the flexural stiffness of the prototype. The struts were made of steel pipes with a uniform outer radius of

10 mm. The specifications of both ground and structure models are shown in Table 2.

Table 2. Specifications of models

	Density(kN/m3)	Water content(%)	Thickness(mm)
Clay	16.7	50.8	330
Silt	18.1	34.5	250
Sand	18.8	27.4	100
Diaphragm wall			10
Barrier piles-wall			1
Side walls of undercrossing			4
Bottom slab of undercrossing			8

The interface between the diaphragm wall and the rigid plate was sealed by elastic sealant to prevent the leakage of fluid from inside while downward vertical movement of the wall was allowable. A rubber slab was installed inside the diaphragm wall at the dredge level to separate the liquid from the soil. And the bottom heave due to the excavation was allowable as well.

After the completion of the set-up, the model ground was re-consolidated under 100g with all hydraulic-locks open and the liquid fully filled till the observed settlement of the undercrossing was stable. Then the excavation and struts installation proceeded as being described above.

3. Results and Discussion

The duration for each test keeping on 100g was around 20 minutes corresponding to approximately 5-month excavation period in real project. The original measured settlement of the undercrossing consisted of two portions. One was induced by excavation while the other was caused by the residual consolidation of the ground model. To eliminate the influence of the latter portion, a test without excavation was conducted to observe the settlement caused purely by consolidation during the same period (Test 4). All the results discussed in the following sections have been rectified by subtracting the extra settlement induced by residual consolidation and presents in prototype scale.

Figure 4 shows the typical settlement time-history curves of the undercrossing induced by excavation (Test 1). It can be found that the settlement occurred after the excavation depth exceeded 5m and then increased gradually as the excavation proceeded. In some monitoring points, the observed value fluctuated when the inner struts were constructed.

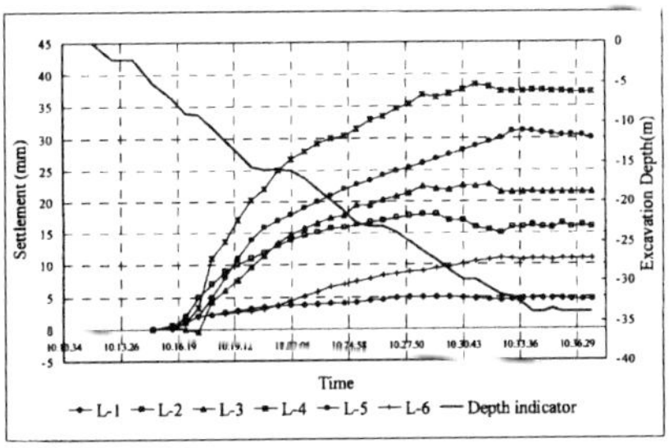

FIG. 4. Typical settlement time-history curves of the undercrossing

In the following part, undercrossing settlement of cross-section 1-1 and cross-section 2-2 are analyzed. (referred to Figure 1)

(1) Cross-section 1-1

Figure5 shows the comparison between the observed settlement of Test 1 and field monitoring of cross-section 1-1 in different phases of excavation. Table 3 shows the details of undercrossing settlement of Test1 and field monitoring of cross-section 1-1 in different phases of excavation.

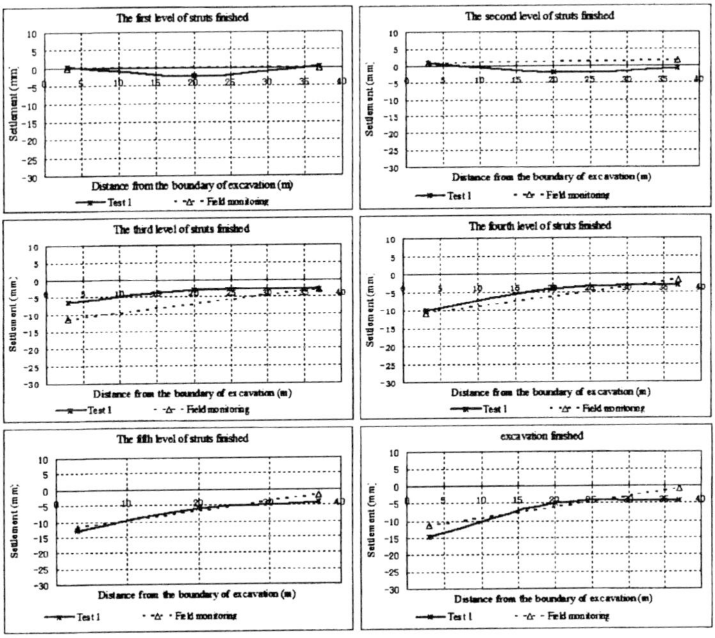

FIG. 5.Undercrossing Settlement curves of Test 1 and field monitoring of cross-section 1-1

Table 3. Details of undercrossing Settlement of Test 1 and field monitoring of cross-section 1-1

	Test 1(mm)			Field monitoring(mm)	
Distance from the excavation	3m	20m	37m	3m	37m
The first level of struts finished	0.3	-2.4	0.3	0.0	0.0
The second level of struts finished	0.9	-2.2	-1.0	0.8	1.4
The third level of struts finished	-6.5	-3.1	-2.8	-11.3	-2.8
The fourth level of struts finished	-10.0	-3.9	-3.0	-10.9	-1.7
The fifth level of struts finished	-12.7	-5.8	-4.0	-11.8	-1.6
Excavation finished	-14.6	-5.2	-4.2	-11.3	-0.9

From Figure5, it can be found that, both in the test and real situation, the undercrossing inclined towards the excavation. Before the third level of struts (-16m) were constructed, the settlement rate in field is quicker than that in the test. After that, the observed settlements of the test and field monitoring tended to be close and reached a maximum of 14.6mm and 11.8mm respectively.

Figure 6 shows the comparison among the observed settlements of Test 1, Test 2 and Test 3 of cross-section 1-1 in different phases of excavation. Table 4 shows the details of undercrossing settlement of Test1, Test2 and Test 3 of cross-section 1-1 in different phases of excavation.

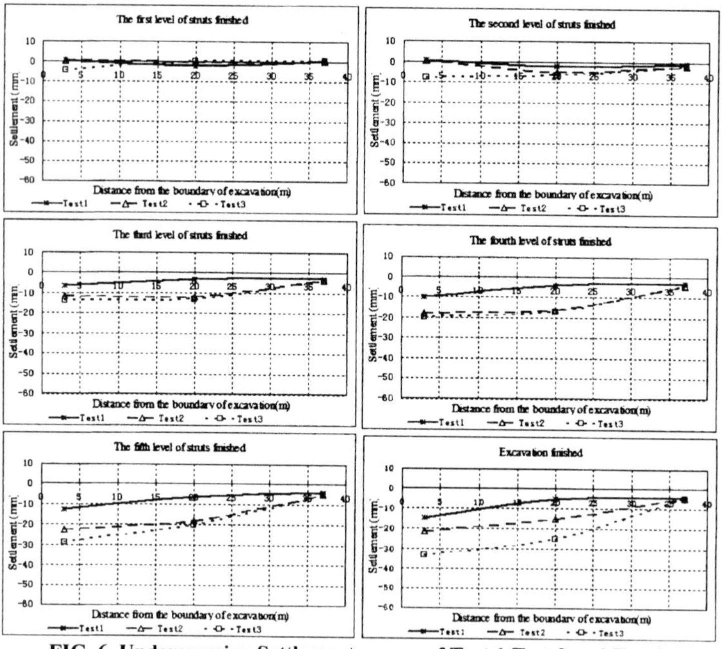

FIG. 6. Undercrossing Settlement curves of Test 1 Test 2 and Test 3 of cross-section 1-1

Table 4. Details of undercrossing Settlement of Test 1 Test 2 and Test 3 of cross-section 1-1

	Test1 (mm)			Test2 (mm)			Test3(mm)		
Distance from the excavation	3m	20m	37m	3m	20m	37m	3m	20m	37m
The first level of struts finished	0.3	-2.4	0.3	0.5	-0.3	-0.1	-4.4	-0.1	0.2
The second level of struts finished	0.9	-2.2	-1.0	0.4	-5.1	-2.0	-7.9	-6.5	-2.0
The third level of struts finished	-6.5	-3.1	-2.8	-11.4	-12.2	-3.5	-13.7	-13.1	-3.5
The fourth level of struts finished	-10.0	-3.9	-3.0	-17.8	-15.9	-4.1	-19.9	-17.0	-4.2
The fifth level of struts finished	-12.7	-5.8	-4.0	-22.3	-18.0	-5.0	-28.6	-20.2	-4.8
Excavation f inished	-14.6	-5.2	-4.2	-21.5	-15.2	-4.4	-33.4	-24.9	-4.3

From Figure 6, it was found that the barrier piles-wall had a good effect to protect the undercrossing. The observed settlement was smallest in Test 1 with 40 meters long piles-wall while the largest settlement was found in Test 3 with no piles-wall at any given phrase. Besides, the settlement rate in both Test1 and Test 2 became stable after the fourth struts(-23m) were finished whereas the settlement in Test 3 kept developing till all the excavation finished. The ultimate settlement in the near-end of the undercrossing in Test 3 was about 150% of that in Test 2 and more than 200% of that in Test 1. The settlement in the far-end appeared to be very close in all the three tests, which indicates that the impact of excavation weakened as the distance increases.

(2) Cross-section 2-2

Figure7 shows the comparison between the observed settlement of Test 1 and field monitoring of cross-section 2-2 in different phases of excavation. Table 5 shows the details of undercrossing settlement of Test1 and field monitoring of cross-section 2-2.

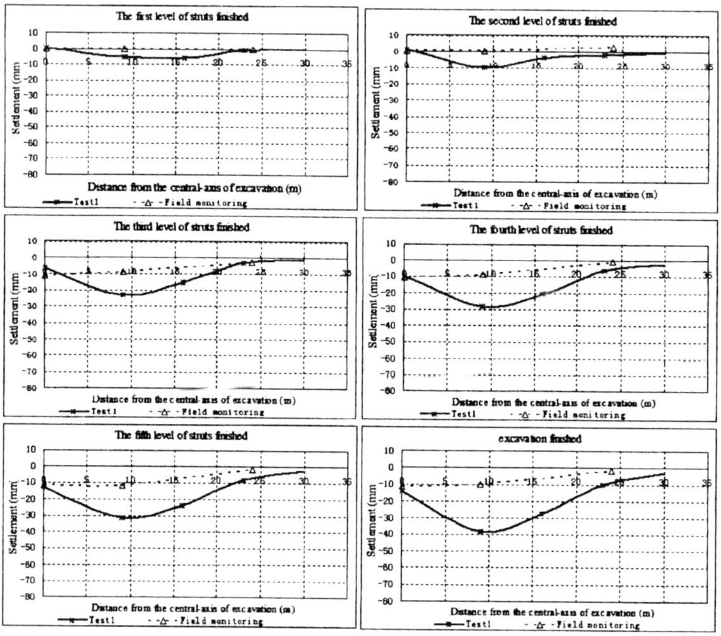

FIG.7. Undercrossing Settlement curves of Test 1 and field monitoring of cross-section 2-2

Table 5. Details of undercrossing Settlement of Test 1 and field monitoring of cross-section 2-2

	Test1(mm)				Field monitoring(mm)		
Distance from the central-axis of excavation	0m	9m	16m	23m	0m	9m	24m
The first level of struts finished	0.3	-5.3	-5.6	-0.5	0.0	0.0	0.0
The second level of struts finished	0.9	-9.7	-4.1	-1.9	0.8	0.8	2.9
The third level of struts finished	-6.5	-23.3	-15.1	-3.2	-11.3	-9.0	-2.7
The fourth level of struts finished	-10.0	-28.3	-20.5	-6.1	-10.9	-8.7	-0.9
The fifth level of struts finished	-12.7	-31.3	-23.8	-7.8	-11.8	-11.7	-1.5
Excavation finished	-14.6	-37.8	-27.7	-10.1	-11.3	-10.3	-2.0

From Figure 7, it was found that in field monitoring, the largest settlement occurred at the central-axis of excavation and it decreased as the distance from the excavation increases. But the differential settlement was not obvious within the range of excavation (less than 10m from the central-axis of excavation). As to Test 1, the largest settlement didn't happen at central-axis but about 9 meters from it, which may due to the friction on the interface between the diaphragm wall model and the rigid plate. The largest settlement in Test 1 was approximately 3 times of that in field, which is probably because of a low density replacing liquid was employed in present study that caused larger bottom heave. The impact range both in field and Test 1 was about 3 times the width of excavation.

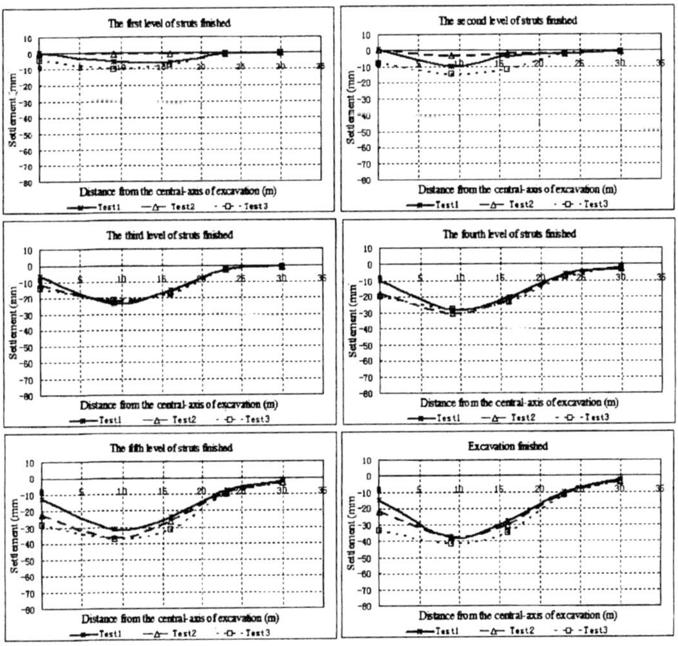

FIG.8. Undercrossing settlement curves of Test 1,Test2 and Test3 of cross-section 2-2

Table 6. Details of undercrossing Settlement of Test 1 Test 2 and Test 3 of cross-section 2-2

	Test1(mm)				Test2(mm)				Test3(mm)			
Distance from the central-axis of excavation	0m	9m	16m	23m	0m	9m	16m	23m	0m	9m	16m	23m
The first level of supports finished	0.3	-5.3	-5.6	-0.5	0.5	-0.6	-0.6	-0.5	-4.4	-10.0	-7.7	-0.5
The second level of supports finished	0.9	-9.7	-4.1	-2.0	0.4	-3.3	-2.1	-1.9	-7.9	-15.0	-12.2	-2.0
The third level of supports finished	-6.5	-23.3	-15.1	-3.1	-11.4	-22.1	-16.0	-3.1	-13.7	-21.3	-17.2	-3.1
The fourth level of supports finished	-10.0	-28.3	-20.5	-6.0	-17.8	-30.3	-22.0	-6.9	-19.9	-27.6	-23.4	-7.9
The fifth level of supports finished	-12.7	-31.3	-23.8	-8.2	-22.4	-36.8	-26.1	-9.2	-28.6	-36.9	-31.4	-10.2
Excavation finished	-14.6	-37.8	-27.7	-9.9	-21.5	-37.3	-30.0	-11.1	-33.4	-41.6	-35.1	-11.0

Figure 8 shows the comparison among the observed settlements of Test 1, Test 2 and Test 3 of cross-section 2-2 in different phases of excavation. Table 6 shows the details of undercrossing settlement of Test1, Test2 and Test 3 of cross-section 2-2 in different phases of excavation.

It can be found that within the range of excavation (less than 10m from the central-axis of excavation), settlement of Test 1 with the longest barrier piles-wall was smallest while settlement of Test 3 with no piles-wall was largest. However, beyond that range, settlements in all the three tests decreased rapidly and were of little difference. It was believed that the impact range of excavation was limited laterally and the induced impact weakened as the distance increases. So the width of barrier piles-wall could be optimized within 2 times the width of the excavation.

4. Conclusions

(1) It was an effective way to simulate deep excavation with multi-level inner struts in high centrifugal field by combiningly using a newly developed hydraulic-control struts installation device and draining-fluid method in this study.

(2) Na2O.3SiO2 is a nontoxic solution that can be used to replacing the soil to be excavated with a density of 1.3-1.5g/ml providing a lateral pressure close to the soil at all depths. However it should be noted that is would cause a larger bottom heave than the real case due to its relative low pressure in vertical direction.

(3) The impact of the excavation on adjacent undercrossing was unneglectable with an impact range of about 3 times the width of the excavation along the direction parallel to the undercrossing. The impact weakened rapidly as the distance from the excavation increases, especially beyond the range of excavation.

(4) Barrier piles-wall was proved to be effective in protecting the undercrossing. The longer the piles-wall, the less settlement occurred. However, there was no significant difference of undercrossing settlement between tests with and without the protection of barrier piles-wall beyond the excavation range. It is suggested that the width of piles-wall could be controlled within 2 times the width of the excavation.

Acknowledgement

The authors would like to express their appreciation to the supports from Tongji Centrifuge Laboratory. They also give their thanks to the 4th division of China Academy of Engineering Physics who jointly invented the simulation device in this research.

Reference

1. Y.M.Yao, S.H.Zhou and Y.C.Li. (2004). "Analysis of Boundary Effect in Centrifuge Model Test". *Season Journal of Mechanics*, 25(2), pp.291-296.
2. J.L.Li, J.P.Wang and G.M.Xu. (2000). "A New Method to Simulate the Excavation in Centrifuge Model Test". *Scientific Research on Water Conservancy and Marine Conveyance*, 4, pp.69-71.
3. J. S. Sharma, M. D. Bolton. (1996). "Centrifuge Modelling of an Embankment on Soft Clay Reinforced with a Geogrid". *Geotextiles and Geomembranes*, 14, pp. 1-17

4. D.J.Richards and W. Powrie. (1998). "Centrifuge model tests on doubly propped embedded retaining walls in overconsolidated kaolin clay". *Geotechnique*, 48(6), pp.833-846.
5. Y.G.Lin, Z.M.Zhou and G.B.Liu. (2001). "Research on the Shelter Effect of Single Pile Settlement". *Tongji University Transaction*, 29(5), pp.520-525.
6. C.Gaudin, J. Garnier, P. Gaudicheau, G. Rault. "Use of a robot for in-flight excavation in front of an embedded wall". *Laboratoire Central des Ponts et Chaussees, Nantes, France.*

Effectiveness of Ground Improvement for a Cut-and-Cover Tunnel with a Backfill Slope based on Finite Element Analysis

Hong Yang, PhD, PE, M.ASCE

Parsons Brinckerhoff, 303 2nd St, 700N, San Francisco, CA 94107, USA; yangh@pbworld.com

ABSTRACT: Two-dimensional finite element analyses were performed using the PLAXIS program to evaluate the effectiveness of ground improvement for a cut-and-cover twin-tunnel structure with a backfill slope. The soils underlying the tunnels include layers of soft clay and loose sand that are subject to consolidation and liquefaction. The soils are to be improved with cement deep soil mixing (CDSM). The analyses show that the ground improvement will greatly reduce the total and differential settlements and eliminate the liquefaction-induced settlement for the tunnel structure. The use of CDSM will also significantly reduce the deformation of the fill slope and increase the global slope stability. The imposed load will essentially be carried by the CDSM walls and transferred to the underlying dense sand layer.

INTRODUCTION

In San Francisco, a new roadway has been proposed to replace the existing Doyle Drive in order to improve the seismic, structural and traffic safety of the roadway, and the proposed Main Post Tunnels will be part of the replacement. The tunnels are located in an area consisting of 8 to 10 m thick loose sand and soft clay which are subject to an unacceptable level of settlements due to consolidation and post-liquefaction reconsolidation. Therefore, ground improvement using the cement deep soil mixing (CDSM) method was considered.

CDSM is an in-situ soil treatment technology using cementitious materials to blend with the soil. CDSM has been widely used for ground treatment and liquefaction mitigation (e.g. Bruce, 2000; Bruce and Bruce, 2003). Pretreatment of the ground by CDSM increases the strength and reduces the compressibility of the soils. The uses of CDSM for liquefaction mitigation include liquefaction prevention, reinforcement of liquefiable soils and pore-water pressure reduction (especially when using a grid pattern) (Bruce, 2000). In this paper, two-dimensional (2D) finite element analyses were performed to model the tunnel behaviors without and with the ground improvement using CDSM, and to evaluate the effectiveness of the CDSM. An advanced soil model, the Hardening Soil model, was used to simulate the soil behaviors in the analyses.

SITE CONDITIONS

The proposed tunnels are a cut-and-cover, two-cell structure of about 305 m in length and typically about 38 m in total width. One side of the tunnels will also form an approximately 2H:1V embankment fill slope to meet the aesthetic requirements.

The tunnels are located in an area of reclaimed tidal marsh characterized by soft marsh deposits (the Young Bay Mud, mainly soft clay), loose marine-deposited sands (the Marine Sand) and hydraulically-placed fill (the Artificial Fill, mainly loose sand) with a total thickness ranging from 8 to 10 m. These soils are underlain by the dense Colma Sand and the stiff Old Bay Clay. The groundwater table is located near the ground surface. A typical transverse cross-section showing the subsurface conditions is presented on Figure 1.

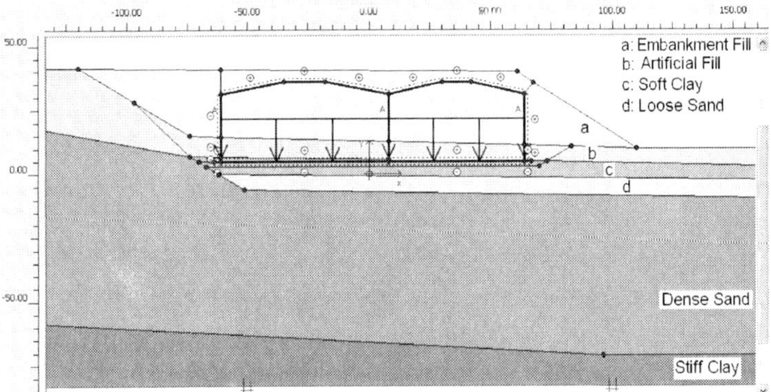

Figure 1. Soil layers and basic model (dimensions in ft; 1 ft = 0.305 m)

METHOD OF ANALYSIS

The study was performed using the finite element program of PLAXIS 2D version 9.02 (Brinkgreve et al., 2008). Two cases of analysis were conducted, the baseline case (i.e. without CDSM) and the CDSM case. The main features of the finite element model are shown on Figure 1. The concrete linings of the tunnels were modeled as "plates" that were composed of beam elements and their properties are summarized in Table 1. The connections between the plates were assumed to be "rigid", i.e. the connection point contains only one rotational degree of freedom.

The soils involved in the analyses include the various existing soils, the proposed backfill and a layer of 0.6 m thick crushed aggregate that will be placed below the tunnel slab. The soil properties are summarized in Tables 2 through 4, which were developed based on the Mohr-Coulomb model for the crushed aggregate and CDSM and the Hardening Soil model for the remaining soils. The Mohr-Coulomb model is an elastic perfectly-plastic model that has been routinely used in geotechnical engineering. The Hardening Soil model is an advanced model for simulating the behavior of different types of soil, both soft and stiff soils. The Hardening Soil model incorporates soil behaviors and features such as stress dependent stiffness and elastic loading and unloading (which are lacking in the Mohr-Coulomb model), and failure according to the Mohr-Coulomb criteria (Brinkgreve et al., 2008). An interface strength reduction factor of 0.30 was used to model the soil-structure interaction considering that a layer of waterproof membrane, which is relatively smooth, will be applied to the outside of the tunnel linings.

The CDSM wall normally consists of series of overlapped cylindrical columns that form a grid pattern. The small "cells" created by CDSM will support the structure and limit liquefaction and lateral spreading. The CDSM was proposed for the area below the tunnel structure and the fill slope. Due to the nature of the 2D plane strain model,

Table 1. Material Properties for Tunnel Lining

Location	Thickness	Axial Stiffness	Flexural Rigidity	Weight	Poisson's Ratio
	d	EA	EI	w	ν
	[m]	[kN/m]	[kNm^2/m]	[kN/m^2]	[-]
Left wall; Right wall	0.91	2.8E+07	1.9E+06	21.5	0.15
Right roof	1.08	3.2E+07	3.1E+06	25.1	0.15
Left roof	1.23	3.6E+07	4.6E+06	28.7	0.15
Slab; Center wall	1.36	4.2E+07	6.5E+06	32.3	0.15

Based on concrete properties of Young's modulus of 30.5E+6 kN/m^2 and density of 23.6 kN/m^3.

Table 2. Soil Densities and Permeabilities

Soil	Material Type	Unsaurated Density	Saurated Density	Horizontal Permeability	Vertical Permeability
		g_unsat	g_sat	k_x	k_y
		[kN/m^3]	[kN/m^3]	[m/s]	[m/s]
Embankment Fill	Drained	18.9	20.4	1.0E-05	1.0E-05
Artificial Fill	Drained	18.2	19.2	5.0E-05	5.0E-05
Young Bay Mud (soft clay)	Undrained	16.2	16.2	4.9E-09	9.9E-10
Marine Sand (loose)	Undrained	17.3	19.6	9.9E-07	9.9E-07
Colma Sand (dense)	Drained	20.7	20.7	4.0E-05	2.0E-05
Old Bay Clay (stiff)	Undrained	18.2	18.2	9.9E-09	9.9E-10
Crushed Aggregate	Drained	22.0	22.0	1.0E-04	1.0E-04
CDSM	Drained	17.3	17.3	1.2E-05	1.2E-05

Table 3. Stiffness and Strength Properties – Hardening Soil Model

Soil	E_{50}ref	E_{oed}ref	E_{ur}ref	m	v_ur	p_ref	K_0nc	c_ref	phi
	[kN/m^2]	[kN/m^2]	[kN/m^2]	[-]	[-]	[kN/m^2]	[-]	[kN/m^2]	[°]
Embankment Fill	114,912	143,640	344,736	0.5	0.250	96	0.486	4.8	35
Artificial Fill	38,304	47,880	110,492	0.5	0.250	96	0.550	4.8	30
Young Bay Mud	1,532	613	7,748	1.0	0.275	96	0.577	9.6	25
Marine Sand	4,788	1,919	23,940	1.0	0.250	96	0.577	9.6	25
Colma Sand	179,550	214,775	538,650	0.5	0.250	96	0.398	0.5	37
Old Bay Clay	1,915	862	7,900	1.0	0.275	96	0.562	9.6	26

Notes: E_{50}ref: secant stiffness in standard drained triaxial test; E_{oed}ref: tangent stiffness for primary oedometer loading; E_{ur}ref: unloading/reloading stiffness; m: power for stress level dependency of stiffness; v_ur: Poisson's ratio for unloading-reloading; p_ref: reference stress for stiffness; $K_o$nc: K_o-value for normal consolidation; c_ref: effective cohesion; phi: effective friction angle. Dilatancy angle is taken as zero for all soils.

Table 4. Stiffness and Strength Properties – Mohr-Coulomb Model

Soil	Young's Modulus	Poisson's Ratio	Cohesion	Friction Angle	Dilatancy Angle
	E_ref	ν	c_ref	phi	ψ
	[kN/m^2]	[-]	[kN/m^2]	[°]	[°]
Crushed Aggregate	287,280	0.250	0.2	40	0
CDSM	248,976	0.300	420	0	0

the CDSM walls in the transverse direction cannot be explicitly modeled. Therefore, only the CDSM walls in the longitudinal direction were modeled. In the model, the CDSM had an equivalent thickness of 0.85 m at 3.87 m spacing, representing an area replacement ratio of 22% in longitudinal direction. The CDSM was extended from below the crushed aggregate to about 0.31 m in the dense sand.

Each of the analyses consisted of six calculation phases (Table 5), which modeled the performance of the tunnels starting from existing ground condition, the various construction stages, elastic deformation, consolidation and post-liquefaction densification. The effect of liquefaction on the untreated ground in both the baseline and CDSM cases was model by applying a 1.0% volumetric strain (phase 6, Table 5), which was based on results of previous liquefaction evaluation for the project site. For the CDSM-treated ground, the likelihood of liquefaction triggering will be greatly reduced since the ground is constrained (e.g. O'Rourke and Goh, 1997); and a 0.05% volumetric strain was assumed to model the effect of liquefaction.

The displacements were obtained for each phase and those for phases 4 through 6 were summed up to obtain the total displacements. Displacements resulted from phases 1 through 3 were considered irrelevant to the tunnel performance as these displacements would occur prior to or during the construction.

Table 5. Phases of Calculation

No.	Phase Description	Remarks
1	Equilibrium	Model existing ground condition.
2	CDSM, excavation and dewatering	No CDSM for baseline case.
3	Activate crushed aggregate and slab	
4	End of construction – activate walls, roofs, backfill and traffic loading on slabs	Model elastic deformation; Traffic load = 250 psf (12 kPa).
5	Consolidation	Only modeled for primary consolidation.
6	Liquefaction – apply volumetric strain to three soils: Artificial Fill, Young Bay Mud and Marine Sand.	Volumetric strain is 1.0% for baseline case; for CDSM case, 0.05% for the three soils between CDSM and 1.0% for the three soils beyond CDSM.

RESULTS AND DISCUSSIONS

Settlements of the Tunnels. The phase and total settlements of the tunnel slab for the baseline case are presented on Figures 2 and 3. The figures show that, without the ground improvement, a significant amount of total settlement (0.43 to 1.02 ft, or 0.131 to 0.311 m) would occur; the consolidation settlement (0.33 to 0.84 ft, or 0.101 to 0.256 m) formed the main portion of the total settlement. The settlement due to elastic deformation and liquefaction was less than 0.06 ft (0.018 m) and 0.13 ft (0.040 m), respectively. The differential settlement of the slab was about 0.59 ft (0.180 m).

For the CDSM case, the phase and total settlements of the tunnel slab are presented on Figure 4, which indicates that, with the ground improvement using CDSM, the total settlement (ranging from 0.07 to 0.11 ft, or 0.021 to 0.034 m) was greatly reduced. The decrease in the total settlement was primarily contributed to the decrease in consolidation settlement that became less than 0.07 ft. The liquefaction settlement (<0.005 ft, or 0.0015 m) was also greatly reduced as compared with the baseline case and became negligible. The elastic settlement (<0.04 ft, or 0.012 m) was only slightly reduced. The differential settlement of the slab was reduced to 0.04 ft (0.012 m), which was much smaller than that of the baseline case.

It is noteworthy that, for both the baseline and CDSM cases, the distributions of the elastic settlement along the slab were similar; the settlements under the three walls were higher than those under the mid spans of the tunnels. This suggests that the tunnel slab of as thick as 4.5 ft (1.372 m) still exhibited some flexibility due to the relatively large width of the tunnels.

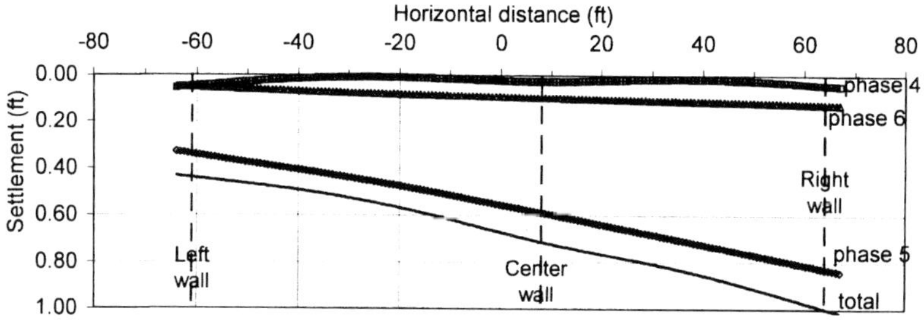

Figure 2. Phase and total settlements of the tunnel slab, baseline case (no CDSM)

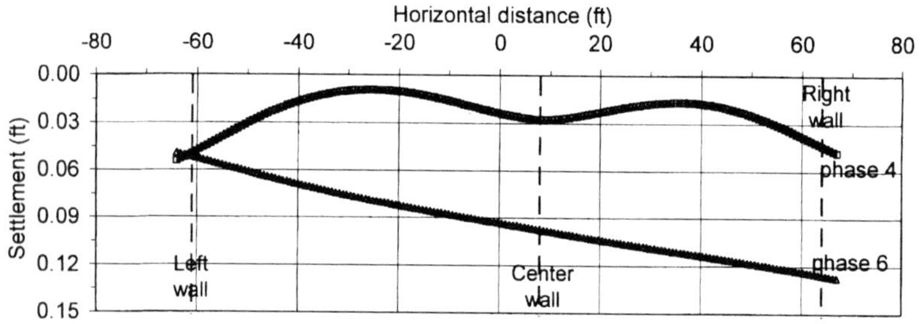

Figure 3. Settlements of the tunnel slab for phases 4 and 6, baseline case (no CDSM)

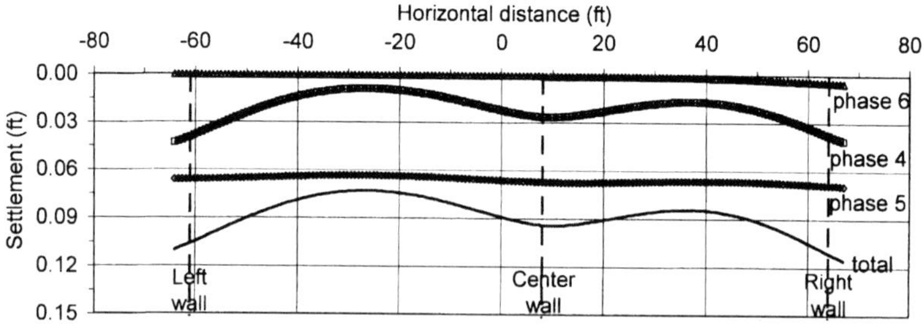

Figure 4. Phase and total settlements of the tunnel slab, CDSM case

Deformation and Stability of the Fill Slope. The deformed fill slope next to the tunnel at end of phase 6 calculation for the baseline and CDSM case is shown on

Figure 5. Although the magnitudes of the extreme total displacement was comparable between the baseline case and CDSM case (1.72 ft versus 0.90 ft, or 0.524 m versus 0.274 m), the characteristics of the deformation were different. In the baseline case, the deformation occurred in the entire slope and extended to the bottom of the loose sand layer (or top of the dense sand layer). There was also an apparently continuous circular surface developed in the soft clay and loose sand that distinguishes the deformed soil portion from the relatively less deformed soil portion. In the CDSM case, on the contrary, the deformation primarily developed in the embankment fill and the Artificial Fill, and only slightly extended to the soft clay layer; there was essentially no deformation in the underlying loose sand layer. The difference in the deformation is also quantitatively illustrated by the horizontal displacement contours as shown on Figure 6. In the baseline case, the horizontal displacement in the Artificial Fill and soft clay was about 1.0 ft (0.305 m) (the maximum value) while it was only 0.2 ft (0.061 m) at the same depth for the CDSM case.

The different behaviors of the slope for the two cases are further demonstrated by the shear strains developed in the slopes (Figure 7). In the baseline case, a continuous shear zone (as indicated by the shear strain concentrations) was developed from the top to the toe of the slope passing through the weak soil layers. In the CDSM case, only several localized shear zones were developed and were not continuous through the entire slope. The continuous shear zone in the slope of the baseline case indicates that global potential failure has developed in the entire slope along the circular slip surface; while such global potential failure has been eliminated when CDSM was provided below the slope, and only localized potential failure occurred. The different behavior of the slope indicates that, the CDSM has effectively restricted the slope deformation along the weak soil layers thus increased the global stability of the slope.

The results also show the CDSM has experienced lateral movement of up to 0.3 ft (0.091 m) (Figures 5(b) and 6(b)). However, considering that the results are based on 2D analysis, the lateral movement of the CDSM (and the soils between/near the CDSM) should be lesser under the actual 3D condition, as the CDSM in transverse direction (which was not modeled in the 2D analysis) would provide additional resistance to the ground movement, thus the slope deformation should be smaller and the slope stability would be further increased.

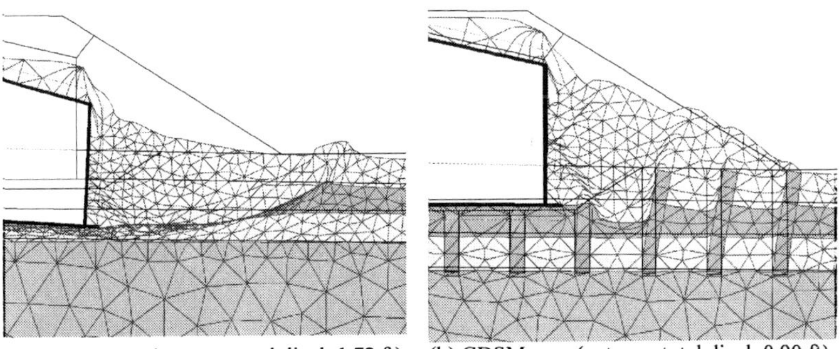

(a) Baseline case (extreme total displ. 1.72 ft) (b) CDSM case (extreme total displ. 0.90 ft)

Figure 5. Deformed mesh, phase 6 (displacement scaled to 10 times)

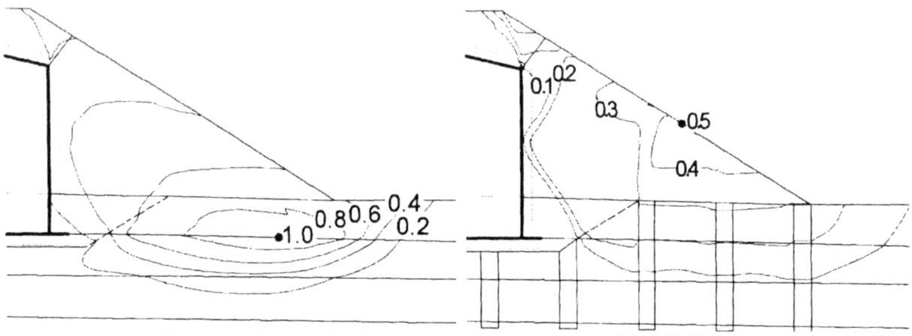

(a) Baseline case (no CDSM) (b) CDSM case

Figure 6. Horizontal displacement contours for the slope (in ft)

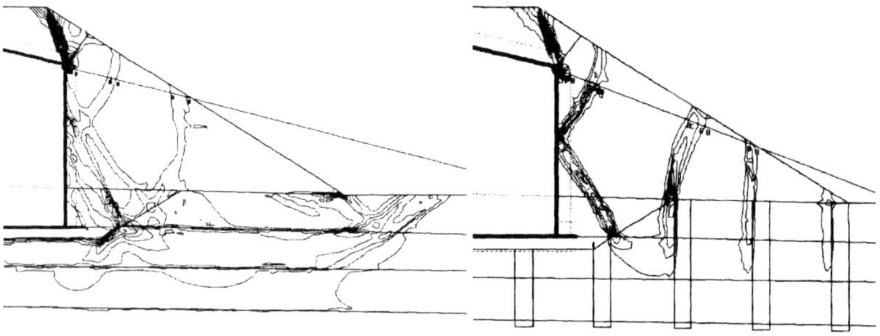

(a) Baseline case (no CDSM) (b) CDSM case

Figure 7. Shear strains developed in the slope

Vertical Stress in CDSM. The distribution of the vertical stress in the ground for the CDSM case is presented on Figure 8. The results show that, in general, the stress in the CDSM was much higher than that in the surrounding soils. The stress along each of the CDSM wall was also relatively uniform in most of the CDSM walls. These observations indicate that the majority of the imposed load was carried by the CDSM and transferred to the underlying bearing layer, the dense sand. These results are somehow expected since the CDSM has much higher stiffness than the surrounding soils. Further, the stresses in the CDSM located below the tunnel walls were much higher than those located in the mid span of the tunnels, which could be attributed to the flexibility of the slab. The extreme vertical stress of 17.2 ksf (826 kPa) in CDSM (located below the left tunnel wall) indicates that the CDSM was reaching its maximum capacity in the localized area, given the approximate unconfined compressive strength of 17.5 ksf (840 kPa) (two times the CDSM cohesion as in Table 4). The stress level in the CDSM should be lower when the 3D condition and the CDSM walls in the transverse direction are taken into account.

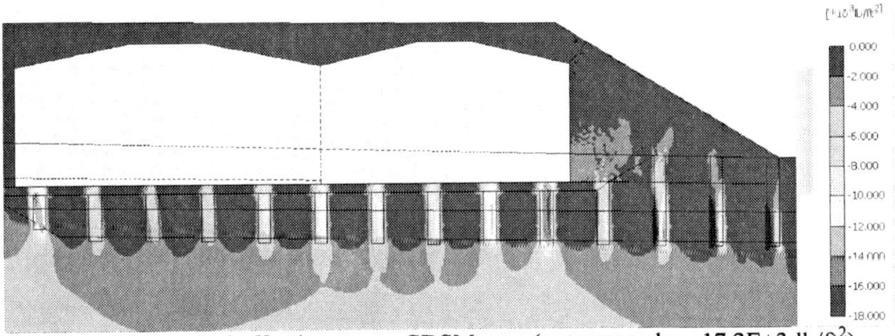

Figure 8. Vertical effective stress, CDSM case (extreme value -17.2E+3 lb/ft^2)

SUMMARY AND CONCLUSIONS

Two-dimensional finite element analyses were performed for a cut-and-cover tunnel with a fill slope. The tunnel foundation soils comprise several layers of weak soils that are subject to consolidation and liquefaction; and ground improvement using CDSM in a grid-pattern was considered. The results of the analyses indicate that the total and differential settlements for the tunnel structure were greatly reduced after the ground improvement; the liquefaction-induced settlement was essentially eliminated due to the cell-structure of the CDSM. The use of the CDSM also reduced the deformation of the fill slope and increased the global slope stability. The imposed load was essentially carried by the CDSM walls and transferred to the underlying competent layer.

ACKNOWLEDGEMENTS

The author greatly appreciates the permission from San Francisco County Transportation Authority (the leading agent of the project) to publish the paper. The author also thanks Professor Juan Pestana of the University of California in Berkeley for his advice on the study. Further, author is grateful to the insightful review comments provided by his colleagues. The points of view expressed in the paper are only those of the author.

REFERENCES

Brinkgreve, R.B.J., Broere, W. and Waterman, D. (2008). PLAXIS 2D Professional Version 9. User's Manual. PLAXIS BV, P.O. Box 572, 2600, AN Delft, the Netherlands.

Bruce, D.A. 2000. An introduction to the deep soil mixing methods as used in geotechnical applications. Report No. FHWA-RD-99-138.

Bruce, D.A. and Bruce M.E.C. 2003. The practitioner's guide to deep mixing. In Grouting and Ground Treatment, ASCE Geotechnical Special Publications, No. 120, 474-488.

O'Rourke, T.D. and Goh, S.H., 1997. Reduction of liquefaction hazards by deep soil mixing. Proceedings, Workshop on Earthquake Engineering Frontiers in Transportation Facilities, NCEER-97-0005, August, 87-103.

Reconstruction of the Temperature Distribution on the vertical Direction of Tunnel in Fire Accidents

Fang Yin-gang[1,2], Zhu He-hua[1,2], Yan Zhi-guo[1,2]

[1]Key Laboratory of Geotechnical and Underground Engineering of the Ministry of Education, Tongji University, Shanghai 200092, China

[2]Department of Geotechnical Engineering, Tongji University, Shanghai 200092, China

ABSTRACT: The aim of the study here was to reconstruct the temperature field of the tunnel in fire accidents. The most important thing during the study was to set up the temperature distribution curve on the vertical direction of tunnel. CFD (Computational Fluid Dynamics) program SMARTFIRE was used to simulate the tunnel fire with different HRR (Heat Release Rate), longitudinal ventilation velocity and number of traffic lanes. According to analyzing the results of simulations, the longitudinal ventilation velocity was considered as the most important factor. Otherwise, the distance from the point of ignition also had a great influence on temperature distribution. Therefore, the temperature distribution curves were set up according to different longitudinal ventilation velocities and distance from the point of ignition.

1 INTRODUCTION

Due to the fast development of the tunnel construction, and the recent catastrophic fire in several tunnels, fire safety in tunnels has become a well-studied topic in the world (Ingason and seco, 2005). It is very important to know the development of tunnel fire. It will be beneficial to passenger evacuation and fire extinguishment if the management staffs get more information in fire accidents.

The aim of the study here was to reconstruct the temperature field of tunnel in fire accidents. The temperature distribution curve on the Z axis direction (the vertical direction of tunnel, shown as in Figure 1) was found. Of cause, the curve contains some unknown parameters. Fiber will be fixed on the crown of tunnels. By using the information collected by the fiber, these unknown parameters can be determined. Then, the temperature distribution on vertical section of tunnel will be known through the combination of temperature distribution curves (on the vertical direction) at any X point. Next, if the temperature on the Y axis direction (Figure 1) is considered as the same, then the temperature field of the tunnel will be obtained. Therefore, the most important thing during the study is to set up the temperature distribution curve on the vertical direction of tunnel.

CFD (Computational Fluid Dynamics) software named SMARTFIRE was used to simulate the tunnel fire with different HRR (Heat Release Rate), longitudinal ventilation velocity and geometry dimensions of tunnels. Through analyzing and fitting the temperature data of simulations, the temperature distribution curve on the vertical direction of tunnel was found.

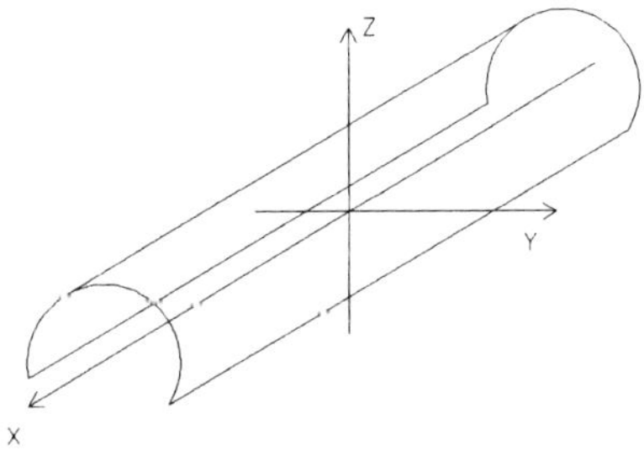

FIG. 1. Simple sketch of tunnel

2 NUMERICAL SIMULATIONS

The model scale tunnel was simulated with the CFD program SMARTFIRE. In Figure 2, a geometrical representation of the model tunnel is shown. The length of the model scale is 600m. And the distance from the point of ignition to the nearside entrance of tunnel is 100m.

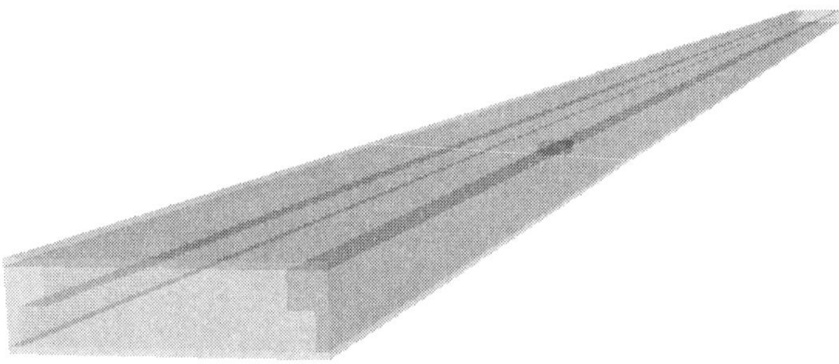

FIG. 2. A geometrical representation of the model tunnel

Tunnel fire was simulated under different fire scenes. There are three variable factors. They are HRR (Heat Release Rate), longitudinal ventilation velocity and geometry dimensions of cross section. And every factor also includes three different levels.

According to different types of vehicle fire, three Heat Release Rates were used in simulations, including 5MW (representing passenger car), 20MW (representing bus) and 50MW (representing heavy goods vehicle) (PIARC, 1999). And the geometry dimensions of fuel sources with different HRR are charged. The geometry dimensions of 5MW fire is length×width×height=4.5m×2.0m×1.5m, the 20MW fire is length ×width×height=8.5m×2.5m×3.0m, and the 50MW fire is length×width×height =6.0m×2.5m ×3.5m.

There are three different longitudinal ventilation velocities, including 0m/s, 1.5m/s and 3.0m/s. At one entrance of tunnel (upstream side), an Inlet was set. It can provide different longitudinal ventilation velocities in simulations. And at the other entrance of tunnel (downstream side), a Vent was set. It can allow free atmospheric exchange.

The research object of the study here was road tunnel. Therefore, the geometry dimensions of cross section were mainly decided by the number of traffic lanes. There are three types of traffic lanes in total, including two traffic lanes, three traffic lanes and four traffic lanes. According to Code for Design of Road Tunnel (JTG D70-2004), the geometry dimension of cross section of two traffic lanes is width×height=11×5m, the geometry dimension of three traffic lanes is width×height=14.5m×5m, and the geometry dimension of four traffic lanes is width×height=18m×5m.

There are three different factors, and every factor also includes three levels. So that, there will be 27 types of different combinations in total. For saving computing time, Orthogonal Design Method was used in this study. The Orthogonal Design will choose representative combinations so as to greatly decrease the number of simulations. According to the Orthogonal Table $L_9(3^4)$ (Li and Hu, 2008), there are 9 types of combinations as shown in Table 1.

Table 1. Scheme of numerical simulation

Factors	HRR	Longitudinal Ventilation Velocity	Number of Traffic Lane
Simulation 1	5MW	0m/s	two traffic lanes
Simulation 2	20 MW	1.5 m/s	two traffic lanes
Simulation 3	50 MW	3 m/s	two traffic lanes
Simulation 4	50 MW	0 m/s	three traffic lanes
Simulation 5	5 MW	1.5 m/s	three traffic lanes
Simulation 6	20 MW	3 m/s	three traffic lanes
Simulation 7	20 MW	0 m/s	four traffic lanes
Simulation 8	50 MW	1.5 m/s	four traffic lanes
Simulation 9	5 MW	3 m/s	four traffic lanes

Gasoline was chosen as the fuel in simulations. The main component of gasoline is Octane (C_8H_{18}). The combustion equation of Octane is as below.

$$C_8H_{18}+12.5O_2 \rightarrow 8CO_2+9H_2O \tag{1}$$

3 RESULTS AND DISCUSSION

The research range of the study here was the downstream side of fire source. Origin was used to analyze and fit the temperature distribution curve. In above three factors, the longitudinal ventilation velocity had the most important influence on temperature distribution in the tunnel. Otherwise, the distance from fire source also had a great influence on temperature distribution. After analyzing the data of simulations, 30m from the fire source was considered as an appropriate critical point. Because the shapes of temperature distribution were relatively close whether within 30m from the fire source or beyond 30m from the fire source. Therefore, the temperature distribution curves were fitted according to different longitudinal ventilation velocities and distance from the fire source.

When the longitudinal ventilation velocity is 0m/s, the temperature distribution curves (on the vertical direction of tunnel) is shown in Figure 3. Functions within and beyond 30m from the fire source are shown as Equation (2) and Equation (3), the correlation coefficients are 0.934 and 0.966, respectively.

$$\frac{T - T_0}{T_{max,t} - T_0} = 1.166e^{-4.025\frac{h}{H}} - 0.061 \quad (2)$$

$$\frac{T - T_0}{T_{max,t} - T_0} = \frac{1}{0.992 + 272.916(\frac{h}{H})^{4.753}} \quad (3)$$

Where, T is the temperature at any time and any point in the tunnel, T_0 is the normal atmosphere temperature, $T_{max,t}$ is the temperature of crown on any distance from the fire source at any time, h is the vertical distance from the measured point to the crown, H is the height of tunnel.

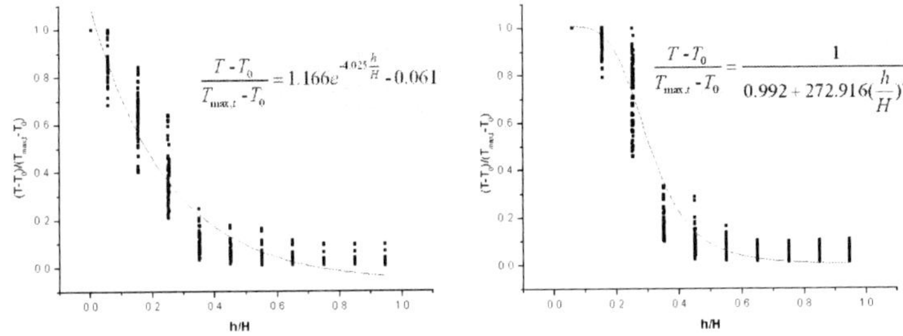

(a) Within 30m from the fire source (b) Beyond 30m from the fire source

FIG. 3. Temperature distribution curve on the vertical direction of tunnel when the longitudinal ventilation velocity is 0m/s

When the longitudinal ventilation velocity is 1.5m/s, the temperature distribution curves (on the vertical direction of tunnel) is shown in Figure 4. Functions within and beyond 30m from the fire source are shown as Equation (4) and Equation (5), the correlation coefficients are 0.966 and 0.977, respectively.

$$\frac{T - T_0}{T_{max,t} - T_0} = 0.640(\frac{h}{H})^2 - 1.434\frac{h}{H} + 1 \quad (4)$$

$$\frac{T - T_0}{T_{max,t} - T_0} = 4.502(\frac{h}{H})^3 - 6.370(\frac{h}{H})^2 + 1.124\frac{h}{H} + 0.941 \quad (5)$$

Where, parameters in Equation (4) and Equation (5) are the same as which in Equation (2) and Equation (3).

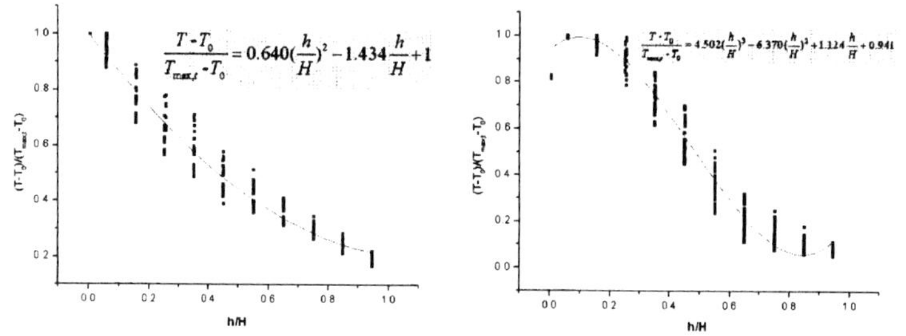

(a) Within 30m from the fire source (b) Beyond 30m from the fire source

FIG. 4. Temperature distribution curve on the vertical direction of tunnel when the longitudinal ventilation velocity is 1.5m/s

When the longitudinal ventilation velocity is 3.0m/s, the temperature distribution curves (on the vertical direction of tunnel) is shown in Figure 5. Functions within and beyond 30m from the fire source are shown as Equation (6) and Equation (7) , respectively, the correlation coefficients are both 0.970.

$$\frac{T-T_0}{T_{max,t}-T_0}=0.563(\frac{h}{H})^2-1.408\frac{h}{H}+1 \tag{6}$$

$$\frac{T-T_0}{T_{max,t}-T_0}=2.860(\frac{h}{H})^3-4.922(\frac{h}{H})^2+1.297\frac{h}{H}+0.916 \tag{7}$$

Where, parameters in Equation (6) and Equation (7) are the same as which in Equation (2) and Equation (3).

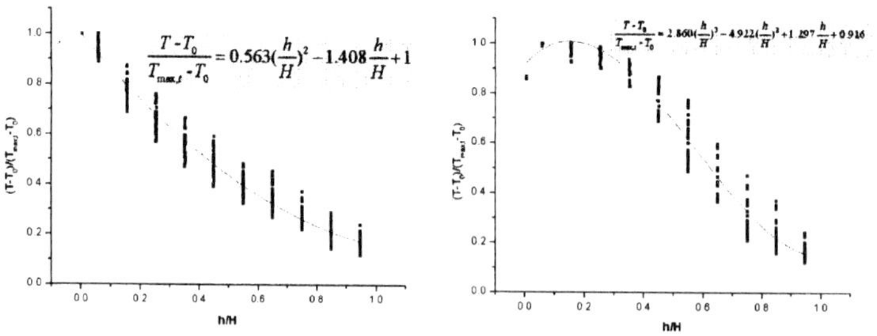

(a) Within 30m from the fire source (b) Beyond 30m from the fire source

FIG. 5. Temperature distribution curve on the vertical direction of tunnel when the longitudinal ventilation velocity is 3.0m/s

Above 6 graphs show that the temperature is higher in areas near the crown. With the reduction of height, temperatures become lower and lower. The maximum temperature is on the crown in areas near the fire source. However, when the distance from fire source is larger, for example beyond 30m, the maximum temperature is not always on the crown but near the crown under longitudinal ventilation. Otherwise, the temperature distribution trend to uniform with the increase of longitudinal ventilation velocity. And the same trend happened when the distance from fire source become larger.

4 CONCLUSIONS

Three factors ware considered in simulations, the longitudinal ventilation velocity had the most important influence on temperature distribution. Meanwhile, the distance from fire source also had a great influence. Through analyzing the results of simulations, 30m from the fire source was considered as an appropriate critical distance. Therefore, the temperature distribution curves were set up according to different longitudinal ventilation velocities and distance from the fire source. Temperature is higher in areas near the crown. With the reduction of height, temperatures become lower and lower. Otherwise, the temperature distribution trend to uniform with the increase of longitudinal ventilation velocity and distance from the fire source.

The temperature distribution curves on the vertical direction of tunnel had been set up. And the temperature of crown can be obtained by fiber fixed on the crown of tunnels. Therefore, the temperature field of any road tunnel in fire accidents can be reconstructed under certain longitudinal ventilation velocity and geometry dimensions of cross section of tunnels.

ACKNOWLEDGMENTS

The authors appreciate the support of the Project (50808137) supported by National Natural Science Foundation of China and Key Project (2006BAJ27B04, 2006BAJ27B05) supported by National Natural Science & Technology Pillar Program.

REFERENCES

Ingason, H. and Seco F. (2005). "Numerical simulation of a model scale tunnel fire test." *SP Fire Technology Report. Sweden*, 2005:47

JTG D70-2004. *Code for design of road tunnel.*

Li, Y.Y. and Hu, C.R. (2008). *Experiment design and data processing.* Beijing: Chemical Industry Press

PIARC Committee on Road Tunnels (C5). (1999). *Fire and Smoke Control in Road Tunnels.*

Determination of stress release coefficient and analysis of influence factors in granular soil tunnel

Zhuang Li[1] and Zhou Shun-hua[2]

[1] Ph.D. student, Department of Civil and Environmental Engineering, Yamaguchi University, Ube, Japan, 755-8611; n006wf@yamaguchi-u.ac.jp
[2] Professor, Key Laboratory of Road and Traffic Engineering of the State Ministry of Education, Tongji University, Shanghai, China, 200092; zhoushh@tongji.edu.cn

ABSTRACT: Formula of stress release coefficient in granular soil tunnel was presented based on the laboratory loading and unloading tests of sand. It is shown that the coefficient only depends on the stress state before excavation and the internal friction resistance. Influences of internal friction angle and excavation shape on the stress release coefficient were analyzed based on the continuum theory. Results show that: stress release behavior is different according to tunnel shapes, and also varies at different parts around tunnel; there are more stress release at the top and bottom than the lateral side of shallow tunnel, and also generally more stress release than in deep tunnel. Stress release will be hindered by increase of friction angle, however, there is no independent relationship between stress release and soil friction angle.

INTRODUCTION

Stress release must be aimed at those objects with initial stress, and can be defined as partial or total 'remove' of initial stress due to of internal or external condition change. In geotechnical engineering, stress release is related to unloading in most cases. However, it is different from unloading. As shown in Fig.1, there will be a new boundary after excavation or remove of load, and thus new load acting at the boundary, which will induce deformation and stress variation in soil. Apparently, stress release is a behavior of soil itself, indicating stress reduced due to deformation. The stress release is generally affected by excavation shape, soil properties and construction method (Zhou 2005, Ren 2005, Ng 2002). The ratio between released stress and the initial stress in the ground is termed as the excavation stress release coefficient, which is noted here as α (Mana 1981). Excavation load is determined by Terzaghi's soil arch formula (1936) or thought as the initial stress multiplied by α, which is usually estimated by displacement back-analysis or simply on engineering experiences for various construction methods, e.g., 10% in shield tunnel, 40~70% in mining tunnel and 100% in opening excavation (Bernat 1998). Residual stresses were found in some in-situ measurements at the base of foundation pit and considered in calculation (Kajigaya 1989). However, it is still a controversial problem about the residual stress in soil. The above review shows that the methods to estimate the released stress induced by excavation are generally empirical. This paper is to study how to determine stress release in granular soil tunnel.

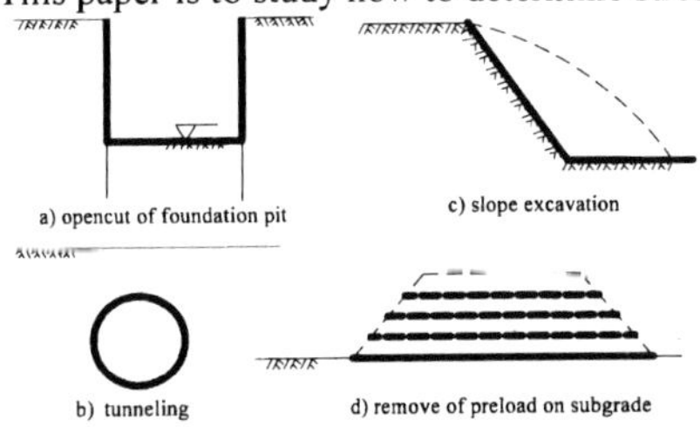

Fig.1. Four main kinds of unloading.

Fig.2. Test model with dead load applied.

STRESS RELEASE COEFFICIENT AND ITS IMPACT PARAMETERS

A quartz sand, with grain size varying from 0.2 to 0.5mm and specific gravity G_s=2.35, is used in the static loading and unloading model test. As shown in Fig.2, dead weight load is applied to the sand model in steel rectangular box, which were prepared by pluviating sand through air by a funnel with a free fall height of 70cm. The stresses developed in sand at different depths are measured using mini soil stress cells. It is found that the stresses developed in sand increase linearly during staged loading, while are different with those during unloading at the same external load level. The stresses in sand have no or little changes at the beginning of unloading; however, significant release of vertical stress may take place when the external load was removed to almost zero. The similar test results were also obtained in sand models in glass cylinder box.

Based on the loading and unloading model test results of sand (Zhuang and Zhou 2009), a model shown in Fig.3 was proposed to describe the stress release behavior of granular materials. This model consists of two traditional elements: the spring element and the St-Venant element. The stress in soil is stored in the spring while the St-Venant element is used to represent the internal frictional resistance σ_f. The stored stress σ will be released only when the removed load σ_u is larger than internal frictional resistance σ_f. The initial stress will also affect the mechanical process of stress release, since the coupling of the stored stress σ and the removed load σ_u dominates the deformation of soil element.

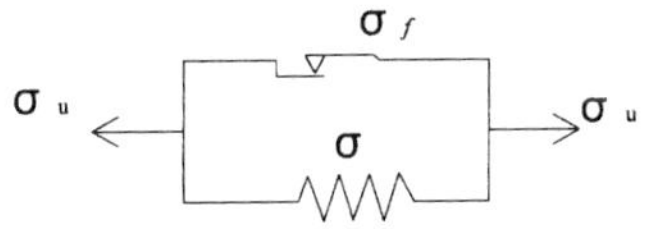

Fig. 3. Stress release model with two cells.

According to the release model in Fig. 3, the stress release coefficient α can be estimated by the following formula as

$$\alpha = \frac{\sigma_r^0 - \sigma_f}{\sigma_r^0} \tag{1}$$

where $\sigma^0{}_r$ is the initial stress prior to excavation and σ_f is the internal frictional resistance of soil. The product of the initial stress and the stress release factor is the required support from the retaining structure to keep the stability of the system. Consequently, it is important to determine the value of α. In this section, we discuss the determination of stress release coefficient using the proposed model for excavations of three different geometries.

The internal frictional resistance σ_f in Eq. (1) can be determined as follows. Since soil by the excavation boundary of the tunnel is in an extension state, σ_f can be considered as the octohedral shear stress τ_8 at the critical state of extension (Zhuang 2009); i.e.

$$\sigma_f = \tau_8 = \frac{\sqrt{2}}{3} q \tag{2}$$

with

$$q = \sigma_\theta - \sigma_r = \frac{2\sin\varphi}{1+\sin\varphi}\sigma_\theta \tag{3}$$

where φ is a representative friction angle of soil, while σ_θ and σ_r are the circumferential and radial stresses respectively. Substitution of Eq. (2-3) into Eq. (1) yields

$$\alpha = 1 - \frac{2\sqrt{2}\sin\varphi}{3(1+\sin\varphi)}\frac{\sigma_\theta}{\sigma_r^0} \tag{4}$$

with σ^0_r is the initial radial stress before excavation.

According to the definition in Eq. (4), the value of α, which is different at different construction stages, varies continuously around the boundary of an excavation. Two main influence factors on α were discussed as follows.

Influence of excavation shape and cover depth

It has been known that the stress field after excavation depends on both the initial stress and geometry of excavation. Let us examine the stress field around three geometries of excavation shown in Fig. 4; namely a circle hole of diameter d, a square hole with side length $2a$, and an elliptical hole with major and minor axis $2a$ and $2b$. Without loss of generality, the ground soil is assumed as homogeneous isotropic material. Compressive stress being noted as positive following soil mechanics, the initial stress field under gravity at the center of the hole is then

$$\begin{cases} \sigma_z = \rho g z \\ \sigma_x = \sigma_y = K_0 \rho g z \end{cases} \tag{5}$$

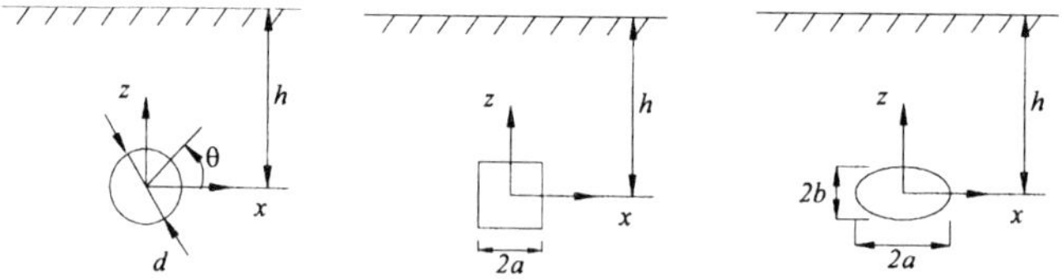

Fig. 4. Three different shapes of tunnel.

where ρ is the density of soil, K_0 is the coefficient of earth pressure at rest, which can be determined as $K_0=\mu/(1-\mu)$ with μ being the Poisson's ratio of soil or by Jacky's equation $K_0=1-\sin\varphi$ for normally consolidated soils.

Through Eq. (5), we can get the initial stresses at the center of the hole, which are noted as $p=\rho gh$ and $q=K_0 p$. According to the solution of plane strain problem in elastic mechanics, by the aid of complex variable function, the stresses around the excavation boundary for the three geometries in Fig. 4 are given as follows:

$$\text{circle:}\quad \sigma_\theta = \frac{1}{1-\mu}p + 2\frac{1-2\mu}{1-\mu}p\cos 2\theta \qquad (6)$$

$$\text{ellipse:}\quad \sigma_\theta = p\cdot\frac{\mu}{1-\mu}\frac{1-m^2+2m-2\cos 2\theta}{1+m^2-2m\cos 2\theta} + p\frac{1-m^2-2m+2\cos 2\theta}{1+m^2-2m\cos 2\theta} \qquad (7)$$

$$\text{square:}\quad \sigma_\theta = p\cdot\frac{\mu}{1-\mu}\frac{4(AC+BD)}{C^2+D^2} + p\frac{4(AC+BD)}{C^2+D^2}\Big|_{\theta-\pi/2} \qquad (8)$$

$$A = 14 - 24\cos 2\theta - 7\cos 4\theta$$
$$B = -24\sin 2\theta - 7\sin 4\theta$$
$$C = 56 + 28\cos 4\theta$$
$$D = 28\sin 4\theta$$

where θ is the angle between the position and horizontal axis as shown in Fig. 4 and is calculated anti-clockwise.

Table 1 summarizes the circumferential stresses at different locations along the perimeter of tunnel result from Eqs. (6)-(8). When the circumferential stresses along the perimeter are determined, the coefficient of stress release can be calculated according to Eq. (4) given the initial geostatic stresses (see Eq. (5)) and friction angle of soil. For example, when assume $\varphi=30°$, $\sigma^0_r|_{\theta=\pi/2}=p$, $\sigma^0_r|_{\theta=\pi/4}=p$, and $\sigma^0_r|_{\theta=0}=q$ according to Fig .4 (here, dimension of tunnel was neglected), the values of α at various locations for different excavation geometries when $\mu= 0.3$(usually for shallow soil tunnels) are listed in Table 2. For the deep tunnel, usually cover depth is several times of tunnel diameter, the initial stresses around tunnel were recognized as the same, thus $\mu=0.5$.

One observes that the amount of stress release is different at various locations on the perimeter of excavation of different geometries. For the shallow circular and ellipse tunnels, stress release coefficient is the biggest at the top and bottom of tunnel, and then at the 45° direction along central axis, and is the smallest at the horizontal sides. For square tunnel, α is the smallest at the 45° direction (corners of tunnel), which subjected to stress concentration. For deep tunnel, stress release is the same around the circular tunnel and shows the same rule with shallow one for ellipse tunnels. All in all, stress release in deep tunnel is lower than in the same shape of shallow tunnel.

Table 1. Circumferential stress along the perimeter of tunnel

Hole shape / Location	circle	ellipse (a=2b,m=1/3)	square
	σ_θ	σ_θ	σ_θ
θ=0, θ=π (lateral sides)	$\frac{3-4\mu}{1-\mu}p$	$\frac{6\mu-5}{1-\mu}p$	$\frac{31-48\mu}{21(1-\mu)}p$
θ=π/4, θ=3π/4 θ=5π/4, θ=7π/4 (45° direction)	$\frac{1}{1-\mu}p$	$\frac{6\mu+1}{5(1-\mu)}p$	$\frac{3}{1-\mu}p$
θ=π/2, θ=3π/2 (vertical)	$\frac{4\mu-1}{1-\mu}p$	$\frac{3\mu-1}{1-\mu}p$	$\frac{48\mu-17}{21(1-\mu)}p$

Table 2. α at different locations with μ= 0.3 (shallow) and μ= 0.5 (deep)

Location \ Hole shape	circle		ellipse (a=2b,m=1/3)		square	
	$\alpha_{\mu=0.3}$	$\alpha_{\mu=0.5}$	$\alpha_{\mu=0.3}$	$\alpha_{\mu=0.5}$	$\alpha_{\mu=0.3}$	$\alpha_{\mu=0.5}$
θ=0, θ=π	0	37.2%	0	0	17.1%	78.9%
θ=π/4, θ=3π/4 θ=5π/4, θ=7π/4	55.1%	37.2%	74.9%	62.3%	0	0
θ=π/2, θ=3π/2	90.9%	37.2%	100%	68.6%	100%	78.9%

Influence of friction angle of soil

According to Eq. (4), stress release coefficient α is also relevant with friction angle besides stress level. Fig. 5 presents the influence of soil friction angle φ on stress release coefficients in shallow and deep tunnel. It is evident that friction angle of soil has significant impact on stress release as one may expect. More specifically, an increase in friction angle of soil tends to decrease stress release. In general, the friction angles of granular soils are higher than cohesive soils, thus one may conclude that the stress release in granular soils tends to be lower than that in clays (particularly saturated soft clays). This conclusion is consistent with observations based on in-situ monitoring by some researchers that "stress in soil under foundation pit base due to excavation was not clear but a part of residual stress remained, which is bigger in sand".

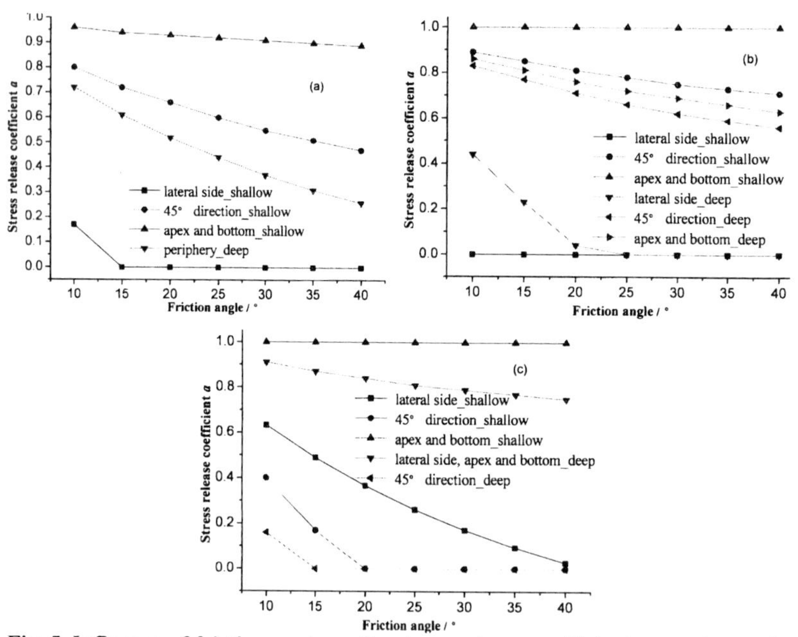

Fig. 5. Influence of friction angle on the stress release coefficient around tunnel: (a) circular tunnel, (b) ellipse tunnel, (c) square tunnel.

Fig.5 also shows that the influence of stress states on the relationship between the coefficient of stress release and the friction angle of soil. For example, α is always 0 when the friction angle varies from 15°~40°at the lateral side of shallow circular opening in Fig. 5(a), which means nearly no stress release; α is always 1 when the friction angle varies from 10°~40°at the apex and bottom of shallow ellipse opening in Fig. 5(b), indicating complete stress release. These are also found in the case of square opening, see in Fig. 5(c). According to Eq. (4), we can know that stress release coefficient $\alpha$is determined by both stress state and friction angle at the same time. There is no independent relationship between stress release and friction angle.

SUMMARY

In this study, stress release coefficients around tunnels with different geometries were calculated based on the continuum analytic solution combined with release behavior of granular soil. The coefficients of stress release vary along the boundaries of excavation, and depend on the geometry of the opening and internal friction of soil. Increase in the friction angle will reduce the stress release in soil under certain stress conditions.

It should be noted that the stresses at the apex and bottom of the tunnel are the same, since the tunnel is considered as a small hole in this study when determining stresses on its boundary and hence the difference in geostatic stress at these two locations may be neglected. Further work is required to investigate the dependency of stress field on principal stress rotation induced by excavation, which may results in different stress releases at the bottom and the apex of a tunnel.

ACKNOWLEDGMENTS

The authors appreciate the support of National Natural Science Foundation of China. (NO. 50878151)

REFERENCES

Zhuang L. (2009). "Loading and unloading behavior and stress release of granular." Tongji University, Doctor Thesis.

Ren, G., Smith, J.V., Tang, J.W. and Xie Y.M. (2005). Underground excavation shape optimization using an evolutionary procedure. *Computers and Geotechnics*, 32: 122-132.

Ng, C.W.W. and Lee, G.T.K. (2002). A three-dimensional parametric study of the use of soil nails for stabilizing tunnel faces. *Computers and Geotechnics*, 29: 673-697.

Zhou, S.H. (1997). Excavation theory. Beijing, China Railway Press.

Mana, A.I., and Clough, G.W. (1981). Prediction of movements for braced cuts in clay. *Journal of Geotechnical Engineering Division*, 107: 759-778.

Masaru Kajigaya, Junichi Ozaki, Kaich Sakamoto and Kazuyuki Hukada. (1989). Behavior and analysis of braced excavation system in soft ground. *Soil and foundation*, 37-5(376): 23-28.

Bernat, S., and Cambou, B. (1998). Soil-structure interaction in shield tunneling in soft soil. Computers and Geotechnics, 22(3/4): 221-242.

Study on Rock Mass Stability effect of High Water Pressure Tunnels by Hydraulic Fracturing Failure

Zongli Li [1] Xiaohui Liu [1] Qingwen Ren [2]

[1]College of Water Conservancy and Architecture Engineering of Northwest A & F University ,Yangling, Shaanxi, 712100; bene@nwsuaf.edu.cn, summerd@yahoo.cn
[2]College of Civil Engineering of Hohai University, Nanjing ,Jiangsu, 210098;renqw@hhu.edu.cn

ABSTRACT: The hydraulic fracturing is one of common failure in hydraulic high water pressure tunnels. The process of hydraulic fracturing of high water pressure tunnel with different initial fracture length in tunnel inner wall is simulated using nonlinear FEM. The couple effect of fracture inner water pressure with surrounding rock mass deformation is considered in the model. The fracture propagation process with increasing of tunnel inner water pressure is analyzed, its effect on stress, displacement and plastic zone of tunnel surrounding rock mass is obtained. The results are compared with that obtained without considering existence of fracture water pressure. There are different results from two analysis method. The results show that only couple analysis can obtain relative authentic stress, displacement and plastic zone of tunnel surrounding rock mass, providing base data for correct stability assessment of surrounding rock mass.

1 INTRODUCTION

According to the requirements in specification for design of hydraulic tunnel, a design scheme must satisfy the lifting top criterion, the hydraulic fracture criterion, the initial minimum principal stress criterion and some others to prevent hydraulic fracturing failure and control leakage effectively. There are lots of fissures and joints in the surrounding rock mass, and many small fractures in the liner for some reasons such as tunnel construction. So even if a design scheme satisfied the requirements of specification, the hydraulic fracturing failure would still occur(Zhang,2005,and Wu,2000).

The hydraulic fracturing of high water pressure tunnel cause not only leakage, but also instability of surrounding rock, and safely running. Without considering of tunnel running and mechanics condition, the hydraulic fracturing problem of a single fracture in surrounding rock mass only from the perspective of fracturing destabilization is analyzed(Huang, 2000; Li,2005; Sheng, 2005) . In this paper, based on the water-filling process in a specific tunnel project, the practical application of the numerical simulation model of hydraulic fracturing of rock with a single fracture (Li, 2007) is applied to the dynamic propagation process of an initial fracture at tunnel wall. In addition, the effects of the fracture on the stress field, the displacement field and the distribution of plastic zone are also studied.

2 NUMERICAL SIMULATION MODEL OF HYDRAULIC FRACTURING OF ROCK WITH A SINGLE FRACTURE

The numerical simulation model of hydraulic fracturing of rock with a single fracture includes three major parts. It is included water movement model of a single fracture, elastic-plastic analysis model of rock mass, and fracture propagation model of a fracture. Each of them includes some sub-models (Li, 2007).

In the model of water movement in a single fracture, the some assumptions are followed. The fracture water is incompressible and satisfies Newton inner friction law. The flow regime in fracture is laminar. The relationship of flow and the fracture opening width agrees with the cubic law. The fracture is in low permeable porous medium, so the leakage of fracture surface can be neglected.

The fracture water movement model can be simplified to one-dimensional problem. Based on the control volume method, the direct calculation formula between the fracture water pressure and the fracture opening width are derived from the flow and mass equation of fracture water. In the analysis process, the advance of fracture water flow and the increase of the inner water pressure are approximately considered as same time. That means, at the end of a period, the fracture water pressure can achieve to the steady state corresponding to a certain inner water pressure. In the model of elastic-plastic analysis of rock mass, the nonlinear finite element method based on continuous medium is adopted. In the model of fracture propagation, the Hillerborg cohesive zone model, and COD criterion improved by damage mechanics theory are applied to. In order to simulate fracture, zero-thickness contact elements are prearranged in the possible propagation direction, so the advance and retreat of fracture length will be skillfully realized through the increase and decrease of actual contact element numbers in iterative computation.

The coupled analysis method is divided into directly and indirectly coupled method in ordinary. Because the problem in this paper exists nonlinear coupled analysis, the indirectly method is selected here. The analysis steps are followed:

(1)The elastic or elasto-plastic rock deformation analysis is carried out based on initial loads and fracture length. The stresses of interface elements are obtained also.

(2)The stiffness factors of interface elements are adjusted according to stress state, and the new whole structure stiffness array is renewed again. Then the analysis is turned into step (1). When the interface state of the first analysis and the secondary is coincidence, then analysis is turned into next step.

(3)The fracture opening width is calculated according to rock deformation results. Then, the fracture's propagation state is judged according to propagation criterion. If fracture is not propagated, then the analysis is turned into next step. If fracture has been propagated, the new tip location, actual fracture length and opening width are calculated, then, the analysis is turned into next step.

(4)The new fracture inner water pressure is calculated according to actual fracture length and opening width, and converted to structure load, and the analysis turns into step (1). When the discrepancy of fracture inner water pressure is smaller than permitting value and the fracture length is not changed, then, the analysis is turned into next step.

(5)The fracture boundary water pressure is increased according to time step, then, the analysis is turned into step (1).

3 CASE STUDIES

3.1 Basic Information

There is an underground pressure tunnel with radius 2 m lined by reinforced concrete. The concrete grade is C30 and lining thickness is 0.8m. The vertical in-situ stress around tunnel is 7.08 MPa, and the horizontal is 3.32 MPa, such as figure 1. In tunnel roof, a single fracture which length is 0.4 m exists due to some other reasons. To simplify the analysis model, the surrounding rock mass is approximately considered as homogeneous. The physical and mechanical parameters of the liner, surrounding rock and fracture are shown in Table 1. In the analysis, the theory of elastic-plastic analysis and Mohr-Coulomb yield criterion are used. For fracture simulation, the zero-thickness contact elements are adopted. The fracture tangential and normal stiffness coefficients, the formulas recommended by reference(Zhu,1998) are applied.

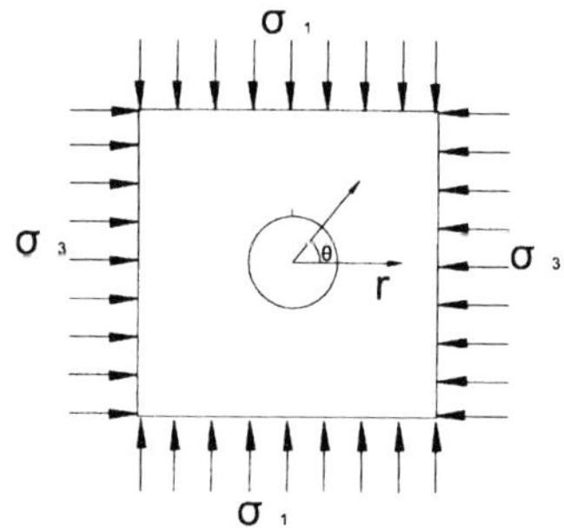

Fig. 1 Analysis example tunnel

Table 1. Physical mechanical Parameters of Rock and Fracture

Materials	**Elastic Modulus (GPa)**	**Poisson Ratio**	**Internal Friction Angle (°)**	**Cohesive Force (MPa)**	**Tensile Strength (MPa)**	**Fracture Toughness (MPa $\sqrt{m}$)**
Concrete	30	1.67	54.903	3.162	2.0	2.5
Rock	60	0.22	46.57	1.30	1.61	1.0
Fracture	**Internal Friction Angle (°)**	**Cohesive Force (MPa)**	**Tensile Strength (MPa)**	**Parameter n**	**Parameter F**	**Air pressure (MPa)**
	20	0	0	1.758	28.3	0.1

According to elastic mechanics theory, the stress of surrounding rock of tunnel is compressive everywhere without inner water pressure. While in water-filling process, pressure water may enter into the fracture, causing the hydraulic fracturing failure. When the water pressure inside fracture is greater than the fracture surface compressive stress, the fracture will tend to opening or even propagation. In order to study the effects of the displacement, stress and plastic zone around the tunnel due to fracture propagation, the water-filling process is selected in analysis. In this process, the inner water pressure increases from 0.0 MPa linearly.

3.2 Analytical scheme

The First one is hydraulic fracturing elastic-plastic couple analysis. The couple effects of fracture inner water pressure with surrounding rock stress and displacement field is considered. The second one is the analysis based on the conventional previous treatment method. The fracture propagation and the gradual increase of tunnel inner water pressure is considered, but not water pressure in the fracture.

3.3 Results analysis

3.3.1 The critical tunnel water pressure at the beginning of fracture propagation

According to elastic analytical solution(XU,1982), if the tunnel wall is elastic and no any fracture in the wall while inner water pressure is 4.88MPa, the hoop stress at tunnel roof is equal to the tensile strength of liner concrete. In the elastic-plastic couple analysis, the tunnel inner water pressure at the beginning of fracture propagation is 1.95MPa while the initial fracture length is 0.4m.While no considering fracture water pressure, it is 2.75MPa. Those results show obviously that initiation pressure for the fracture propagation of the couple analysis is lower than that of not considering the fracture water pressures.

3.3.2 The influence of hydraulic fracturing on stress field around tunnel

The existences of fractures, as well as fracture water pressure can influence the stress field of surrounding rock mass. From the analysis results, it can be seen that when the length of fracture is small, the effects of stress field is limited to small region. While at the part far away from the fracture, the stress field is close to that of the no fracture condition. Based on other results, with the fracture propagation, the influence region expands gradually. However, when without considering the fracture water pressure, the existence of fracture make the hoop stress became smaller and the radial stress larger near the fracture tip.

3.3.3 The influence of hydraulic fracturing on displacement field around tunnel

In displacement field calculation, the actual running condition of the tunnel is considered. After the deduction of the displacement originated from initial in-situ stress, the changing laws of the displacements of the there control points at the roof, the bottom and the right side of tunnel inner wall with the fracture propagation are analyzed. Those specific results are shown in Figure 2 and Figure 3.

As can be seen from Figure 2, in couple analysis method, with the development of fracture propagation, the displacements of tunnel roof and bottom develop toward inside, while the displacement of the side wall develops toward outside, indicating that the rock at tunnel roof tends to be loosing and instability. The phenomena indicate the fracture propagation and opening at tunnel roof. As can be seen from Figure 3, in the elastic-plastic analysis without considering the fracture water pressure, on the contrary

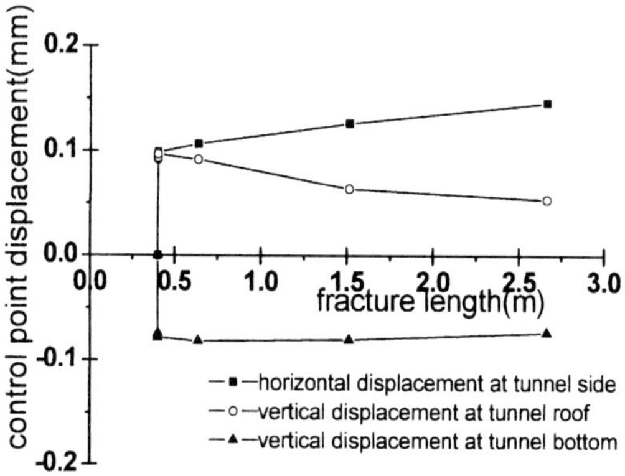

Fig.2 Changing law of control point displacements of coupled analysis as fracture propagation

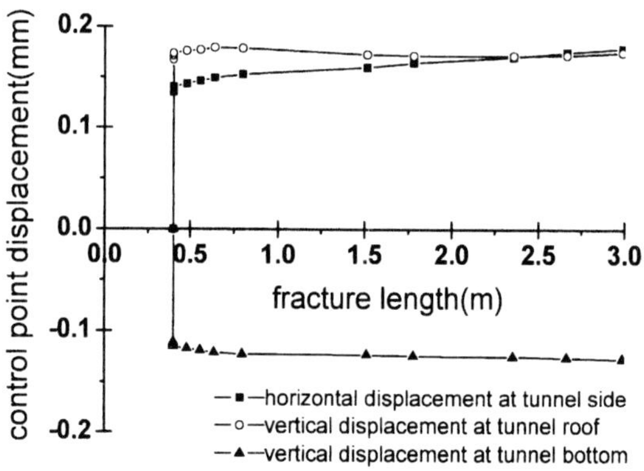

Fig.3 Changing law of control point displacements of no water in fracture as fracture propagation

of couple analysis, the displacements of tunnel roof and bottom develop toward outside, while the displacement of the side wall develops toward inside. Thus, if not considering the fracture water pressure, the stability assessment of rock mass will cause incorrect conclusion.

3.3.4 The influence of hydraulic fracturing on plastic zone around tunnel

Due to high in-situ stress of analysis tunnel, the plastic zone has yet appeared in local region of tunnel inner wall even if there's no inner water pressure in the tunnel. With the increase of inner water pressure, this plastic zone disappears gradually, and a small plastic zone appears in the fracture tip finally. In couple analysis, the fracture tip plastic zone is larger than that of no considering fracture water pressure.

4 CONCLUSIONS

(1) For the high water pressure tunnel with initial fractures in rock mass, the hydraulic fracturing couple analysis can reflect the actual fracture propagation process and obtain relative authentic stress field, displacement field, the distribution of plastic zone around tunnel. Based on couple analysis, the correct assessment of surrounding rock mass stability can be obtained.

(2)The couple analysis results of hydraulic fracturing indicate that once the fracture appeared in high water pressure tunnel, a further fracture propagation is apt to happen. The longer the fracture is, the easier propagation is. When the fracture length is small, the influence region of stress and displacement field is less. With the development of fracture propagation, the influence region will increase gradually.

ACKNOWLEDGMENTS

The authors appreciate the support of the Natural Science Foundation of China under grant No.50779057, and the National Basic Research Program of China ("973" Program) under Grant No.2007CB714104.

REFERENCES

Huang Runqiu, Wang Xianneng, Chen Longsheng(2000). “Hydro-splitting off analysis on underground water in deep-lying tunnel and its effect on water gushing out.” Chinese Journal of Rock Mechanics and Engineering, Vol. 19(5):573-576

Li Zongli, Zhang Hongchao, Ren Qingwen, et, al(2005). “Analysis of hydraulic fracturing and calculation of critical internal water pressure of rock fracture.”Rock and Soil Mechanics,Vol.26(8): 1216-1220

Li Zongli, Wang Yahong, Ren Qingwen(2007). “Numerical simulation model of hydraulic fracturing of rock with a single fracture under natural hydraulic power.” Chinese Journal of Rock Mechanics and Engineering,Vol. 27(4): 7227-733

Sheng Jinchang, Zhao Jian, Su Baoyu(2005). “Analysis of Hydraulic fracturing in Hydraulic Tunnels under high water pressure.” Chinese Journal of Rock Mechanics and Engineering, Vol.24(7):1226-1230

Wu Zhiqin(2000). “Initial filling and emptying tests on headrace system of Guangzhou pumped storage power station phase.” Water Resources and Hydropower Engineering,(4):48-51

XU Zhilun(1982). “Elastic Mechanics (the second edition)”. Beijing: China Higher Education Press

Zhu Bofang(1998). “The Finite element method theory and application (the second edition).” Beijing: China Waterpower Press

Zhang Youtian (2005). “Rock hydraulics and engineering.” Beijing: China Waterpower Press

Application of Single Pass Tunnel Lining with Steel Fibre Reinforced Shotcrete on the Ventilation Shaft of Mount Motian Tunnel

L. J. Su[1], X. K. Xing[2], Z. P. Song[3†], H. J. Liao[4] and S. Y. Wang[5]

[1] Corresponding author, School of Civil Engineering, Xi'an University of Architecture & Technology, Xi'an Shaanxi 710055, PRC; Tel: +86-29-8220-7823, Email: sulijun1976@163.com
[2] Civil Engineering Department, Guilin University of Technology, Guilin 541004, PRC
[3] School of Civil Engineering, Xi'an University of Architecture & Technology, Xi'an 710055, PRC
[4] Corresponding author, Department of Civil Engineering, Xi'an Jiaotong University, Xi'an Shaanxi 710049, PRC; Email: hjliao@mail.xjtu.edu.cn
[5] Centre for Geotechnical and Materials Modelling, Civil, Surveying and Environmental Engineering, the University of Newcastle, Callaghan, NSW 2308, Australia

ABSTRACT: The ventilation shaft of Mount Motian tunnel is an inclined shaft with a dip angle of 24°21'48". Investigations were conducted to evaluate the feasibility of applying single pass tunnel lining (SPTL) of steel fibre reinforced shotcrete (SFRS) to the ventilation shaft as a permanent lining. Concrete mix ratio tests were carried out to find out a proper mix ratio under which the optimum strength and degree of slump can be achieved. Tests were performed on specimens cut from sprayed shotcrete samples for flexural and cleavage tensile strength of the shotcrete. Results of the tests showed that the inclusion of steel fibre largely increased the flexural and tensile behaviour of the shotcrete. The shaft is in construction now and the performance of the completed sections is good.

INTRODUCTION

The function of a tunnel lining is to absorb the deformation of the surrounding rocks and utilize their self-support capability. Fibre reinforced concrete displays good performance in anti-cracking and absorbing deformation. Therefore it is widely used in pavements, bridges and tunnel linings.

The mechanical bahaviour of fibre reinforced concrete have been widely investigated internationally. Ding and Kusterle(1999 and 2000) found that steel fibres can enhance the energy absorption and the ductility. Haktanir et al. (2007) found that the average three-edge-bearing strength and crack size of steel-fibre concrete pipes having steel fibres at a dosage of 25 kg/m^3 turned out to be 82% greater and 47% smaller than those of plain concrete pipes, and 6% greater and 15% smaller than those

† The first tree authors contribute equally to this paper.

of reinforced-concrete pipes, respectively. Tan and Saha (2005) found that the addition of steel fibers to concrete was effective in containing long-term deflection and restraining crack widening. Other type of fibre such as polypropylene fibre is also used to increase the anti-cracking capability of concrete (Zhu et al. 2004 and Sivakuma and Manu 2007). SPTL of SFRS has been used as permanent tunnel lining since 1980s in Europe and America. In China, fibre reinforced shotcrete is beginning to be used in tunnels and electrical constructions (Ren and She 2005 and Yu 2006). Successful experiences accumulated in these applications are very useful for future applications.

As the dip angle of the ventilation shaft is too large, the construction of the secondary moulded lining is difficult. After reviewing the investigations of fibre reinforced concrete and applications of SFRS in tunnel lining, the contractee and contractor agreed to use SPTL of SFRS as the permanent lining. Tests for both physical and mechanical properties of the SFRS were conducted and the results are presented in this paper. Displacements and stress of the SPTL were monitored during construction. Ventilation capability of the tunnel was evaluated to make sure it will not decrease compared with the original supporting plan.

STRENGTH TESTS ON SPRAYED SHOTCRETE SAMPLES

The cement used here is a type of 42.5R ordinary Portland cement. The coarse aggregate is crushed limestone with particle size from 6.0 mm to 10.0 mm and fine aggregate is a mix of 70% crushed sands and 30% river sands. Quick-setting additive was used to help in developing a higher initial strength of the shotcrete. Silica fume was used to fulfill the requirement of strength, permeability and degree of slump of the shotcrete at the same time. Concrete mix ratio tests were carried out to determine mix ratio for the five 20 m long testing sections. The mix ratios for the five sections are listed in Table 1. The percentage of water reducing agent and quick-setting additive in Table 1 is the percent of the total mass of cement and silica fume. Using the above mix ratio, the obtained degree of slump was around 180 mm which is acceptable.

Sprayed shotcrete slabs were cut into 100mm×100mm×100mm cubic specimens for compressive and tensile strength tests and 100mm×100mm×400mm beam specimens for flexural strength test. The test results are summarized in Table 2.

Table 1. Concrete Mix Ratio for the Testing Sections

No.	Steel fibre			Water	Cement	Coarse aggregate	Water reducing agent	Quick-setting additive	Sands	Silica fume
	Type	Length-diameter ratio	Mass (kg/m^3)	(kg)	(kg)	(kg)	(%)	(%)	(kg)	(kg)
1	Cold-drawn	60	40	225.6	441.6	598.8	0.9	7	1112	38.4
2	Incised	60	45	225.6	441.6	597.1	0.9	7	1109	38.4
3	Incised	60	40	225.6	441.6	598.8	0.9	7	1112	38.4
4	Incised	60	40	225.6	441.6	514.6	0.9	7	1201	38.4
5	Incised	60	45	225.6	441.6	597.1	0.9	7	1109	38.4

Table 2. Measured Strength Parameters for the SFRC

Testing section	Average strength				
	Flexural (MPa)	Compressive (MPa)		Tensile (MPa)	
		14d	28d	14d	28d
1	8.21	41.89	45.24	4.29	5.45
2	7.64	31.08	39.84	2.55	2.86
3	8.06	31.23	41.92	2.70	3.46
4	7.93	36.72	43.82	2.57	3.11
5	8.12	35.81	50.86	3.53	4.33

. The inclusion of steel fibre largely improved the ductility of the shotcrete although the increase in compressive strength of the shotcrete is not very apparent. FIG. 1 shows the compressive stress-strain curves for SFRS and plain shotcrete. It can be observed that the peak strain for the SFRS is larger than that of the plain shotcrete. After the peak stress, the stress for the plain shotcrete decreases abruptly while that for the SFRS decreases slowly to a residual stress. Cracks and small spallings are found on the specimen but the main parts of it are still held together. This indicates that the inclusion of steel fibre changed the failure mode of concrete from brittle failure to ductile failure.

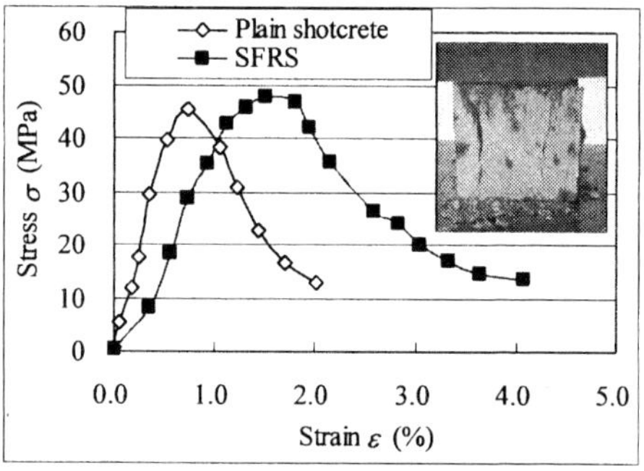

FIG. 1. Compressive stress-strain curves for SFRS and Plain shotcrete.

The tensile strength of plain concrete is around 2.0 MPa. The tensile strength of SFRS increased 43% to 172.5% compared with that of plain concrete. The flexural strength of plain concrete is normally about 5 MPa and that of the SFRS increased 52.8% to 64.2% compared with this value. From Table 2, it can be observed that the strengths of the shotcrete reinforced by cold-drawn steel fibres are the highest ones. For the wave-form incised steel fibre reinforced shotcrete, the strength increases with the increase in the percentage of steel fibre. The strengths of the shotcrete reinforced by end-hooked type incised steel fibre are the lowest ones even a higher percentage was added. Considering the performance and price of the steel fibre, the wave-form incised steel fibre was recommended to be used for the remaining part of the ventilation shaft.

SFRS was sprayed on rock blocks to prepare specimens for bond resistance tests on shotcrete-rock interface. After 7-days curing, the sprayed samples were cut into 100mm cube specimens for cleavage testing on the shotcrete-rock interface. Test results are listed in Table 3. The results indicate that the measured bond resistances satisfy the requirement in Specifications for bolt-shotcrete support (GB 50086-2001).

Permeability tests were carried out and the obtained average leakage pressure is 1.2 MPa which satisfies the requirement of the corresponding tunnel lining standard.

Table 3. Bond Resistance on Shotcrete-rock Interface

Rock mass classification*	Cleavage force (kN)	Bond resistance (MPa)	Minimum requirement (MPa)
□	10.23	1.02	0.5
□	16.85	1.69	0.5
□	23.63	2.36	0.5
IV	22.39	2.24	0.8
IV	21.90	2.19	0.8

*according to the "Standard for engineering classification of rock masses" (GB 50218 94)

MONITORING DISPLACEMENT AND STRESS OF THE SPTL

During construction of the SPTL, stress and displacement of the lining were monitored. FIG. 2 shows the locations for the stress monitoring. Displacements between the crest and left and right feet of the side wall were monitored. FIG. 3 shows the Displacement and its rate between left and right feet of the side wall for one of the testing sections. The recorded maximum displacement rate was 0.51 mm/day and the average displacement rate was 0.098 mm/day. After about 7 days of the construction, the increase in displacement was minimized and the displacement rate tended to zero which indicate the deformation tended to be stable.

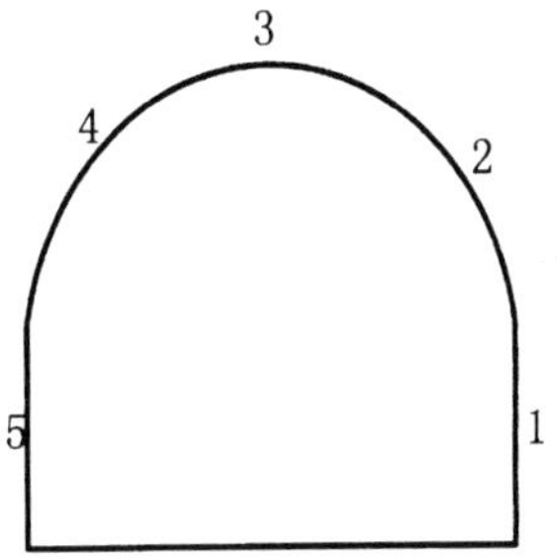

FIG. 2. Locations for stress monitoring

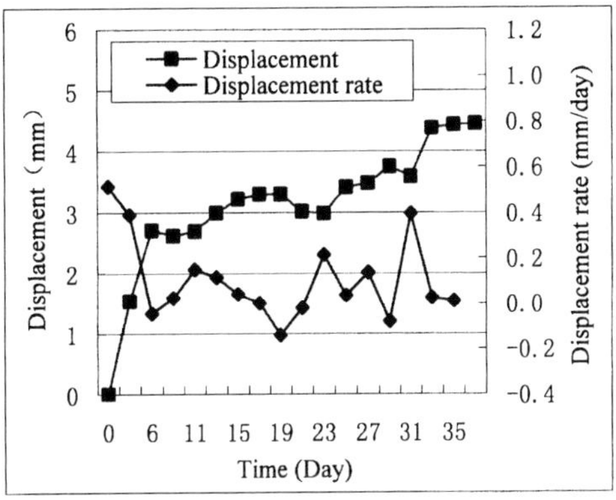

FIG. 3. Displacement and its rate between left and right feet of the side wall

FIG. 4 shows the stresses in the SPTL. It is observed that the measured stresses were small. The largest compressive stress was recorded at the crest which was 0.50 MPa. With the increase in time, the measured stresses firstly increased and then tended to be stable. It can be observed that the incremental rate of the stress tended to zero at the end of the curve. This indicated that the stress in the lining was very small and tended to be stable in a short time, which was consistent with the displacement monitoring results.

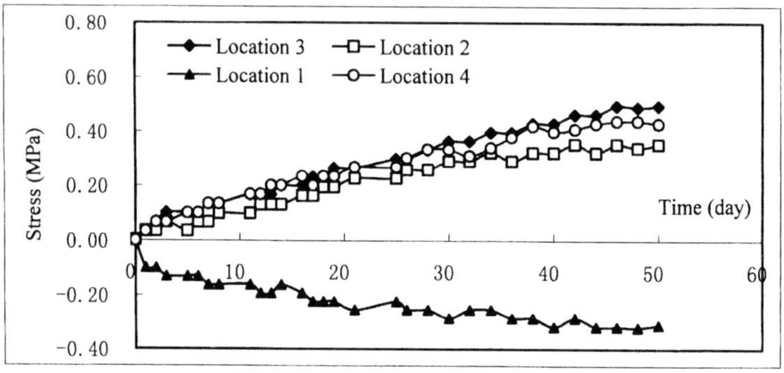

FIG. 4. Stresses in the SPTL

VERIFICATION OF VENTILATION CAPACITY

As a ventilation shaft, it is the most important to ensure the ventilation capacity of it. The SPTL should not reduce the ventilation capacity of the shaft compared with the secondary moudeld lining in the initial design. Compared with the initial design, the cross-section area of the tunnel would increase which will benefit the ventilation capacity. However, the surface roughness of the lining would also increase which is unfavorable for the ventilation. Considering these two effects, the ventilation capacity of the shaft with the SPTL was calculated and compared with that of the initial design.

The frictional resistance of the tunnel surface to the ventilation h_f is expressed as:

$$h_f = \lambda \gamma L U Q^2 / 8 g S^3 \tag{1}$$

where λ is the coefficient of frictional loss of the tunnel surface, γ is the unit weight of air, L is the length and U and S the cross-section perimeter and area of the tunnel, Q is the amount of the ventilation, and g is the gravity acceleration.

Therefore the ratio between the frictional resistances of the SPTL and cast-in-place concrete lining can be expressed as:

$$h_f = \left(\frac{\lambda_1 \gamma L U_1 Q^2 / 8 g S_1^{\ 3}}{\lambda \gamma L U Q^2 / 8 g S^3} \right) = \left(\frac{\lambda_1 U_1 / S_1^{\ 3}}{\lambda U / S^3} \right) \tag{2}$$

where, λ_1, U_1 and S_1 and λ, U and S are corresponding parameters for the SPTL and cast-in-place concrete lining, respectively.

Parameters for calculation of the ratio between the frictional resistances of the SPTL and moulded cast-in-place concrete lining are listed in Table 4. The λ for the moulded lining was obtained from the "Standard for ventilation and illumination design of highway tunnels" (JTJ026.1-1999) and that for the SPTL was calculated based on a measured average roughness of 100 mm. Using the listed parameters, the calculated ratios of frictional resistances for both the exhaust and blowing shafts are 1.01 and 0.98 respectively. This indicates that the ventilation capacity of the shaft with SPTL is almost the same as that for the initial design.

Table 4. Ventilation Parameters for the Initial Design and SPTL

		λ	U (m)	S (m^2)	Ratio
Exhaust shaft	SPTL	0.048	20.89	27.8	1.01
	Initial design	0.025	19.57	24.4	
Blowing shaft	SPTL	0.048	19.84	25.2	0.98
	Initial design	0.025	18.52	21.9	

CONCLUSIONS

The inclusion of steel fibre changed the compressive failure mode of concrete from brittle failure to ductile failure and increased its tensile and flexural strengths. Both the bond resistance on shotcrete-rock interface and permeability of the SFRS satisfied corresponding standards. These indicated that the SPTL of SFRS can be used as the permanent lining of the ventilation shaft. This was proved by measured small and stable values of displacement and stress during construction. The ventilation capacity of the tunnel with SPTL permanent lining is almost the same as that for the initial design. All of these indicate that the application of SPTL of SFRS as the permanent lining of the ventilation shaft of Mount Motian tunnel in testing sections was successful and can be used for the remaining part of the shaft.

REFERENCES

Ding Y. N. and Wolfgang Kusterle (1999). "Comparative study of steel fibre-reinforced concrete and steel mesh-reinforced concrete at early ages in panel tests." *Cement and Concrete Research*, (29): 1827–1834.

Ding Y. N. and Wolfgang Kusterle (2000). "Compressive stress-strain relationship of steel fibre-reinforced concrete at early age." *Cement and Concrete Research*, (30): 1573-1579.

Haktanir, Tefaruk, Kamuran Ari, Fatih Altun and Okan Karahan (2007). "A comparative experimental investigation of concrete, reinforced-concrete and steel-fibre concrete pipes underthree-edge-bearing test." *Construction and Building Materials*, (21): 1702–1708.

Ren S. X. and She Y. H. (2005). "Application of polypropylene tiny fiber shotcrete in tunnel construction. *Modern Tunnelling Technolog*, Vol. 42(1): 26-29. (in Chinese)

Sivakumar A. and Manu Santhanam (2007). "Mechanical properties of high strength concrete reinforced with metallic and non-metallic fibres." *Cement & Concrete Composites*, Vol. 29(8): 603-608.

Tan K. H. and Mithun Kumar Saha (2005). "Ten-Year Study on Steel Fiber-Reinforced Concrete Beams Under Sustained Loads." *ACI Structural Journal*, Vol. 102(3): 472-480.

Yu C. X. (2006). "Application of steel fiber wet-shot concrete technique in tunnel construction." *Shanxi Architecture*, Vol. 32(10): 5-136. (in Chinese)

Zhu Y. Q., Liu Y., Liu Z. C. and Bian Y. H. (2004). "Experiment on behaviour of polypropylene-fiber-net Shotcrete and lining." *Chinese Journal of Rock Mechanics and Engineering*, Vol. 23(19): 3376～3380. (in Chinese)

Numerical Analyses and Elasto-plastic Behavior Study on Surrounding Rock Mass of the Underground Caverns in a Hydropower Station during Deep Excavations

Yong Li[1], Weishen Zhu[2] and Linfeng Sun[3]

[1] Geotechnical & Structural Engineering Research Center, Shandong University, Jinan, P.R. China, 250061; andyli.liyong@yahoo.com.cn
[2] Geotechnical & Structural Engineering Research Center, Shandong University, Jinan, P.R. China, 250061; zhuw@sdu.edu.cn
[3] Fuzhou Investigation and Surveying Institute, Fuzhou, P.R. China, 350000; 19786771@qq.com

ABSTRACT: For the time being in China, aggressive development of the western hydroelectric resources is being performed. Taking the Shuangjiangkou (SJK) Hydropower Station as an engineering background, the elasto-plastic behavior is studied on the surrounding rock mass of the underground caverns, including the main powerhouse, the transformer house, the surge chamber, the busbar chamber, etc. in the whole process of deep excavations. A 3D numerical simulation is performed using $FLAC^{3D}$. Four typical yield criteria which are Mohr-Coulomb (MC) criterion, Drucker-Prager model for the outer adjustment (DPO), Drucker-Prager model for the inner adjustment (DPI) and Zienkiewicz-Pande criterion (ZP) are all utilized in the elasto-plastic behavior study on the surrounding rock mass of the underground caverns. Among of the criteria, ZP criterion is written in the VC++7.1 compiling environment and the other three criteria are included in $FLAC^{3D}$. The displacement field of the surrounding rock mass are obtained and compared under the four criteria in the process of excavations. The plastic zones are also obtained and compared in this paper. It is concluded that the results obtained through MC and DPI criteria are much greater than those obtained by the other two criteria, and the results obtained though ZP criterion are close to the practical engineering results.

INTRODUCTION

Understanding the mechanical behaviors of discontinuous rock masses is important for the development and utilization of deep underground engineering such as hydropower stations and petroleum reservoirs (Jiang et al., 2009). Analysis of stress and displacement around the underground caverns excavated in isotropic rock masses has been one of the fundamental problems in geotechnical engineering (Lee et al., 2008). According to the existing literature reviews (Alonso et al., 2003), elasto-plastic approaches seem to be the most popular in studying the stability of underground caverns.

In the past, MC yield criterion was the most common in the elasto-plastic analysis of rock masses, due to its simplicity. However, a large number of experimental observations show that the strength envelope for most of rock-like materials is not linear. Therefore, the other nonlinear yield criteria were also studied just like DPO, DPI, ZP, Hoek-Brown (HB) (Hoek and Brown, 1980) and generalized HB (Hoek et al., 2002). Therefore, in this paper, MC, DPO, DPI and ZP were all utilized in the elasto-plastic behavior study of the surrounding rock mass in the underground caverns of SJK Hydropower Station.

YIELD CRITERIA

FIG. 1 shows yield curves of different criteria in the π plane.

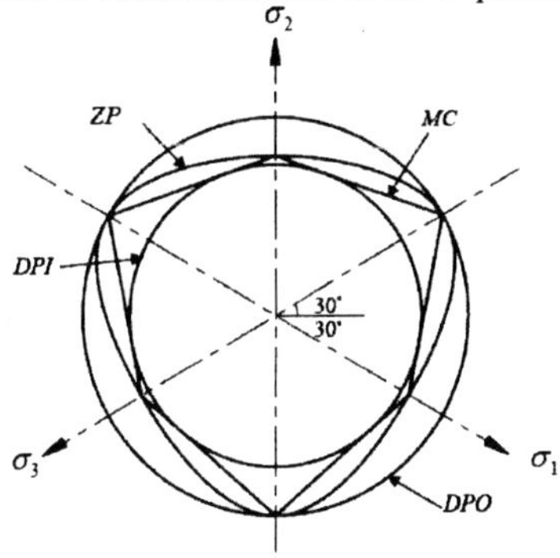

FIG. 1. Yield curves of different criteria in the π plane

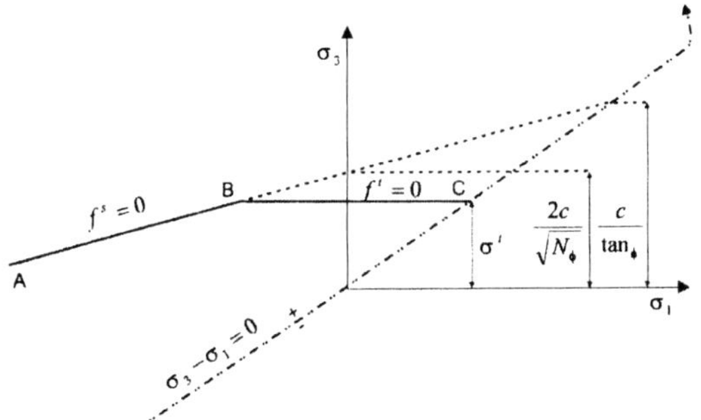

FIG. 2. MC criterion in FLAC^{3D}

1) MC yield criterion

The failure criterion used in FLAC^{3D} model is composite MC criterion with tension cutoff. In labeling the three principal stresses so that

$$\sigma_1 \le \sigma_2 \le \sigma_3 \tag{1}$$

This criterion can be represented in the plane (σ_1, σ_3) as illustrated in FIG. 2. The failure envelope $f(\sigma_1,\sigma_3)=0$ is defined from point A to B by the MC failure criterion $f^s=0$ with the following

$$f^s=\sigma_1-\sigma_3N_\varphi+2c\sqrt{N_\varphi} \tag{2}$$

where, $N_\varphi=\dfrac{1+\sin\varphi}{1-\sin\varphi}$, φ represents the friction angle.

2) DP yield criterion

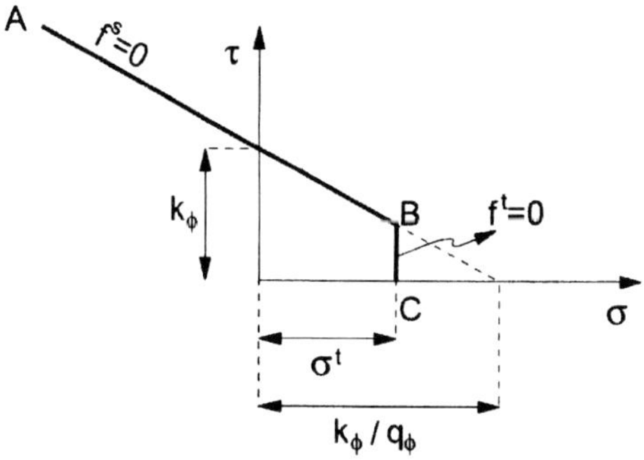

FIG. 3. DP criterion in FLAC^{3D}

The failure criterion used for this FLAC^{3D} model is a composite DP criterion with tension cutoff as sketched in the (τ,σ) representation of FIG. 3. The failure envelope $f(\tau,\sigma)=0$ is defined, from point A to B on the figure, by the DP failure criterion $f^s=0$, with

$$f_s=\tau+q_\varphi\sigma_m-k_\varphi \tag{3}$$

where q_φ and k_φ are positive material constants. And q_φ and k_φ would be different values under the conditions of DPO and DPI.

3) ZP yield criterion

The expression of ZP yield criterion can be defined as

$$f_s=\beta\,\sigma_m{}^2+\alpha_1\,\sigma_m-k+\overline{\sigma_+}^2=0$$

where $\overline{\sigma}_+=\dfrac{\sqrt{J_2}}{g(\theta_\sigma)}$, and $g(\theta_\sigma)=\dfrac{2K'}{(1+K')-(1-K')\sin 3\theta_\sigma}$, $K'=\dfrac{3-\sin\varphi}{3+\sin\varphi}$.

PROJECT DESCRIPTIONS

Shuangjiangkou Hydropower Station is located on the upper reaches of the Dadu River, Sichuan Province, China. It has a maximum output of 2000MW (500 MW×4 units) and is the largest hydropower station in the Dadu River basin. It is located 2-6 km downstream of the confluence of the Zumuzu and the Zhuosijia rivers. The stream valley belongs to the entrenched meander valley. The mountains at both sides of the river are so high and the slope height beside the river reaches up to more than 1000 meters. The natural slope inclinations are 35°-50° at the left bank and 45°-60° at the right bank. The rock masses are in a drastic unloading state towards the free surface. The mountains stretch along the slope inclinations and reach up to the altitude of 3000 m. The whole valley looks like a nearly symmetrical V-shaped valley below the altitude of 3000 m. Hence, the peak must have an effect on the initial *in situ* stress field near the underground cavern group. From the *in situ* stress measurement, the maximum principal stress near the underground cavern group is up to 38 MPa. The high *in situ* stress is absolutely rare for hydropower stations in China.

NUMERICAL DESCRIPTIONS

FIG. 4(a) shows the 3D numerical model of the cavern group. FIG. 4(b) shows the three dimensions of the cavern complex and the excavation steps. In the numerical analysis, the $FLAC^{3D}$ numerical method was adopted for the underground cavern complex considering the boundary effects, the two generating sets area, including the main power house, the transformer house, the surge chamber, the busbar chamber, et al. The elements for linings and the surrounding rock masses were the eight-node isoparametric elements. The bolts were modeled by using the bolts elements. The whole numerical model has 20754 nodes and 96662 elements. FISH is a programming language embedded within $FLAC^{3D}$ that enables us to define new variables and functions.

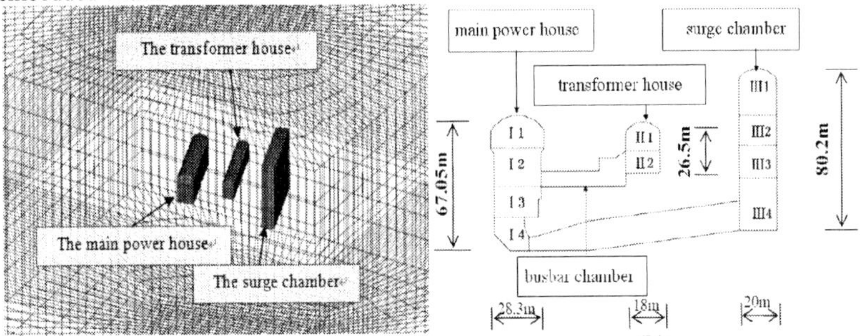

(a) **(b)**

FIG. 4. 3D numerical model and excavation steps of the cavern group

Due to some irregular excavating shapes in the in-situ excavations, the whole excavating subsequences have been simplified in the numerical simulation. The whole excavation subsequences can be divided into 20 circles and 102 steps. The excavation footage of every step is 10m. The excavation sequences are Ⅰ1,Ⅱ1,Ⅲ1,Ⅰ2,Ⅱ2,Ⅲ2,Ⅰ3,Ⅲ3,Ⅰ4 and Ⅲ4 (see FIG. 4(b)).

NUMERICAL SIMULATION AND RESULTS ANALYSES

The real embedded depth of the cavern complex is approximately 600m. According to the in-situ stress field, the initial horizontal stresses are almost 1.5 times the vertical stresses near the caverns so that the coefficient of horizontal earth pressure was determined to be $K_0 = 1.5$. Table 1 shows the physico-mechanical parameters of the rock mass.

Table 1.The physico-mechanical parameters of the rock mass

Rock mass density	Elastic modulus	Cohesion	Friction angle
26.5 KN/m^3	10.83GPa	2MPa	40.36°
Compressive strength	**Tensile strength**	**Poisson's ratio**	
80MPa	4MPa	0.25	

FIGS. 5-8 show the displacement contours after excavations using the four yield criteria. FIGS. 9-12 show the plastic zones after excavations using the four yield criteria. From the figures below, it is concluded that $D_{DPI} > D_{MC} > D_{ZP} > D_{DPO}$, here, D represents the displacement and the subscripts represent the used yield criteria. In the same way, it is concluded that $PA_{DPI} > PA_{MC} > PA_{ZP} > PA_{DPO}$, and PA means the plastic zone areas after excavations.

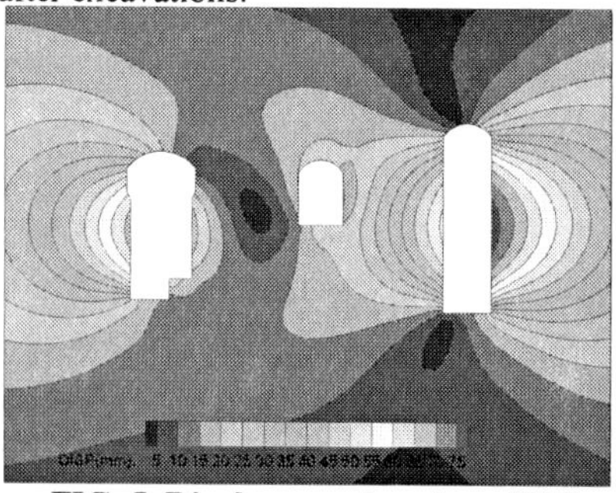

FIG. 5. Displacement contour-DPO

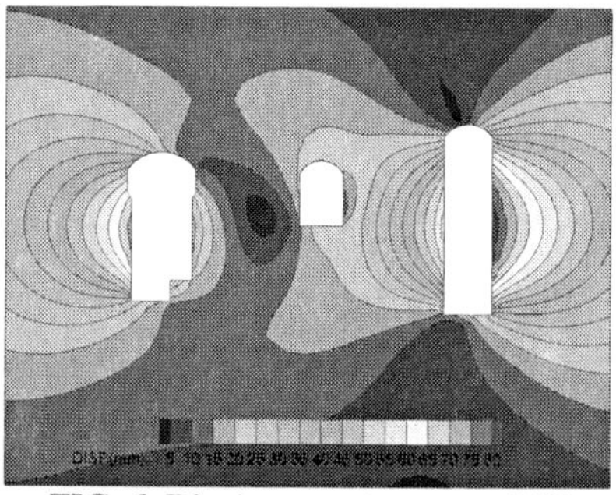

FIG. 6. Displacement contour-MC

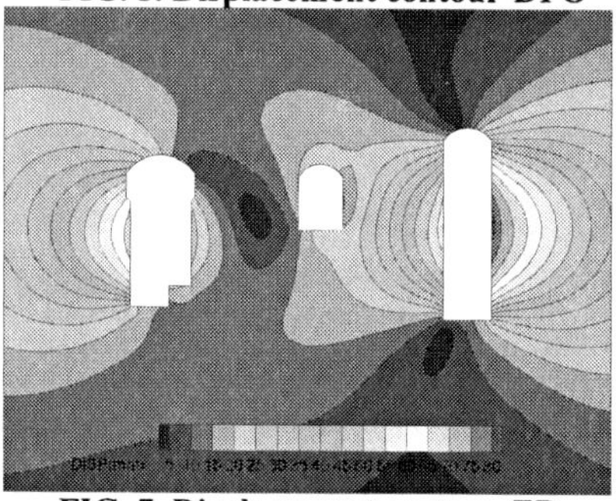

FIG. 7. Displacement contour-ZP

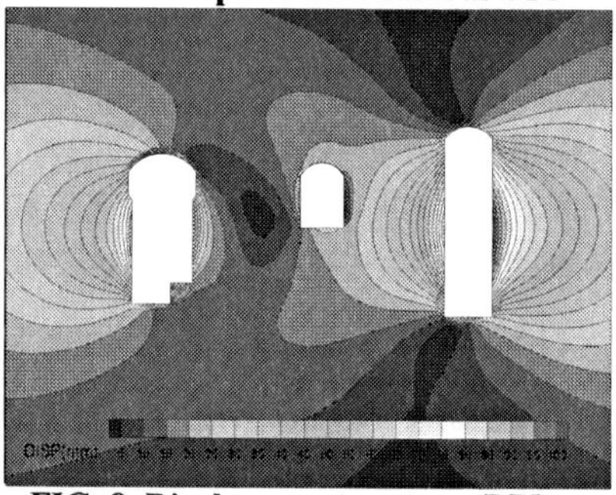

FIG. 8. Displacement contour-DPI

CONCLUSIONS

Four typical failure criteria are utilized to study the stability of the surrounding rock mass of the underground cavern group in SJK Hydropower Station. The displacement field and plastic zones of the surrounding rock mass are obtained and compared under the four criteria in the process of excavations. It is concluded that the results obtained under MC criterion are much greater than those obtained by the other three criteria, and the results obtained though ZP criterion are close to the practical engineering results.

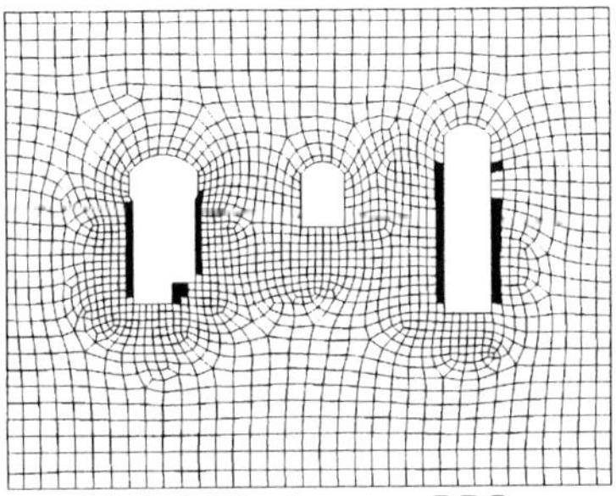

FIG. 9. Plastic zones-DPO

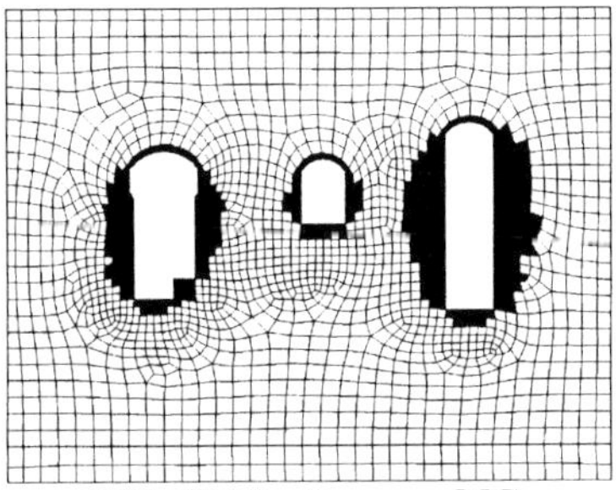

FIG. 10. Plastic zones-MC

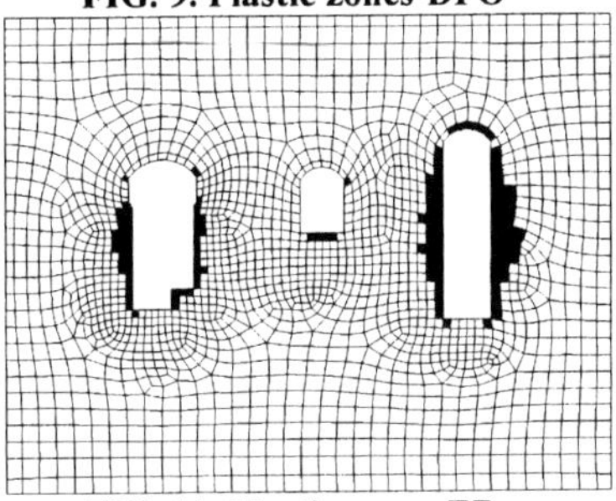

FIG. 11. Plastic zones-ZP

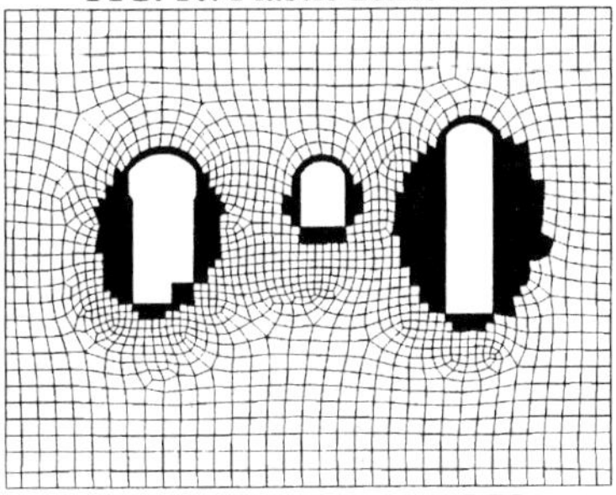

FIG. 12. Plastic zones-DPI

ACKNOWLEDGMENTS

The work was financially supported by the Key Project of Natural Science Foundation of P. R. China (No. 90510019).

REFERENCES

Alonso, E., Alejano, L.R., Varas, F., Fdez-Manin, G., and Carranza-Torres, C. (2003). "Ground response curves for rock masses exhibiting strain softening behavior." *Int. J. Numer. Anal. Meth. Geomech.*, Vol. 27: 1153-1185.

Hoek, E. and Brown, E.T. (1980). "Underground Excavation in Rock." Institution of Mining and Metallurgy, London.

Hoek, E., Carranza-Torres, C.T., and Corkum, B. (2002). Hoek-Brown failure criterion-2002 edition. In: *Proceedings of the 5th North American Rock Mechanics Symposium and 17th Tunnelling Association of Canada Conference*, Toronto: 267-273.

Jiang, Y., Li B. and Yamashita Y. (2009). "Simulation of cracking near a large underground cavern in a discontinuous rock mass using the expanded distinct element method." *International Journal of Rock Mechanics & Mining Science*, Vol. 46: 97-106.

Lee, Y.K. and Pietruszczak, S. (2008). "A new numerical procedure for elasto-plastic analysis of a circular opening excavated in a strain-softening rock mass." *Tunnelling and Underground Space Technology*, Vol. 23: 588-599.

Study on Mechanism of Simultaneous Backfilling Grouting for Shield Tunneling in Soft Soils

Zhiren Dai[1], Yun Bai[2], Fangle Peng[3], Shaoming Liao[4]

[1]Department of Geotechnical Engineering, Tongji University, 1239 Siping Road, Shanghai 200092, P.R. China; dzrzss@126.com.
[2]Department of Geotechnical Engineering, Tongji University, 1239 Siping Road, Shanghai 200092, P.R. China; baiy@public1.sta.net.cn.
3Department of Geotechnical Engineering, Tongji University, 1239 Siping Road, Shanghai 200092, P.R. China; pengfangle@tongji.edu.cn
[4]Department of Geotechnical Engineering, Tongji University, 1239 Siping Road, Shanghai 200092, P.R. China; liaosm@126.com.

ABSTRACT: The shield tail void formed in unit time in tunneling is simplified as a three-dimensional annular space with shied advancing at certain speed. As the filling process can not be observed in-situ, it is very difficult to simulate it in detail. Taking the cementitious grouts used in grouting injection as Newton fluid and Bingham fluid respectively, based on the typical four grouting holes design, the distribution of grouts pressure in cross section is obtained with the diffusion model of grouts pressure and the principle of superposition. With further consideration of grouts infiltration into the surrounding strata and its hardening process, the distribution of grouts pressure in longitudinal section is obtained. Therefore, the diffusion model of grouts pressure for simultaneous backfilling grouting in shield tunneling is founded. Finally, the comparison between theoretical result and the site observation is made.

1. INTRODUCTION

As for the modern, pressurized face shield machine, ground deformation is the main concern encountered during construction, and grouting through the tunnel rings is the key solution. At present, simultaneous backfilling grouting technology has been studied by many researchers. However, most of their studies are concentrated on the physical and mechanical characters of grouts (Zhao 2008), its penetration and diffusion into the surrounding strata (Yang 2005), the corresponding pumping technology, and ground settlement induced by the close of tail void (Ou and Cherng 1995). Due to the process of grouts diffusion in tail void is very difficult to be observed or modeled (Nomoto et al 1999), few corresponding studies have been carried out. A sort of lining piezometer wa adopted by *Hashimoto et al* (2004) to monitor grouts pressure after the tunnel boring machine in Japan, but the initial diffusion process of grouts pressure was not studied. Grouts pressure is measured by piezometers near grouting holes in construction, but not in tail void.

The flow of grouts in tail void is depended on its rheological property and the shield gap size. Therefore, as for the geometry space of shield tail void, based on the study of rock crack grouting by predecessors (Ruan 2005), especially on the diffusion law of Newton fluid and Bingham fluid, taking grouts pressure as the main parameter, and the filling process is divided into two relatively independent processes, the diffusion model of grouts pressure for simultaneous backfilling grouting is established.

2. ASSUMPTIONS AND PRECONDITIONS

In order to establish the diffusion model, some assumptions are made as follows:

- Isotropic homogeneous incompressible and stable fluid, large viscosity, dynamic shear stress of Bingham fluid remains unchanged during grouting.
- The diffusion velocity in tail void is slow and unchanged, the mutual penetration of grouts and surrounding strata is ignored, and the flow is laminar.
- The volume of tail void is fixed in unit time with certain tunneling speed, and this part of void is filled by grouts, the influence of the earlier injected grouts will not been considered.
- During the filling process of cross-section grouting, grouts viscosity remains unchanged, and its change is mainly seen in the change of grouts pressure in longitudinal section.

In this paper, the inert and harden grouts are treated as Newton fluid and Bingham fluid, the typical four grouting holes design is assumed, the grouts pressure is taken as the controlling factor, and the flux of each grouting hole is not a variation value.

3. DERIVATION OF GROUTS PRESSURE FORMULA

Assuming that the thickness of tail void is δ, tunneling speed is v, width of tail void formed in unit time is S ($S=v$), so the corresponding volume formed in unit time is $V=\pi(D^2-d^2)\times(S/4)\times k$ (D and d are outer diameters of shield tailskin and tunnel lining respectively, k is the grouting ratio). The upside grouting hole is located at $\theta=45^0$ (the positive direction of Y-axis is at $\theta=0^0$, clockwise rotation, Fig.1), grouting pressure is p_s, grouting flux to the upward and downward is q_1 and q_2. The corresponding parameters about downside grouting hole is $\theta=135^0$, p_x, q_1' and q_2'.

Taking the tunnel center as coordinate origin, vertical axis as Z axis, Cartesian coordinate system is established in the tunnel cross section plane. The shape of tail void and the arrangement of grouting holes are symmetric to vertical axis.

3.1 Derivation of grouts pressure formula in tunnel cross section

Compared to Newton fluid, the flow of Bingham fluid has to overcome the static shear stress τ_0 (Zheng 2005). Most of fluid belongs to Bingham fluid, especially the grouts with high density and thickness recently developed (Shanghai Tunnel Engineering Co., Ltd. 2008). So, taking Bingham fluid as the main research object, and the derived formulae will be more close to reality.

(1) The filling path of grouts flow is from $\theta = 45^0$ to the upward

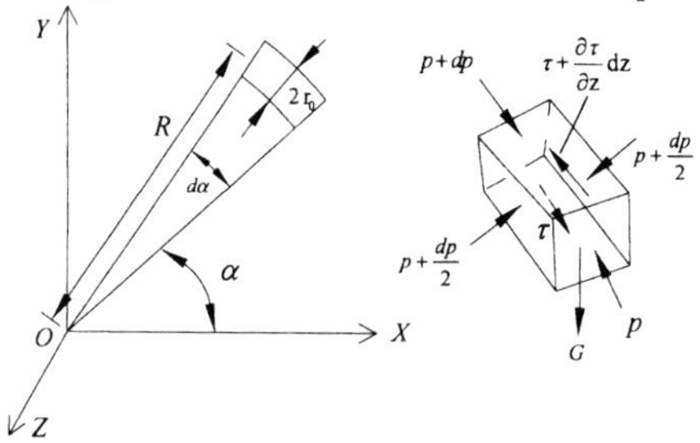

Fig.1 Forces on fluid microelement from $\theta = 45^0$ to the upward and corresponding coordinate system

Grouts flows from $\alpha = 45^0$ to $\alpha = 90^0$, angle starts from the positive X-axis, counterclockwise rotation (Li 2009).

Taking any single micro-fluid from the flow area for instance, just as shown in Fig.1, flow velocity in the coaxial thin layer keeps unchanged along the flow line because of its laminar flow, so shear stress on the left and right side equals zero ($\tau = \tau_0 + \mu\ \gamma = \tau_0 + \mu \cdot du/dz$, $d\tau/dR = 0$). The influence of inertial force is ignored, and sums of all the projective forces along the flow line equal zero. So, we can get the equilibrium equation as follow:

$$p(2r_0)dz\cos\frac{d\alpha}{2} - (p+dp)(2r_0)dz\cos\frac{d\alpha}{2} + (\frac{\partial\tau}{\partial z}dz)\frac{2r_0\left[(R-r_0)d\alpha + (R+r_0)d\alpha\right]}{2} - \rho g\frac{2r_0\left[(R-r_0)d\alpha + (R+r_0)d\alpha\right]}{2}dz\cos(\alpha + \frac{d\alpha}{2}) = 0 \quad (1)$$

So, $$p = p_s + \rho g R\left(\sin\pi/4 - \sin\alpha\right) - A\ R\ (\pi/4 - \alpha) \quad (2)$$

Where, $$A = \left(A_1 + A_2\right)^{\frac{1}{3}}/\left(S\delta^3\right) + A_3^{\ 2}/\left[S\delta^3\left(A_1 + A_2\right)^{\frac{1}{3}}\right] - A_3/\left(S\delta^3\right) \quad (3)$$

Where, $$A_1 = \tau_0^{\ 3}S^3\delta^6 - 64\mu^3 q_1^{\ 3} - 48\mu^2 q_1^{\ 2}S\tau_0\delta^2 - 12\mu q_1 S^2\tau_0^{\ 2}\delta^4 \quad (4)$$

$$A_2 = 4\tau_0\left[-\mu q_1\left(\tau_0/S\right)\left(16\mu^2 q_1^{\ 2} + 12\mu q_1 S\tau_0\delta^2 + 3S^2\tau_0^{\ 2}\delta^4\right)\right]^{\frac{1}{2}} \times S^2\delta^3 \quad (5)$$

$$A_3 = 4\mu q_1 + S\tau_0\delta^2 \quad (6)$$

As for Newton fluid, $\tau_0 = 0$, $\delta_0 = 0$, then $$A = -12\mu q_1/\left(S\delta^3\right) \quad (7)$$

So, $p = p_s + \rho g R\left(\sin\pi/4 - \sin\alpha\right) + 12\mu q_1 R/\left(S\delta^3\right)\left(\pi/4\text{-}\alpha\right)$ (8)

Where, $\pi/4 \le \alpha \le \pi/2$, $0 \le \theta \le \pi/4$ (9)

(2) The filling path of grouts flow is from $\theta = 45^0$ to the downward

Grouts flow from $\alpha = 45^0$ to $\alpha = 180^0$, angle starts from the positive Z-axis, clockwise rotation.

Similarly available, $p = p_s + \rho g R\left(\cos\pi/4 - \cos\alpha\right) \text{-} AR\left(\pi/4\text{-}\alpha\right)$ (10)

Newton fluid: $p = p_s + \rho g R\left(\cos\pi/4 - \cos\alpha\right) + 12\mu q_2 R/\left(S\delta^3\right)\left(\pi/4 - \alpha\right)$ (11)

Where, $\pi/4 \le \alpha \le \pi$, $\pi/4 \le \theta \le \pi$ (12)

(3) The filling path of grouts flow is from $\theta = 135^0$ to the upward

Grouts flow from $\alpha = 45^0$ to $\alpha = 180^0$, angle starts from the negative Z-axis, counterclockwise rotation.

Similarly available, $p = p_x \text{-} \rho g R\left(\cos\pi/4 - \cos\alpha\right) \text{-} A\ R\left(\pi/4\text{-}\alpha\right)$ (13)

Newton fluid: $p = p_x \text{-} \rho g R\left(\cos\pi/4 - \cos\alpha\right) + 12\mu q_1{'} R/\left(S\delta^3\right)\left(\pi/4 - \alpha\right)$ (14)

Where, $\pi/4 \le \alpha \le \pi$, $0 \le \theta \le 3\pi/4$ (15)

(4) The filling path of grouts flow is from $\theta = 135^0$ to the downward

Grouts flow from $\alpha = 45^0$ to $\alpha = 90^0$, angle starts from the positive X-axis, clockwise rotation.

Similarly available, $p = p_x \text{-} \rho g R\left(\sin\pi/4 - \sin\alpha\right) \text{-} AR\left(\pi/4\text{-}\alpha\right)$ (16)

Newton fluid: $p = p_x \text{-} \rho g R\left(\sin\pi/4 - \sin\alpha\right) + 12\mu q_2{'} R/\left(S\delta^3\right)\left(\pi/4 - \alpha\right)$ (17)

Where, $\pi/4 \le \alpha \le \pi/2$, $3\pi/4 \le \theta \le \pi$ (18)

3.2 Derivation of grouts pressure formula in longitudinal tunnel section

The grouts pressure in tail void will decrease with its hardening process and the infiltration into surrounding strata. As for the soft soil, grouts keep in liquid state in the range of about five lining rings away from shield tail, and the grouts pressure should be considered in this range.

(1) Influence of grouts hardening process

The chemical reaction will take place after grouts are mixed, and rheological parameters of any cement grouts are time-dependent. The relation between rheological parameters and time can be described in exponential function, $\mu = \mu_0 e^{\lambda_1 t}$ and $\tau = \tau_0 e^{\lambda_2 t}$. Coefficient λ_1 and λ_2 can be determined by measured data. Viscosity is time-dependent in the process of grouts diffusion, but the static shear stress remains almost unchanged.

(2) Grouts infiltration into surrounding strata

There are three main interaction ways between grouts and surrounding strata, i.e. infiltration, compaction and fracturing grouting (Cooperative group of Geotechnical grouting theory and engineering examples 2001). And this paper only considers the infiltration process resulting in the decrease of grouts pressure.

The corresponding deformation in strata will occur because of tunneling. Assuming that the width of this disturbance area is equal to the width l_s of the annular space out of tail void, and the soil after deformation is still in elastic stage. Taking the micro-element in vertical plane as research object, the length of the micro-element is the sum of l_s and δ. ε_t is the time-dependent strain. According to generalized Hooke's Law (Song and Cai 1998), we can get the loss of pressure due to infiltration.

$$\Delta p = E\,\delta \cdot \varepsilon_t / l_s \times (1-\upsilon)/(1-\upsilon-2\upsilon^2) \qquad (19)$$

3) Grouts pressure formula in longitudinal tunnel section

Taking Newton fluid for instance, the ultimate distribution of grouts pressure in longitudinal section is obtained.

1) From $\theta=45^0$ to the upward ($\pi/4 \le \alpha \le \pi/2$, $0 \le \theta \le \pi/4$)

$$p = p_s + \rho g R(\sin\pi/4 - \sin\alpha) + (12\mu_0 e^{\lambda_1 t} q_1 R / S\delta^3) \times (\pi/4\text{-}\alpha) - E\, C_1\, C_2 \qquad (20)$$

2) From $\theta=45^0$ to the downward ($\pi/4 \le \alpha \le \pi$, $\pi/4 \le \theta \le \pi$)

$$p = p_s + \rho g R\,(\cos\pi/4 - \cos\alpha) + (12\mu_0 e^{\lambda_1 t} q_2 R / S\delta^3) \times (\pi/4\text{-}\alpha) - E\, C_1\, C_2 \qquad (21)$$

3) From $\theta=135^0$ to the upward ($\pi/4 \le \alpha \le \pi$, $0 \le \theta \le 3\pi/4$)

$$p = p_x - \rho g R\,(\cos\pi/4 - \cos\alpha) + \left(12\mu_0 e^{\lambda_1 t} q_1' R / S\delta^3\right) \times (\pi/4\text{-}\alpha) - E\, C_1\, C_2 \qquad (22)$$

4) From $\theta=135^0$ to the downward ($\pi/4 \le \alpha \le \pi/2$, $3\pi/4 \le \theta \le \pi$)

$$p = p_x - \rho g R\,(\sin\pi/4 - \sin\alpha) + \left(12\mu_0 e^{\lambda_1 t} q_2' R / S\delta^3\right) \times (\pi/4\text{-}\alpha) - E\, C_1\, C_2 \qquad (23)$$

Where, $C_1 = \delta \cdot \varepsilon_t / l_s$, $C_2 = (1-\upsilon)/(1-\upsilon-2\upsilon^2)$

From the grouts pressure distribution in longitudinal section, we know that it is relevant to many factors. When all the other factors are fixed, grouts pressure diffusion depends on time. Grouts pressure on lining and surrounding strata near shield tail will decrease with time lapse until reach its balance with surrounding pore pressure.

4 ANALYSIS EXAMPLE AND ENGINEERING APPLICATION

The buried depth of a metro shield tunnel in Shanghai is 13m, strata from up to down are miscellaneous fill, silt soft soil, stiff clay and sand. D=6340mm, R =3.27m, d=6200mm, ν =2cm/min, S =0.02m, δ =0.10m. p_s =0.16MPa, p_x =0.23MPa, W/C =0.65, τ_0 =90 Pa , μ =28.1 $\times e^{0.0165t}$ $Pa.s$, C_1 = $\delta \cdot \varepsilon_t / l_s$ =0.015, $E = \beta E_S$ =1.39MPa, ρ=1500kg/m^3. We assume that grouts flux at each grouting hole is equal, flux to the downward and upward from each grouting hole is equal, and the injection ratio is 150%. Then, we can get the specific distribution of grouts flux:

$$q_1 = q_2' = q/4 = 1.24\times10^{-4} m^3/s\,,\quad q_1' = q_2 = 3q/4 = 3.71\times10^{-4} m^3/s \qquad (24)$$

4.1 Grouts pressure in tunnel cross section

From Eq.(8), (11), (14) and (17), pressure distribution is calculated as follow.

a) From θ=45^0 to the upward (θ=0, $\alpha=\pi/2$; $\theta=\pi/4$, $\alpha=\pi/4$)

$\theta=\pi/4$, p=0.16MPa; $\theta=\pi/6$, p=0.152MPa; θ=0, p=0.146MPa.

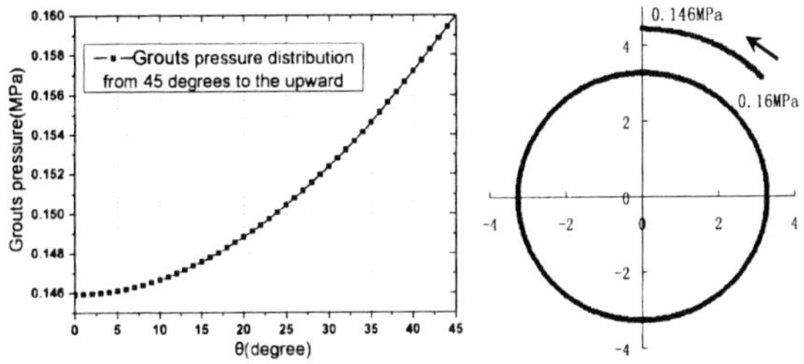

Fig.2 Pressure distribution from θ=45^0 to the upward

b) From θ=45^0 to the downward ($\theta=\pi/4$, $\alpha=\pi/4$; $\theta=\pi$, $\alpha=\pi$)

$\theta=\pi/4$, p=0.16MPa; $\theta=\pi/3$, p=0.17MPa; $\theta=\pi/2$, p=0.194MPa; θ=3$\pi/4$, p=0.228MPa; $\theta=\pi$, p=0.242MPa.

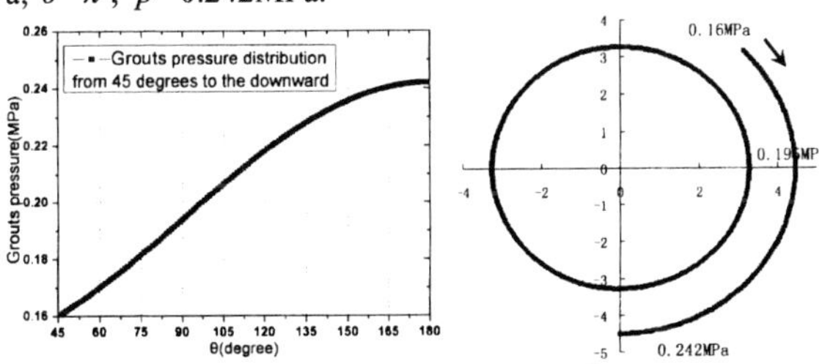

Fig.3 Pressure distribution from θ=45^0 to the downward

c) From θ=135^0 to the upward (θ=3$\pi/4$, $\alpha=\pi/4$; θ=0, $\alpha=\pi$)

θ = 3$\pi/4$, p =0.23MPa; θ = 2$\pi/3$, p =0.22MPa; θ = $\pi/2$, p =0.196MPa; $\theta=\pi/4$, p=0.162MPa; θ=0, p= 0.148MPa.

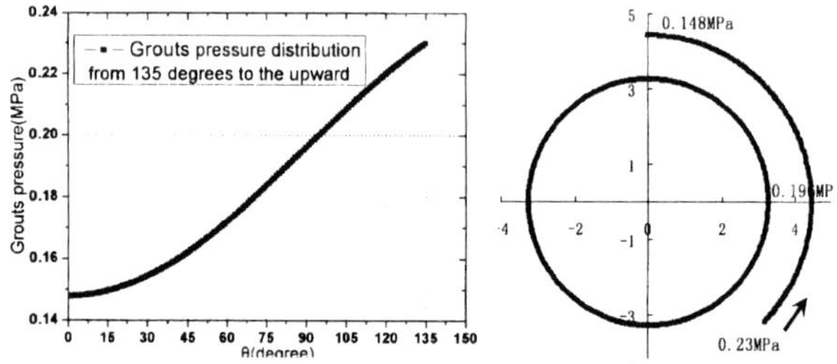

Fig.4 Pressure distribution from θ=135^0 to the upward

d) From θ=135^0 to the downward ($\theta=3\pi/4$, $\alpha=\pi/4$; $\theta=\pi$, $\alpha=\pi/2$)
$\theta=3\pi/4$, p=0.23MPa; $\theta=5\pi/6$, p=0.238MPa; $\theta=\pi$, p=0.244MPa.

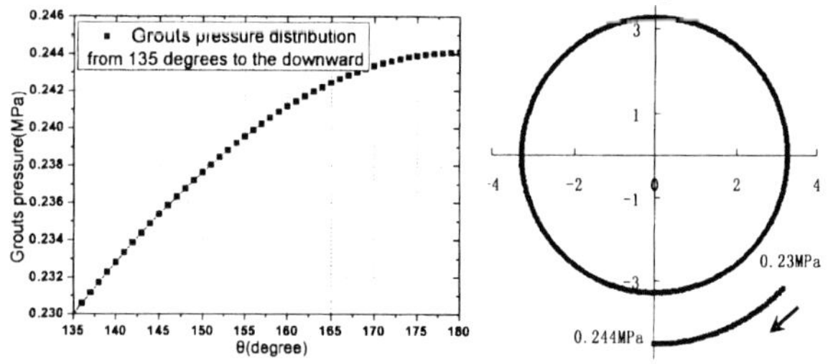

Fig.5 Pressure distribution from θ=135^0 to the downward

Make superposition of grouts pressure at the same location, we can get the final pressure distribution in cross section.

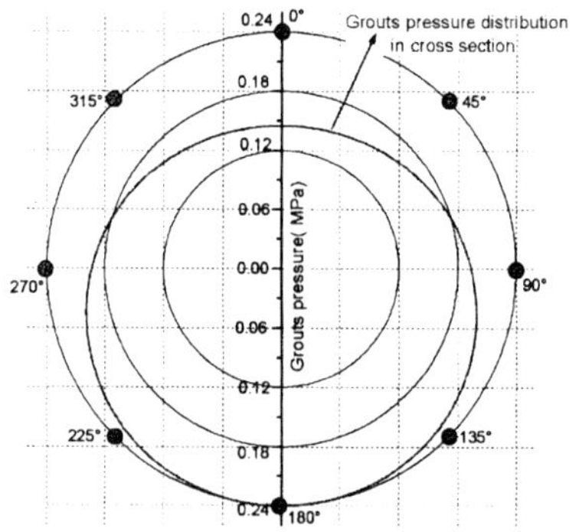

Fig.6 Final distributions of grouts pressure in cross section.

From the final pressure distributions in cross section, we know that the maximum pressure at vault reaches 0.147MPa, the arch bottom is 0.241MPa, and the calculated results are in good agreement with the actual situation in soft soil in Shanghai.

4.2 Grouts pressure distribution in longitudinal section.

For the segment ring with 1200mm width, the duration of shield driving is about 60 min/ring. Considering the transportation of muck, segments, cement grouts, and the shield maintenance, the progress of tunneling is about 10 rings/day. Taking the grouting hole at $\theta = 45^0$ for example, using Eq.(20)~(23), the final pressure distribution in longitudinal section is calculated.

$$p = (p_s + p_x)/2 - \rho g R/\sqrt{2} + 3\pi \times \left(28.1 \times e^{0.0165t}\right) q_1 R / S\delta^3 - E\, C_1\, C_2 \tag{25}$$

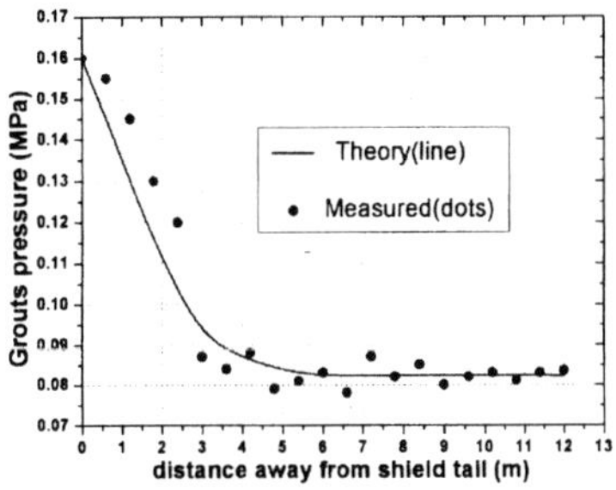

Fig.7 Final distributions of grouts pressure in longitudinal section at $\theta = 45^0$

As Fig.7 shows, grouts pressure decreases rapidly in the range of three or four rings from shield tail. Corresponding data is from 0.16MPa to 0.08MPa. Grouts pressure becomes stable at the fifth ring away from shield tail in longitudinal section. So, calculated results match the general rules of pressure dissipation. From further observation, we know that theoretical values are a little smaller than the measured values. That is because of the time factor for pressure diffusion, which is not completely considered in the example calculation. It is obvious that the further away from shield tail, the greater the gouts pressure loss.

5. CONCLUSIONS

(1) The cross section and longitudinal section can be divided into two relatively independent processes, according to the grouts pressure distribution in simultaneous backfilling grouting in shield tunnel.

(2) Based on the typical four grouting holes design, grouts pressure formula in cross section is derived. Considering the hardening process and infiltration into surrounding strata, grouts pressure formula in longitudinal section is founded. The engineering example shows that the proposed equation obtained matches the actual situation well.

ACKNOWLEDGMENTS

This work was supported by funds from Grant 2006BAJ27B02-02 from National Key Technology R&D Program and Shanghai Leading Academic Discipline Project (B308).

REFREENCES

Cooperative group of Geotechnical grouting theory and engineering examples. (2001). "Geotechnical grouting theory and engineering examples". *Science press*. 53~95. (in Chinese)

Hashimoto, T., Brinkman, J., Konda, T., Kano, Y., and Fedema. (2004). "Simultaneous backfill grouting, pressure development in construction phase and in the long-term", *Proceeding of the 30th ITA-A ITES World Tunnel Congress*, Singapore, 52~59.

Li, Z.M. (2009). "Control theory and technology of simultaneous backfilling grouting for shield close to crossing structures". *Master's dissertation*. Tongji University. (in Chinese)

Nomoto, T., Imamura, S., Hagiwara, T., Kusakabe, O., and Fujii, N. (1999). "Shield tunnel construction in centrifuge". *Jounral of Geotechnical and Geo-enviornmental Engineering*, ASCE, Vol.125, No.4, PP.289-300.

Ou, C.Y. and Cherng, J.C. (1995). "Effect of void closure on ground movement during shield tunneling in sandy soil". *Geotechnical Engineering*, Vol.26, No.1, pp.173~32.

Ruan, W.J. (2005). "Spreading design of grouting in rock mass fissures based on time-dependent behavior of viscosity of cement-based grouts". *Chinese Journal of Rock Mechanics and Engineering*. Vol.24 No.15. 2709-2714. (in Chinese)

Shanghai Tunnel Engineering Co., Ltd. (2008). "Study and application of single-component slurry, instead of two-component slurry, in simultaneous backfilling grouting in slurry balance shield". Shanghai. (in Chinese)

Song, Z.K. and Cai, W.A. (1998). "Material mechanics". *Tongji University press*. 110~116. (in Chinese)

Yang, X.Z. (2005). "Study on grouting diffusion theory and experiments under static or dynamic loading". *Doctor's dissertation*. Center South University. (in Chinese)

Zhao, T.S. (2008). "Study on mortar material and application of tail void grouting of slurry shield tunnel". *Master's dissertation*. Tongji University. (in Chinese)

Zheng, Y.H. (2005). "Research on grouts and controllable method of grouting in rock mass of the fissures". *Doctor's dissertation*. Jilin University. (in Chinese)

Study on Shiziyang Tunnel Engineering Geology and Shield Tunneling

Xinan Yang [1], Yongqin Yao[2], Ying Zhang[3], Peixu Ye[4]
[1]Proffessor, Department of Urban Track and Railway Engineering, Tongji University, Shanghai 200092, China; xyang@tongji.edu.cn
[2]Chief Engineer, China Railway 12th Bureau Group Corporation, Taiyuan 030024, China; yqyao@yahoo.com
[3]Post Graduate, Department of Urban Track and Railway Engineering, Tongji University, Shanghai, 200092, China; lydiabeibei@126.com
[4]Post Graduate, Department of Urban Track and Railway Engineering, Tongji University, Shanghai, 200092, China; 2010ypx@tongji.edu.cn

ABSTRACT: The Shiziyang tunnel is in passenger dedicated rail line connecting Guangzhou, Shenzhen and Hong Kong. It will be the longest underwater tunnel and have the highest technical level in China. This article describes the tunnel's design and construction schemes, analyzes and summarizes the characteristics of this tunnel, in terms of its geological and hydro geological formations; and defines the strata graphic classification. According to sub-formation and characteristics of the tunnel, this article discusses the problems that may arise in the construction of the tunnel in sub-strata, declares the key points in technology of shield tunneling in three major aspects: soft ground, the upper-soft and lower-hard mixed ground and hard rock (medium hard rock) ground.

KEY WORDS: railway tunnel; underwater tunnel; Shield; mixed ground; construction

1 INTRODUCTION

Currently, the Shiziyang underwater railway tunnel which connecting Guangzhou, Shenzhen and Hong Kong in the passenger dedicated line is being constructed. Also some other highway underwater tunnels are being constructed. A large number of underwater tunnels in coastal areas and urban traffic systems need to be designed and planed and constructed [1].

The passenger dedicated line of the Shiziyang tunnel in the area of the Pearl River Delta is located in the interval between Dongyong Station and Humen Station. The tunnel's length is 10800m. It is the key project of the Guangzhou-Shenzhen-Hongkong passenger dedicated line[2,3]. Shiziyang tunnel is a large-section underwater tunnel. It passes through mixed strata which has frequent changes in litho logy with large differences in physical and mechanical properties. In the strata there are wide variations of weathered bedrock interface, intensive distribution of fault broken rock, obvious differences in water content. Thus Shiziyang tunnel is challengeable and has some risks in engineering.

This article is to describe the tunnel's design and construction projects, analyze and summarize the characteristics of this tunnel in terms of its geological and hydro geological formation, divide the tunnel in sections by its sub-strata and analyze the characteristics of each stratum graphic section; and then define the stratum graphic classifications. Moreover, based on geological data, this article will show the type of shield and knife set, the configuration strategy of the cutter aperture ratio for the cutterhead sets. In addition, in accordance with sub-formation and tunnel characteristics, this article will discuss the problems that may arise in the construction of the tunnel through the sub-strata, declare the key points in technology of shield tunneling in three major aspects: soft ground, upper-soft and lower-hard mixed ground, hard rock (medium hard rock) ground. The parameters for the key shield will also be described. Great practical significances for the construction of underwater tunnels and shield tunneling technology in mixed ground strata will be also shown in this article.

2 OVERVIEW OF SHIZIYANG TUNNEL

2.1 Tunnel design

The Shiziyang tunnel is a rated passenger rail line. Its main line is a double track line with a maximum speed of 350km/hr, using the structure of dipolar-single line. The spacing between rail lines are 17~22m, the minimum curve radius is 7000m; the maximum gradient is 2 percent.

The Shiziyang tunnel's passenger dedicated rail line, is located in the interval between Dongyong Station and Humen Station, in Guangzhou-Shenzhen-Hong Kong area. The tunnel's length is 10800m. This Tunnel is the longest underwater tunnel with most stringent specification for an underwater tunnel in China. Moreover, it is the key project of the Guangzhou-Shenzhen-Hong Kong passenger dedicated railway, using the design of dipolar-single shield tunnel.

2.2 Tunnel construction project.

The tunnel construction project requires construction of "Two working wells, four shields to dock in the land". In shield section, number 4 slurry shield is used for TBM during construction [3]. After Dongyong station, the railway has a sloped gradient of 20 ‰ for 2950m downhill to the tunnel entrance, as per department DIK32+600, and pass through Xiaohuli, Xiaohuli island, Shazaili and finally reach to Shazai island. The downhill slope gradient when passing through Shazai island is 3‰ for 3750m until entering the Shiziyang deepwater channel. The tunnel then changes to a positive gradient of 3.5‰ for 1700m as it climbs through Shiziyang channel and the 6th berths of Shatian area in Humen port. Finally the tunnel climbs at a gradient of 20‰ from the Humen Port then passes through the bonded warehouses and goes out at the East side of the highway under the supervision of the Humen Port authority. Each end of the tunnel uses the same gradient of 20‰ to enter the tunnel to minimize the length of shield tunneling in the soft-and-hard rock mixed strata. Two working wells are located respective for the import and export tunnels between the open tunnel and shield tunnel. The altitude of the lowest point in the tunnel track is -61.56m, 64.7m away from the once-climax.

3 GEOLOGICAL ANALYSIS OF SHIZIYANG TUNNEL'S FORMATION

3.1 Vertical stratum graphic distribution

Stratum graphic sequences in the Shiziyang tunnel occur in layers from top to bottom are; artificial fill, a sedimentary layer of marine (including silted clay layer, sludge layer, silt soil layer, silted soil layer, fine sand layer, medium sand layer, coarse sand layer, and gravel layer), alluvium (including clayey layer, silt soil layer, fine sand layer, medium sand layer, coarse gravel sand layer, and gravel layer), residual soil, bedrock (including fully weathered rock, strong weathering rock, and weak weathered rock).

A quaternary stratum graphy is multi-layered, and it changes frequently. It has low intensity poor self-stable ability. The saturated uniaxial compression strength of the strong weathering zone is only 0.70～14.70MPa, while in the weak weathered zone , its value is 3.10～78.70MPa. This data shows that there are strong differences between soft and hard rocks. The quartz content in the strong weathering zone is 46.70%, and in the weak weathered zone is 40.74%, showing that the value of quartz content in the bedrock is high, which can definitely damage the tool significantly, cause excessive wear-out, and slow down the progress of the TBM, due to increased cutter replacements.

3.2 Horizontal stratigraphic distribution

Based on the formation classification and intensity of the tunnel section, the formation of the shield tunneling can be divided into three major categories: soft, upper-soft and lower-hard mixed and hard rocky (medium hard rocky). Figure 1 show the typical profile of three types of formation.

The length of shield tunneling in bedrock is 6540m, which is 70.0% of the tunneling length in total. Its length in mixed ground is 2000m, which is 21.4%of the whole tunneling length. Its length in Quaternary cover formation is 800m, which is 8.6% of the total tunneling length.

The mixed ground where soft formation is under hard rock forms in 3 layers:

(1) This mixed ground has soil on the upper part of the formation, while rock on the lower part is shored up with low soil to rock mixed ground;

(2) This mixed ground has rock on the upper part of the formation, while soil on the lower part is shored up with low rock to soil mixed ground;

(3) This mixed ground has soft rock on the upper part of the formation, hard rock on the lower part and is shored up with soft-hard rock mixed ground.

The features of mixed ground in Guangzhou area are as follows: Many changes in geological conditions, a number of different formations in the tunnel, the most common formation being Semi-soil and Semi-rock, e.g. Figure1(b). The composite stratum graphic formation is strongly heterogeneous, and has many changes in rock weathering and the uniaxial compressive strength. Sand and rock may appear at the same elevation or even in the same section, as a result of the dramatic changes of formation along the tunnel axis.

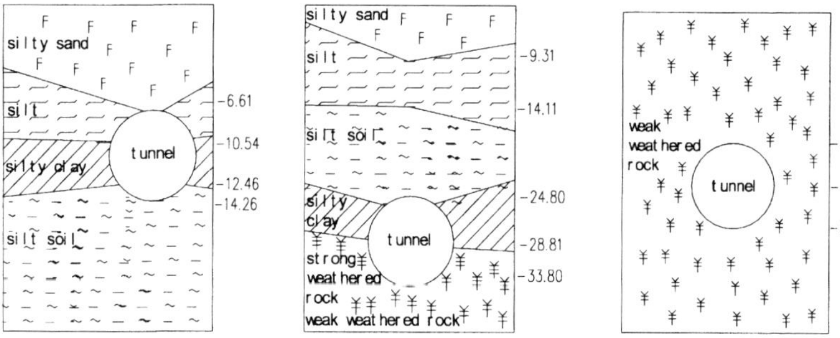

(a) (b) (c)

Fig.1 The typical formation profile of Shiziyang tunnel
(a) soft ground (b) mixed ground (c) hard rock ground

3.3 Hydrogeology

The Shiziyang tunnel is across the plain area of the Pearl River Delta. In this area a sludge layer of Quaternary marine-ferruginous (terrigenous) faces, Quaternary alluvial clay, organic and silted clay, silted clay, silted clay layer of Quaternary residual soil and completed weathered argillaceous siltstone form a part of the relatively impermeable layer. Silted sand or medium sand of Quaternary alluvium, coarse sand, silt, Organic silt, Coarse sand layer with gravel and round gravel layer belong to medium or strongly permeable stratum, in which the groundwater is porously confined water. Strongly weathered and weak weathered argillaceous siltstone aquacade the groundwater and the bedrock is fissure water with pressure afford-ability.

The major surface water crossing the tunnel line is Xiaohuli, Shazaili and Shiziyang. All of them are the branch of the Pearl River, and are navigable waterways. They are supplied by upstream water and discharge to the lower reaches of the Pearl River estuary.

4 SHIELD TUNNELING TECHNOLOGY OF SHIZIYANG TUNNEL

The features of Shiziyang tunnel are sub-strata, mixed ground, long hard rock and complex hydro-geological conditions, which will have significant impact on the efficiency of the construction. Therefore, the formation and geological factors for the shield tunneling must be considered and appropriate technical schemes and measures must be taken [4, 5].

4.1 Cutter head design and tool preparing

In order to meet the tunneling requirements in weathered rock, hard-soft heterogeneous ground and hard rock ground in the Shiziyang tunnel, a slurry shield TBM and panel-type cutter head must be used, see Fig.2. Disc cutters (see Fig.3) and scrapers can be installed at the same time. To attain the structural strength required for the TBM cutters to successfully tackle the tunneling, the following issues have to be defined or determined: the rock support, the cutters' layout, the maximum cutter opening ratio, the cutter opening ratio in the center part of the cutter head (a critical determination). And the 31 % cutter opening ratio of Shiziyang tunnel should be reached.

In the upper-soft and lower-hard mixed ground, the number of disc cutters and scrapers used should be directly contingent upon the formation. If 6 disc cutters in the import bid section of right line tunnel are allocated, the tool will be worn out seriously, and the shield tunneling will be very difficult. If 25 disc cutters are allocated in the left line, the tunneling process increases significantly and tool life also increases. As in any cutting environment, speed, feed, depth of cut must be harmonized to the material being removed. Accordingly, in the mixed ground, the key problem can be seen as the reasonable allocation of tool configuration and soil formation [6,7].

Fig.2 Shield in the Shiziyang tunnel

Fig.3 Disc cutters

4.2 Technical points of sectional construction

4.2.1 Shallow covering ground

When the shield tunneling goes through shallow covering ground, the shield speed and attitude adjustment should also be taken into account in addition to adjusting the incision water pressure and grouting. Since soil settlement is impacted by the tunneling speed and attitude control, the speed should be slowed down to 20~25mm/min when the shield tunneling goes through the shallow covering ground. This means that the tunneling time for one ring is about 60~80min in order to decrease the disturbance of soil. Simultaneously, the shield attitude should not be changed greatly or frequently [6,7].

4.2.2 The upper-soft an lower-hard mixed ground

Problems that will arise when the shield tunneling goes into the upper-soft and lower-hard mixed ground are as follows:

(1) Easily lead to over break, resulting in the loss of the upper soil and the ground subsidence;

(2) The force on the Jack will be unbalanced when the shield tunneling goes through the upper soft lower hard mixed ground, so that the direction of shield is difficult to control and can easily lead to the rise of the shield.

(3) When the tool suddenly shifts from the soft clay to hard rock, the intensity differences of the two formations will have a strong reverse impact force. Tool abscission or tool breakage can occur and can also cause shield body rotation.

According to these issues mentioned above, the measures which should be taken during the construction process are as the follows:

(1) If the upper soil layer is very soft (such as around Mileage DIK33+930), the incision water pressure should be used to maintain stability in order to prevent the collapse of the upper soil. The grouting ratio should be adjusted to reduce the initial setting time and the grouting pressure should also be reasonably grasped. All these considerations are devised to optimize the matching parameters for grouting capacity, advancement speed and other construction parameters.

(2) When the shield tunneling goes from the soft soil into the junction of soft and hard rock (such as around Mileage DIK34+400), the advancing speed should be slowed down, the oil pressure of all regions should be adjusted so that the action points of the resultant forces of the jack can be changed in order to avoid the rise of the shield. When the shield tunneling goes from the hard rock ground into the soft ground, the relative gradient of the shield and design axis should be adjusted, the differential oil pressure of all regions should be adjusted so that the action points of the resultant force of the jack can be changed. And the advancing speed should be increased to avoid the "kowtow" of the shield.

(3) When the uniaxial compressive strength of the lower part of hard rock is beyond 30MPa (such as around Mileage DIK38+500), scraper should be changed by hob, especially the peripheral surface of cutter head. The incision water pressure may be appropriately increased and smaller effective thrust for low-speed tunneling should be. If the knife for over breaking can be used, the lower part of hard rock can be applied with expanding excavation and the rise of the shield can be avoided.

4.2.3 Hard rock ground

(1) Single-blade disc cutter should be used to break the rock and the time and frequency of tool changing should be reduced in order to improve the construction efficiency. The stress states of the tool and cutter head must be constantly monitored during the tunneling process in order to avoid overloading.

(2) The cutting tools should be replaced on time to maintain tunneling efficiency and to avoid the cutter head damage.

(3) The tool should be changed in the cabin. The cutter head's telescopic device

should be switched on in order to reduce the changing times, The tunneling cycle and optimizing performance in the utilization of all the equipment components.

(4) When tunneling in the hard rock ground, the shield's stable device should be switched on to reduce the shield's vibration and prevent the transfinite torsion. That will make the segment forced stably and ensure the tunnel's forming quality. Contemporarily, the segment can be protected and the shield will be prevented from deformation.

(5) One should use the open model of tunneling and select the parameters following the high-speed low torque principle to increase the tunneling speed.

5 CONCLUSIONS

(1) Shield tunneling of the Shiziyang tunnel goes through the three major formations: soft ground, upper-soft and lower-hard mixed ground and hard rock (medium hard rock) ground; and each can be divided into several categories. The lengths of the formations are 6800m, 1237m, 1240m respectively and occupy 73.3%, 13.3%, 13.4% of the total tunneling length.

(2) The Shiziyang tunnel is in a complicated geological area. A slurry shield TBM with the diameter of 11.182m and panel-style cutter head with the 31% aperture ratio should be used. When the soft soil is the main ingredient of the geological formation, the tool should be outfitted mainly with scrapers and supplemented by appropriate hobs. When hard rock is the main ingredient of the geological formation, the tool should be outfitted mainly with disc cutters and supplemented by appropriate scrapers.

(3) The key points in technology of shield tunneling are declared in three major aspects: soft ground, the upper-soft and lower-hard mixed ground and hard rock (medium hard rock) ground, and the parameters to control shield are also described.

REFERENCES

[1] Xinan, Yang, Jingchuan Sun and Fanjiang Meng(2001). "Bridge or Submerged Tunnel?" *Journal of Xuzhou Institute of Architectural Technology*, Vol.1(1):31-34.

[2] Guoqiang Pan(2006). "Study in the crossing of the Pearl River connecting Guangzhou-Shenzhen-Hong Kong Passenger Dedicated Line." *Railway Standard Design*. (10):16-18.

[3] Kairong Hong(2007). "Analysis on technical difficulties of super-long underwater shield-bored tunnels with large cross-sections." *Tunnel Construction*. 27(6):1-3, 18.

[4] Clough G Wayne(1993). "EPB shield tunneling in mixed face conditions." *Journal of Geotechnical Engineering*, 119(10): 1640-1656.

[5] China Railway 12 Bureau Group Co.Ltd., Tongji University(2009). "Attitude control and docking technology in mixed ground of Shiziyang shield tunneling". GuangZhou.

[6] Zhongsheng, Tan, Kairong, Hong and Jianglin Wan(2006). "Shield driven tunneling in the complex layer." *Engineering science*. 8(12): 92-96.

[7] Institute of site workers(2008). "Shield surveys • Design • Construction". Beijing: China Building Industry Press.130-139.

Numerical analysis of the thermo-hydromechancial behaviour of underground storages in hard rock

Y. Jia, H. B. Bian, G. Duveau, and J. F. Shao

Laboratory of Mechanics of Lille, Polytech'Lille, Cité scientifique, 59655 Villeneuve d'Ascq, France
Email: Yun.jia@polytech-lille.fr

ABSTRACT: Understanding the thermal impact on the hydro-mechanical behavior of the host rock is one important issue in the design of underground storages for the high level radioactive wastes. Increasing temperature inevitably affects the rock mass stability with the generation of overpressure and associated mechanical affects. Consequently, it is essential that the designer of an underground storage has appropriate knowledge of the coupled thermal and hydromechanical behaviour of the rock mass. In this paper, an in situ thermal-hydromechanical test is studied. This test consists to heat the host rock by one heating element of 3 meters length with two successive supplied heating powers: 275W and 915W. In the test, the temperature in the storage was increased up to 100° C with a monitor system equipped approximately 50 temperature, pore water pressure and deformation/displacement sensors in order to investigate the response of argillites in different heating phases. The Mohr-Coulomb constitutive model and the FLAC3D numerical code are used for prediction and interpretation of the measurements. The evolutions of temperature, pore-water pressure and strain during heating phases and cooling phase have been simulated and compared with in situ measurements. The good concordance between the numerical simulations and measurement data allow us to identify the in situ THM parameters of argillite and understand the coupled phenomena observed in the experiment.

INTRODUCTION

Unlike rock mechanical tunnel design in general, one of the key issues in designing underground storage for the high level radioactive wastes is understanding the thermal effect of radioactive decay on the hydromechanical behavior of the host rock. Increasing temperature of saturated rock affects its volume change and hydraulic behavior (Abuel-Naga et al., 2007). As a result, the rock mass stability is inevitably influenced with the induced volume change and undesired high pore pressure. This problem is a coupled thermo-hydromechanical (THM) one.

Previous research in the area of underground storage for high radioactive wastes has mainly dealt with the temperature dependence of rock behavior on the sample scale

(Baldi et al. 1988; Hueckel and Baldi,1990; Selvadurai, 1996; Su, 1998; Sultan et al, 2002; Zhang et al., 2007). By using the laboratory values as input to numerical models, the underground storage is simulated. However, the perturbation of geologic system caused by the decay heat of high radioactive wastes introduces various highly coupled thermo-hydromechanical processes around the underground storage (Tsang et al., 2000). To ensure the efficiency of numerical prediction, it is necessary to perform analogical field studied. To this end, several in situ heating tests have been carried out in different counties (Gens et al. 1998; Chijimatsu et al., 2001; Datta et al., 2004; Alonso et al., 2005; Guo and Dixon, 2006; Jia et al., 2007). These tests consist in simulating the heat generation of a high level radioactive canister and measuring the evolution of temperature, water pressure, displacement and stresses around the thermal element.

Among these tests, in order to characterize the thermal properties and the coupled thermo-hydromechanical processes in the argillites formation, an in situ heating experiment, called TER, was carried out in Meuse/Haute Marne Underground Research Laboratory (M/HM URL) by French Radioactive Waste Management Agency (Andra).The objective of this study is to investigate the response of host rock in different heating phases of TER experiment and characterize thermal and thermo-hydromechanical properties of rock by inverse analysis of in situ measurement. The good agreement obtained between the numerical simulation and experimental data helps us to enhance the understanding of the coupled thermo-hydromechanical processes in the rock and then ensure the stability of underground storage.

Presentation of the in situ experiment

The M/HM underground research laboratory, in which the TER experiment was conducted, is construed in one layer of Callovo-Oxfordian argillites formation located on the boundary between the Meuse and Haute-Marne department. The studied argillaceous formation was laid down between 158 and 152 million years ago. It originates from the sedimentation of detritic materials (clay minerals amounting to up to 60% by mass and fine quartz) and carbonates in ancient oceans (Andra, 2005). The mineralogical composition of the Callovo-Oxfordian argillites confers them relatively the high strength with a simple compressive strength at the range of 20MPa to 35MPa and a limited deformability varying from 3 000 to 5 000 MPa.

The TER experiment consists of a heater borehole (TER1101) and 9 observation boreholes (Fig. 1). The heater borehole is drilled horizontally form the TER "Carrure" to the depth of 10m in the direction of maximum horizontal stress (in the direction of N65°E). In addition, TER1101 is equipped with a heating element and temperature sensors in order to control the homogeneity of temperature field within the probe. The TER experiment was started in July 2005 with the drilling of an extensometer borehole (TER1301) perpendicular to the GKE drift. When the GEX drift was excavated, the drilling and instrumentation of water pressure sensors boreholes was performed in October 2005 with the purpose of gaining enough time for the stabilization of pore pressure towards the initial pressure state. After having obtained a rapid pressure recovery, the three temperature measuring boreholes were instrumented in early December 2006 just one week before the creation of heater borehole (TER1101). The first heating period was switched on 11 January 2006. In this stage, two steps of regulated power injected, which are 277 watts during 20 days and 925 watts during 24 days, were applied. Due to the packer membrane failure, the heating power was voluntary switched off in February 2006. In this paper, the experiment results until 6th Septembre 2006 will be analyzed in this study.

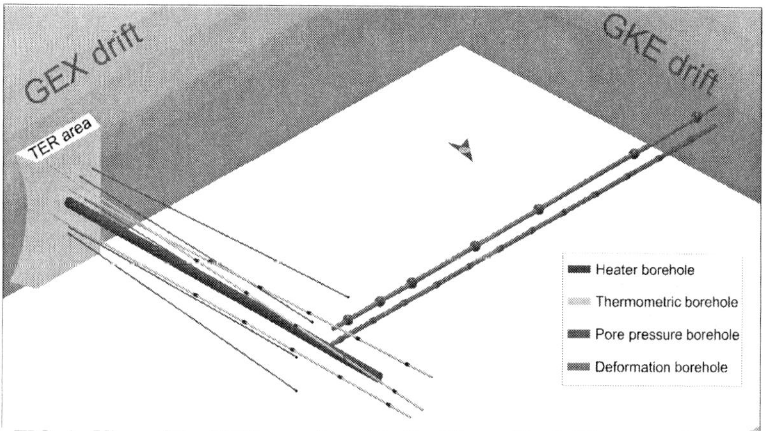

FIG. 1. View of the TER experiment in the Meuse Haute Marne underground research laboratory

3D numerical model description of TER experiment

A rigorous analysis of TER experiment is a difficult task due to the anisotropy of material properties, the anisotropy of stress field and the presence of the adjacent gallery as well as its instrument system, etc.... It is therefore necessary to realize three-dimensional modeling accounting for the anisotropic material's properties and stress field. FLAC3D is a finite different program which is robust enough to deal with the issues mentioned above. In the numerical model, the rock is therefore assumed to be initially saturated by water with initial pore pressure of 5MPa and the initial temperature of 22°C. Furthermore, because the heater borehole is located in the direction of the major horizontal stress , the values of stresses are: $:s_{xx}^{0} = s_{zz}^{0} = s_h = -12.7MPa$, $s_{yy}^{0} = s_H = 1.3s_h = -16.5MPa$. In order to avoid the distrubtion of boundary condition, the dimension of geometry model (30m*30m*20m) is used as large as possible. The heating borehole is considered as a circular borehole with a radius of 0.073m. For the reason of symmetry, only one quarter of domain is studied.

The loading path is composed of three steps: excavation, heating and cooling. During the excavation, the radial stress and liquid pressure are unloaded from the initial values to the atmospheric pressure. After that, the wall of borehole is kept free of mechanical force and is impermeable during 13 hours. Then, the borehole is heated for 20 days using the heating power 275W, followed by a second heating phase during 22 days with a higher heating power 915W. Finally, a cooling period of 200 days is considered. In the last phase, the heater borehole keeps free of heating flow. The boundary conditions during the heating phases and cooling phase are given in Fig. 2. The first figure represents a horizontal plane across the centre of heat element and the second figure gives an axisymmetric vertical section of the heater borehole. The symmetric surfaces, the ground surface and the inner boundary condition of the heater borehole are considered to be impermeable and adiabatic except the heater, where different constant heat flux are imposed according to experimental schedule. In addition, the normal displacement is blocked in the symmetric surfaces and the ground surface. Constant water pore pressure, temperature and stress, equal to their initial values, are imposed in the three other boundaries.

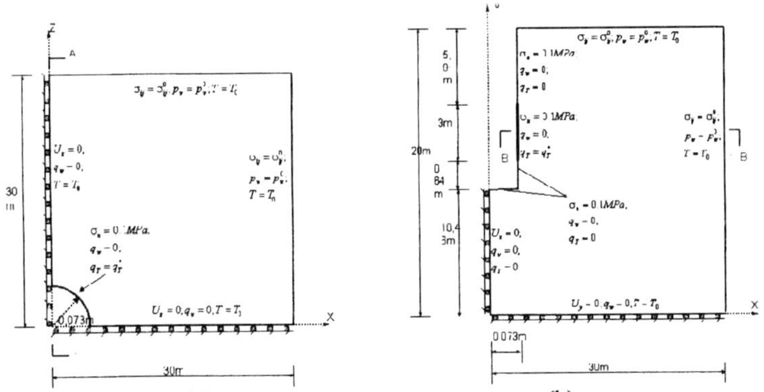

(a) (b)

FIG. 2. The element grids used in the continuum model. (a) The horizontal plan across the heater center; (b) the axi-symmetric vertical section of the heater borehole

According to the experimental investigations obtained by different partners of Andra (Andra, 2005), the basic mechanical behavior of Callovo-Oxfordian argillite exhibit significant plastic deformation coupled with induced damage by microcracks. In addition, the increase of temperature has no significant influence on the mechanical behavior of material until the temperature arrives at 100°C. As the maximum temperature in the TER experiment is less than 100 °C, the influence of temperature on the material properties is then neglected in this study. However, the mechanical behavior is strongly pressure dependent. Under higher confining pressure, all the microcracks are closed and the frictional sliding along cracks becomes more difficult. Therefore, plastic deformation becomes the dominant mechanism with respect to damage by growth of microcracks. In the underground laboratory, the argillites are essentially under high compression. Consequently, for the TER experiment realised at the depth of 490m, the rock mass can be described by an elastoplastic continuum. According to the available of mechanical constructive model in FLAC3D, a Mohr-Coulomb criterion with two non associative flow rules is used for the description of mechanical behavior of argillites under compression and tensile conditions. The value of these parameters can be determined by using the uniaxial compression and extension strength of material. The coupling THM parameters of Callovo-Oxfordian argillites are presented in the Table 1.

Table 1. Typical values of THM parameters for the Callovo-Oxfordian argillite

	Parameter	Used value
Mechanical parameters	Volumetric mass, ρ_m (Kg/m^3)	2700
	Initial porosity, ϕ(%)	15
	Drained Young's modulus E(MPa)	4000
	Drained Poisson's ration ν	0.3
Thermal parameters	Drained linear thermal dilation, α_b(/C)	1.28e-5
	Equivalent thermal conductivity, λ(w/m/K)	$\lambda_{\perp} = 1.3$ $\lambda_{//} = 2.0$
	Special heat capacity, C_p (J/Kg/C^0)	1005
Hydraulic parameters	Initial permeability, k_0(m^2)	5.0E-20
	Biot coefficient, b	0.6
	Biot modulus $M(Mpa)$	6000
	Dynamic viscosity of water μ_w $(Pa.s)$	1.0E-3

Numerical simulation of experiment

The principle aim of this work is the interpretation of thermal effect on the underground storage. In the numerical results, the initial time is assumed to correspond to the start of the first heating phase. Temperature, water pore pressure and strain evolutions at different measured points during the TER experiment are simulated and analyzed in this section.

The variation of temperature on the heater surface is presented in Fig. 3a while the results at different measured points in the rock are compared in Fig. 3b. In general, the numerical simulation agrees very well with the measured data. The increase of temperature in the heating phase and stabilization of temperature at the end of each heating phase is well simulated. In addition, the sharp increase and reduce of temperature with the change of heating power is also correctly reproduced. The anisotropic distribution of temperature in the argillite formation is also good illustrated at two points TER1402TEM01 and TER1405TEM01 which are located at the same distance from the heater surface show that.

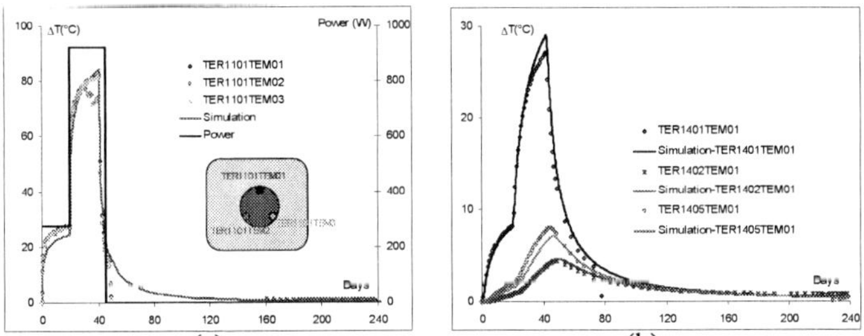

(a) (b)

FIG. 3. Evolution of temperature (a) at the heater/tubing interface borehole; (b) different measure points in the rock

The evolutions of pore pressure during heating and cooling phase at some measured points are also simulated (Fig. 4). In the numerical simulations, these fowling phenomena are well snapped: the increase at the beginning of heating phase, the decrease with the stabilisation of temperature, and the drop due to the switch off the heating power. In conclusion, there are some quantitative scare between the numerical predictions and the experiment data, but the general tendency of the pore pressure evolution is correctly reproduced.

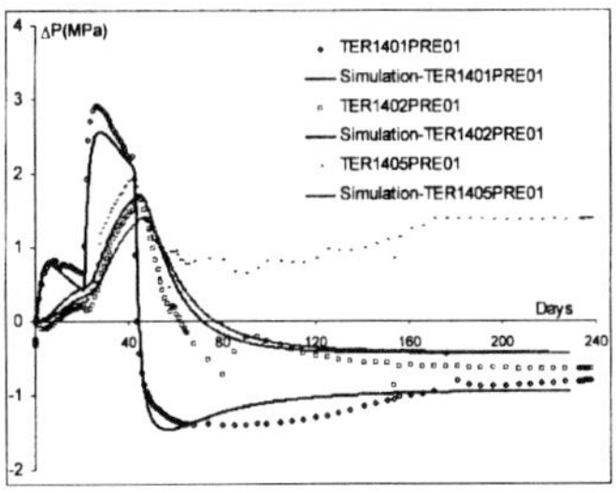

FIG. 4. Variation of water pressure at different measure points in the rock

Evolution of rock deformation is also studied here. Compared to temperature and pore pressure, the evaluation of rock deformation during the experiment is more difficult to capture from the point of view of both experimental measuring and numerical simulations. More important scatter has been obtained in measured data. Fig. 5a et Fig.5b show the numerical predictions of axial strains (relative displacement) along the borehole TER1302. Even if comparison between numerical modelling and experimental data exhibits some difference, the general trends of the deformation have been correctly reproduced. Therefore the numerical modelling proposed here seems to quite correctly describe the thermo-hydromechanical coupling process in the TER experiment. On the other hand, the used numerical model is validated.

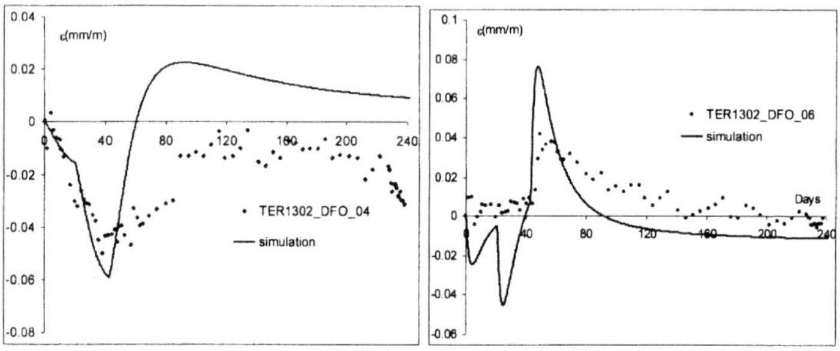

(a) (b)

FIG. 5. Variation of strain at measure point (a) TER1302_DFO_06; (b) TER1302_DFO_06

CONCLUSIONS

A 3D numerical modelling of an in situ heating experiment has been presented in this work. A Mohr-Coulomb model has been proposed for the description of poro-mechanical behavior of saturated argillite. The material resistances in compression and traction have been used for calibrating the model's parameters. The numerical simulation of TER experiment obtained by the coupled THM modelling shows a good agreement with the measured data for temperature and pore pressure variations. Although more important quantitative scatters have been obtained for the rock deformation, the general qualitative trend has been correctly reproduced. In spite of simplifications made, the 3D model seems to provide reasonable predictions of the main features observed in the in situ experiment. However, three dimensional analyses taking into account geometric anisotropy could improve numerical modelling, in particular, the prediction of rock deformation. Furthermore, it is also useful to point out that the damage induced material softening can enhance the onset of strain localization, which leads to the spurious mesh dependency of numerical modelling. It is necessary to introduce an elastoplastic damage model in FLAC3D. This feature has not been considered in the present work and needs to be studied in a future work.

ACKNOWLEDGMENTS

This work was partially supported by the ANDRA (French Agence Nationale de Gestion des Déchets Radioactifs), which is gratefully acknowledged.

REFERENCES

Abuel-Naga H.M., Bergado D.T., Bouazza A., and Ramana G.V., 2007. Volume change behaviour of saturated clays under drained heating conditions: experimental results and constitutive modelling. Canadian Geotechnical Journal 44(8), 942–956

Alonso E.E., Alcoverro J., Coste F., Malinsky L., Merrien-Soukatchoff V., 2005. The FEBEX benchmark test: case definition and comparison of modelling approaches. International Journal of Rock Mechanics and Mining Sciences. 42(5-6), 611-638.

Andra, 2005. Dossier 2005, Référentiel du site de Meuse/Haute-Marne. Rapport Andra (in French), n° C.RP.ADS.04.0022

Baldi G., Hueckel T., Pellegrini R., 1988. Thermal Volume Changes of the Mineral-Water System in Low Porosity Clay Soil, Canadian Geotechnical Journal, 25, 4, 807-825,

Chiarelli A.S., 2000. Experimental investigation and constitutive modeling of coupled elastoplastic damage in hard claystones. Doctoral thesis (in French), University of Lille

Chiarelli A.S., Shao J.F., Hoteit N., 2003. Modelling of elastoplastic damage behaviour of a claystone. International Journal of Plasticity 19, 23-45

Chijimatsu M., Fujita T., Sugita Y., Amemiya K., Kobayashi A., 2001. Field experiment, results and THM behavior in the Kamaishi mine experiment. International Journal of Rock Mechanics and Mining Sciences 38 (1), 67-78

Datta R., Barr D., Boyle W., 2004. Measuring thermal, hydrological, previous termmechanical, next term and chemical responses in the Yucca Mountain drift scale test. GeoProc2003, Stockholm, Sweden, October 2003.

Escoffier S., 2002. Caractérisation expérimentale du comportement hydromécanique des argilites de Meuse/Haute-Marne. Doctoral thesis (in French), INPL

Ewen J., Thomas H. R.,1989. Heating unsaturated medium sand. Geotechnique 39, 455–470.

Gens A., Garcia-Molina A. J., Olivella S., Alonso E.E.,Huertas F.,1998. Ananlysis of a full scale in situ test simulating respository condition. International Journal for Numerical and Analytical Methods in Geomechanics 22, 515-548.

GUO R.P., DIXON D., 2006. Thermohydromechanical simulations of the natural cooling stage of the Tunnel Sealing Experiment. Engineering geology 85(3-4), 313-331

Hueckel T.,Baldi G., 1990. Termoplastic Behavior of Saturated Clays: an Experimental constitutive Study, Journal of Geotechnical Engineering 116 (12), 1778-1796.

Jia Y., Wileweau Y., Su K., Duveau G., Shao J.F., 2007.Thermo-hydro-mechanical modelling of an in situ heating experiment. Geotechnique 57 (10), 845-855.

Laloui L., Modaressi H., 1994. Effets de la thermo-plasticité des argiles sur le comportement des puits de stockage, 1st International Congress on Environmental Geotechnics, Edmonton, Canada.

Su K., Bounenni A., 1998. Couplage entre la perméabilité et l'endommagement dans les matériaux des sites du Gard, de l'Est et de la Vienne. Rapport Andra (in French), n° B.RP.0G3S.98.021/A.

Sultan N., Delage P. , Cui Y. J.,2002. Temperature effects on the volume change behaviour of Boom clay. Engineering Geology 64 (2-3), 135-145.

Tsang C.F., Tephansson O. S, Hudson J.A., 2000. A discussion of thermo-hydro-mechanical (THM) processes associated with nuclear waste repositories. International Journal of Rock Mechanics and Mining Sciences 37, 397-402.

Zhang C.L. , Rothfuchs T., Su K.,Hoteit N., 2007. Experimental study of the thermo-hydromechanical behaviour of indurated clays. Physics and Chemistry of the Earth, 32, 957-965

Study on the Influence of Mix Proportioning on Cemented Mortar Engineering Properties for Tail Void Grouting of Shield Tunnel

Yuewang HAN[1,2], Wei ZHU[3], Quanwei LIU[4], Xiaochun ZHONG[2]

[1]Zhengzhou Rail Transit Co., LTD, Zhengzhou, China; hanyuewang@163.com
[2] College of Civil Engineering, Hohai University, Nanjing, China
[3] College of Environmental Science and Technology, Hohai University, Nanjing, China
[4] Construction Headquarters of Yangtze River Tunnel of Nanjing, China Railway Shisiju Group Corporation, China

ABSTRACT: The application of cemented mortar in tail void grouting of shield tunnel construction is becoming increasingly popular. However, the effect of mix ratios on cemented mortar performance is not well understood, and grout engineering properties cannot meet the demand of shield tunnel construction. Pumping difficulty and pipe blockage were occurred during backfill grouting process, and large stratum settlement was induced while shield passed. Determination method of mix proportioning for desired grout engineering properties is unreasonable, mix proportioning cannot be adjusted quickly with shield construction conditions change. Orthogonal experimental design method is used to study on grout engineering properties. The qualitative and quantitative relationships between mix ratios and grout engineering properties were obtained by variance analysis and regression analysis of experimental results respectively. Finally, mix ratios were calculated for desired grout engineering properties by goal programming method, and the rationality of goal programming results was examined by trial. Experimental analysis results show that the influence of water-binder ratio on all grout engineering properties is mostly obvious, and water-binder ratio is key impact factor. Bleeding rate and segregation degree reduce with the increasing of bentonite-water ratio, so bentonite is beneficial to increase the physical stability of cemented mortar. Grout volume shrinkage is mainly induced by drainage consolidation, so grout shrinkage rate decrease with the enhancing of mortar water retention ability and the decreasing of binder-sand ratio. Cemented hydration products were dispersed by the deflocculating action of water-reducing retarder to prolong the setting time of mortar and decrease the early age strength of mortar. It was illustrated that combination of orthogonal experimental design, regression analysis and goal programming is an effective method to determine mix proportioning quickly that is suitable to shield construction environments.

1. INTRODUCTION

The application of shield tunneling has become one of most popular methods with the construction of urban tunnels and river-crossing tunnels in China. Settlement induced by shield tunneling is one of most concerned problems in engineering fields. In order to predict ground movement by tunneling-induced, gap parameter was introduced to describe the equivalent ground loss of elastoplastic deformation induced by surrounding soil intruding into cutting face and tail void. Analytic solution of ground movement was derived based on elastoplastic theory. The predicted results show that final ground movement tends to increase with the increasing of equivalent ground loss (**Loganathan, 1998**; **Rowe, 1983**). Field monitoring of settlement Tianjin No.1 Metro Line was implemented. The monitoring profile was deployed at the whole section of silty clay. Ground movement above tunnel induced by tunneling occurred immediately while shield tail passed. Ground movement increases with shield further tunneling. Finally, ground movement induced by the tail void closure occupied about 50~70% of total settlement (**JIANG Xin-liang, 2005**). Obviously, ground loss induced by tail void closure is an important aspect to cause ground movement.

Backfill grouting is an important means to minimize ground loss, which pump grout into tail void with certain pressure. In order to simulate the influence of shield tunnel on ground movement, a three-dimensional finite element simulation model was presented. All relevant components of the construction process (face support pressure, tail void grouting pressure, hydraulic jack trusts, self-weight of TBM, etc.) were taken into account. The sensitivity analysis results showed that settlement decrease as the increasing of face support pressure and grouting pressure (**Kasper, 2004**). The strength of grout increases with cement hydration while grout was pumped into tail void. The influence of grout material strength characteristic on surface settlement was investigated numerically. The tail void grout was described as an elastic, fully saturated porous material. The hydration of grout was taken into account by applying time-dependent Young's modulus. The different material strength increasing rate was simulated for stratum deformation. The early age stiffness and strength development of grout were expected to be important, whereas the final stiffness and characteristics were expected to be of minor importance while grout reached similar or higher values than surrounding soil (**Kasper, 2006a; Kasper, 2006b**). Grout cannot coagulate to solid state due to long cemented mortar setting time, and then grout will stay in liquid state for relatively long time. The time-dependent stiffness model only can be used to simulate solid stage of grout. Therefore, it is necessary to take the drainage consolidation shrinkage property of liquid stage into account for settlement prediction. Grout drainage consolidation properties were tested by self-made oedometer in sand soil. Consolidation curves were tended to be stable in a shorter time under 0.3MPa pressure, and shrinkage rate up to about 20%. The shear strength of grout increased as consolidation shrinkage proceeding (**Bezuijen, 2003**). In order to investigate the effect of backfill grouting on soil deformation, stress release method was adopted to analyze the soil stresses state. Squeezing action of backfill grouting pressure on surrounding soil was pointed out, which cause excess pore pressure increase and effective stress decrease. With the movement of soil toward tail void, stratum stress release and excess pore pressure dissipation which induce soil consolidation and settlement (**Sakajo, 1996**). Ground stress release is governed by grout pressure dissipation under combined grout and soil conditions. However, grout pressure dissipation is determined by grout consolidation shrinkage properties. Field monitoring of grout pressure dissipation was apply to Sophia tunnel tail void grouting. The subsoil of monitoring profile is almost sand. While grout pumped into tail void, grout pressure begins to dissipate with consolidation, and final grout pressure decreases to the value very close to pore pressure in soil. Permeability coefficient of soil is a key parameter to influence on grout consolidation. Lower permeability coefficient may be helpful to delay grout pressure dissipation, and reduce the value of grout consolidation shrinkage (**Bezuijen, 2004; Bezuijen, 2006**). Therefore, grout engineering properties are described by fluidity, segregation degree, bleeding rate, setting time, consolidation shrinkage rate, and unconfined compressive strength (UCS), which has an important influence on tail void grouting operation efficiency and ground settlement control.

Grout engineering properties are unsuitable to shield tunneling conditions (grouting strategy, ground conditions, shield advance distance, etc.), which will lead to worse effect of backfill grouting. It is mainly manifested at the following two aspects. Firstly, poor grout physical stability and high segregation cause more condensed grout to deposit at lower part pipe during injecting process. As time going, grouting pipe will be plugged. Grout is hard to be pumped due to excessive fluidity loss caused by shorter setting time. Grouting pressure is lower than earth and water pressure lead to excessive ground stress release. Lower grouting rate leads to fill the tail void incompletely. On the other hand, after injection completion, excessive grout drainage consolidation shrinkage due to high bleeding rate leads to much ground loss before coagulation. Lower early age cementation strength of grout is no helpful to segment stability. Therefore, the effect of tail void grouting is governed by grout engineering properties directly.

Grout engineering properties are directly determined by mix proportioning. The desired grout engineering properties should be adjusted to meet the demand of ground conditions and construction environment with shield tunneling. Therefore, how to adjust mix proportioning immediately based on the relationships of engineering properties and mix ratios, which has important significance for enhancing settlement control effect of tail void grouting.

The main purpose of present work is to investigate the effects of common used admixtures (cement, fly ash, sand, bentonite and water-reducing retarder) on the grout engineering properties, and find out a reasonable mix proportioning determination method. The relationships between mix ratios and engineering properties were obtained by orthogonal experimental design, variance analysis, and regression analysis. Finally, it was proved that goal programming method is an effective method for determining mix proportioning of cemented mortar.

2. Material and testing methods

2.1. Materials

Generally, mix materials of cemented mortar are provided locally close to project site. Although mix proportioning is same, grout engineering properties are different from admixture types. The materials used commonly include ordinary Portland cement (Strength degree is 32.5, and basic performance parameters were given in Table 1), two-grade fly ash (Basic performance parameters were given in Table 2), sand, Na-bentonite (Main component is montmorillonite), and HLC-NAF water-reducing retarder (Water reducing ratio is 24%). Main chemical constitutions of cement, fly ash and bentonite were given in Table 3.

Table 1 Main performance index of cement

Normal consitency (%)	Setting time (h:min)		28d strength (MPa)	
	Initial	Final	Flexural strength	Compressive strength
25.0	2:35	3:50	8.1	38.4

Table 2 Main performance index of Fly ash

Fineness (%)	45μmSieve residue (%)	Burning amount (%)	Water requirement (%)
25.6	14.6	1.53	101

Table 3 Properties of cement, fly ash and bentonite

	SiO_2	Al_2O_3	Fe_2O_3	CaO	MgO	SO_3
Cement (%)	21.50	5.80	2.70	63.00	2.09	3.55
Fly ash (%)	52.70	30.50	4.32	6.38	1.46	0.72
Bentonite (%)	70.76	14.93	1.89	0.96	1.77	-

2.2. Testing methods

The test procedures are performed referring to National Regulation (**JGJ70-90, 1991**). Grout admixtures were prepared based on orthogonal experimental design groups, and casted in stirring apparatus at one time. The blender was constantly working while all of these materials stirred well-mixed, and stirring process continued for 5 minutes. Testing procedures of engineering properties are listed in detail as follows:

- Fluidity: Cemented mortar fluidity detector is used. After jump table flopped 25 times, the diameter of grout flowed over is mortar fluidity.
- Segregation degree: Firstly, mortar consistency is measured by consistency

detector, and marked as A, then putting mortar in delamination degree barrel. Mortar in lower part of barrel is taken out after 30 minutes, which is agitated for 2 minutes. Secondly, mortar consistency is measured again, and marked as B. The value of A minus B is segregation degree of cemented mortar.

- Bleeding rate: Final bleeding of freshly mixed grout is evaluated in a glass graduated cylinder of 100ml. The volume percentage of bleed water and whole grout is bleeding rate after 3 hours.
- Setting time: Freshly mixed grout is standing for several hours. Penetration resistance is measured at 30 minutes intervals by setting time apparatus of cemented mortar. The time corresponding to penetration resistance reached 0.5MPa is setting time of cemented mortar.
- Shrinkage rate: One dimensional consolidation of grout is tested by self-made oedometer under 0.3MPa consolidation pressure. The volume percentage of final maximum consolidation deformation and total grout volume is defined as shrinkage rate.
- Unconfined compression strength for 1 day curing age (1-day UCS): The cubic sample of 70.7mm×70.7mm ×70.7mm is made. Unconfined compression strength of 1 day curing age is measured on Hydraulic Universal Testing Machine.

2.3. Experimental design

Mix proportioning of cemented mortar is established by five ratios of six admixtures. Five ratios are water-binder ratio (W/G), blinder-sand ratio (G/S), bentonite-water ratio (M/W), flyash-cement ratio (F/C), and additive-blinder ratio (A/G) respectively. Water-binder ratio is the mass ratio of water and cementing material (cement and fly ash). Blinder-sand ratio is the mass ratio of cementing material (cement and fly ash) and sand. Bentonite-water ratio is the mass ratio of bentonite and water. Flyash-cement ratio is the mass ratio of fly ash and cement. Additive-blinder ratio is the mass ratio of water-reducing retarder and cementing material (cement and fly ash). Four levels are assigned to each ratios were listed in Table 4.

Orthogonal experimental design is an important method of fractional factorial design, which chooses some representative groups to test, which is applied widely in scientific research fields (**FANG Kai-tai, 2001**). According to five-factors and four-levels, standard orthogonal table of $L_{16}(4^5)$ was adopted.

Table 4 Test factors used and their levels

Level	Factors and their levels				
	W/G	**G/S**	**M/W**	**F/C**	**A/G**
1	0.7	0.6	0.05	1.0	0.000
2	0.8	0.7	0.10	1.5	0.010
3	0.9	0.8	0.15	2.0	0.015
4	1.0	0.9	0.20	2.5	0.020

3. Results and discussion

3.1. Experimental results

Grout engineering properties were tested according to 16 groups of orthogonal experimental designed. Table 5 shows the experimental results of grout engineering properties. The relationships of engineering properties and mix ratios are not obtained directly through orthogonal experimental results, which must be analyzed by means of mathematical statistics methods.

Table 5 Experimental results of grout for each trial

Trial numbe	Before injection			After injection			
	Initial fluidity cm	8-hour fluidity cm	Segregation degree mm	Bleeding rate %	Setting time h	Shrinkage rate %	1-day UCS KPa
1	19	11.5	1	1	8	4.95	315
2	23.9	19.4	2	2	10.5	8.81	217
3	20.8	15.1	1	1.5	18	8.65	159
4	18.8	14.4	1	0.5	23.5	5.62	58
5	25.8	19.8	1	2	17	7.3	95
6	33	32	4	5	12	8.25	4
7	14.8	10.5	1	0.5	10.5	7.65	303
8	22	12	1	1	13.5	5.48	260
9	19.5	14	1	2	15.5	9.51	179
10	19.3	11	1	2.5	11.5	10.82	110
11	30.5	29.5	3	3	23	13.21	44
12	29.3	26	2	2	14.5	13.26	224
13	19.5	16	1	3	16	11.94	84
14	25.5	22.5	2	4.5	19.5	15.28	79
15	30	24	3	6	25	15.9	40
16	31	30	4	8	20.5	25.12	16

3.2. Discussion

3.2.1 Analytical method of experimental results

The quantitative relationships must be obtained by means of regression analysis method based on experimental results. Water-binder ratio, blinder-sand ratio, bentonite-water ratio, flyash-cement ratio, and agent-blinder ratio were defined as x_1, x_2, x_3, x_4, and x_5 respectively. The second-order polynomial, given by Eq.(1), was used to fit experimental results. This equation was used very popular in scientific optimum design fields (**TANG Qi-yi, 2006**; **Fall, 2008**). The functional relationships between mix ratios and engineering properties were obtained by regression function of statistical analysis software (DPS Data Processing System).

Eq. (1) was shown below

$$y = b_0 + \sum_{k=1}^{5} b_k x_k + \sum_{j=1}^{5}\sum_{k=j}^{5} b_{jk} x_j x_k + \varepsilon \qquad (1)$$

Where y is the predicted responses (i.e. fluidity, bleeding rate), b_0 is the intercept term, b_k, b_{jk} are the constant regression coefficients for the linear terms, and the cross-product terms respectively. The x_k variables represent the normalized values of each of the input parameters that influence the response y, the cross-term $x_j x_k$

represents the first-order interactions between x_j and x_k. Finally, ε is the associated random error reflecting the combined effects of variables not included in the models. It is assumed that the additive error ε is normally distributed with a mean of zero and standard deviation.

The admixtures quantitative effect were analyzed based on the functional relationships of mix ratios and engineering properties limited in the range from lower to upper boundaries of each factors. Analysis results of main impact factors were listed in detail as follows.

3.2.2 Relationship of fluidity and mix ratios

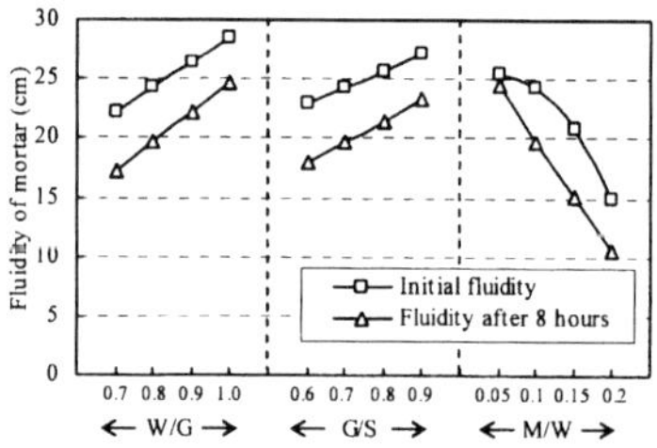

Fig. 1 Response graphs for fluidity

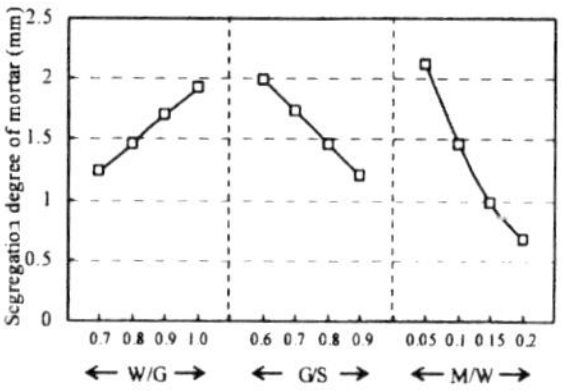

Fig. 2 Response graphs for segregation

Eq. (2) and Eq. (3) were obtained by regression analysis of initial fluidity and 8 hours fluidity respectively. M/W ratio, W/G ratio and G/S ratio are important effect factors for fluidity from variance ranges. Fig. 1 illustrates the effects of W/G ratio, G/S ratio and M/W ratio on mortar fluidity. With the increasing of M/W ratio, initial fluidity presents a non-linear decrease tendency whereas 8h fluidity presents a linear decrease tendency. Inversely, grout fluidity tends to increase with increasing of W/G ratio and G/S ratio. Grout fluidity is obviously depended on the amount of free-water in mortar. The more free-water, the better flow performance may be obtained. High water absorption characteristic of bentonite decreases the amount of free-water, and causes grout fluidity decrease. But, the particularly lubricating property of bentonite is beneficial to enhance the pumpability of cemented mortar. Cement and fly ash belong to fine materials with high specific surface area. Therefore, sand particles are packaged by flocculent structures of fine materials formed, and then grout flow performance is enhanced by the reduction of friction force between sand particles.

$$F_{0h} = 11 - 482.6x_3^2 + 17.5x_1x_2 + 84.1x_1x_3 - 1618.9x_3x_5 + 209.5x_4x_5 \quad (2)$$

$$F_{8h} = 8.3 - 83.3x_3 + 19x_1x_2 + 3.1x_1x_4 + 486.9x_1x_5 \quad (3)$$

3.2.3. Relationship of segregation and mix ratios

Eq. (4) presents the relationship of segregation and mix ratios by means of regression analysis. M/W ratio, W/G ratio and G/S ratio are important effect factors for segregation degree from variance ranges. Fig. 2 shows that segregation degree of mortar varies with three mix ratios increasing. Segregation degree of mortar reduces linearly with the increasing of M/W ratio and G/S ratio. But with W/G ratio increasing, segregation degree tends to increase linearly. Segregation degree reflects the phenomenon of particles relative displacement. The slower of solid particles sediment speed, the smaller of segregation degree. Cement, fly ash and bentonite are fine materials, and sand is coarse granule. Sand granules are packaged by the flocculent structures induced by fine materials. The buoyancy force of sand granules gained from flocculent structures causes the lower sediment speed of sand. The thickening effect and suspension action increase with free-water reducing due to high water absorption of bentonite, which leads to mortar segregation degree decreasing.

$$L = -0.2 + 33.7x_3^2 + 0.6x_4^2 + 3.3x_1x_2 - 12x_3x_4 \quad (4)$$

3.2.4. Relationship of bleeding and mix ratios

Based on regression analysis of experimental bleeding results, Eq. (5) was obtained. It can be seen from variance ranges that the important effect factors on final bleeding rate are W/G ratio, M/W ratio and F/C ratio. Fig. 3 shows the main effect relationships of bleeding rate and mix ratios. Final bleeding rate reduces with the increasing of M/W ratio, and the decreasing of W/G ratio and F/C ratio. Bleeding represents the phenomenon of solid particles subsidence or free-water floating upward solid. So, the excess free-water existed in grout, the more seriously bleeding. Because of the high water absorbability of bentonite, and cement particles are more easily to adsorb free-water than fly ash, bleeding rate can be decreased by the increasing of M/W ratio and F/C ratio.

$$B = 21.6 - 60.2x_1 + 16.4x_3 + 42x_1^2 - 4.2x_2^2 + 4.6x_2x_4 - 19.2x_3x_4 \quad (5)$$

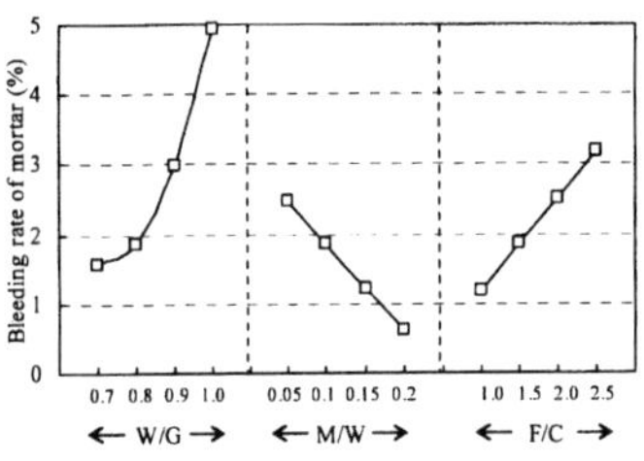

Fig. 3 Response graphs for Bleeding

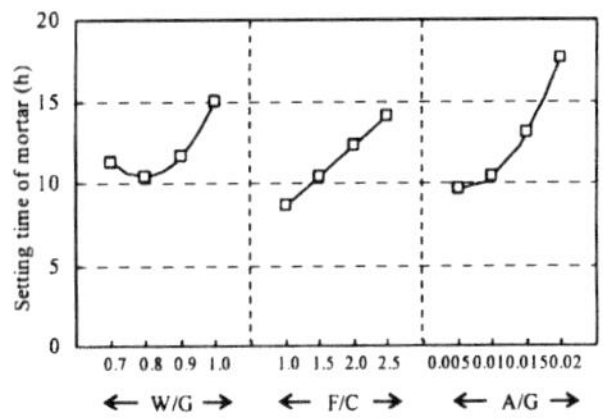

Fig. 4 Response graphs for setting time

3.2.5. Relationship of setting time and mix ratios

Through the regression analysis of setting time, Eq. (6) can be fitted. It can be seen from the values of variance ranges that W/G ratio, A/G ratio and F/C ratio are most influential factors for setting time. Fig. 4 illustrates the relationships of setting time and W/G ratio, A/G ratio and F/C ratio. With the increasing of W/G ratio, F/C ratio and A/G ratio, setting time of mortar was prolonged. The process of coagulation is a part of cement hydration. The hydration activity of cement is higher than fly ash, which causes shorter setting time with the decrease of F/C rato. With the development of cement hydration, cemented flocculent structures begin to connect each other, the fluidity of grout gradually lose. The mortar finally coagulates to solid state which has certain bonding strength. Water-reducing retarder is a polymer surface active agent which can disperse cemented flocculent products. The hydration products are prevented to connect each other, coagulating and hardening speeds are slowed down. But the retarding effect is not significant while less addition.

$$C = 65.3 - 162.2x_1 + 102.7x_1^2 + 36513.1x_5^2 + 8.8x_2x_4 - 251.9x_4x_5 \quad (6)$$

3.2.6. Relationship of shrinkage and mix ratios

Based on regression analysis of experimental consolidation shrinkage rate results, Eq. (7) was obtained. It can be seen from variance ranges that the influence of W/G ratio, G/S ratio and M/W ratio on shrinkage rate are significant. Fig. 5 shows the main effect curves of shrinkage rate and mix ratios. Shrinkage of mortar under compressive pressure is mainly induced by drainage before coagulated. Therefore, excess free-water existed in grout and poor grout water retention must lead to large consolidation shrinkage. Grain stress gradually increases during compression process as the contacting of fine aggregate which minimizes the compressive deformation. Shrinkage rate tends to increase with the increasing of W/G ratio and G/S ratio, and with the decrease of M/W ratio.

$$S = 77.9 - 202.6x_1 + 130x_1^2 + 19.9x_1x_2 - 33.1x_2x_3 \quad (7)$$

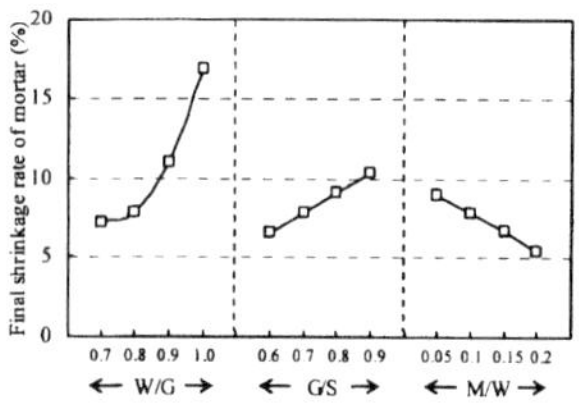

Fig. 5 Response graphs for shrinkage

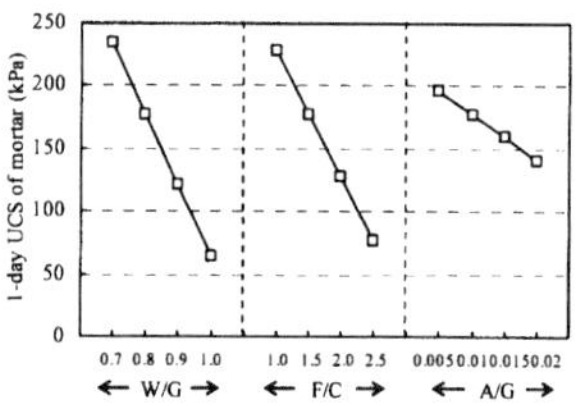

Fig. 6 Response graphs for 1-day UCS

3.2.7. Relationship of UCS and mix ratios

Through the regression analysis of 1-day UCS, Eq. (8) can be obtained. It can be seen from the values of variance ranges that F/C ratio, W/G ratio and A/G ratio are mostly influential factors for 1-day UCS of grout. Fig. 6 shows the relationships of 1-day UCS and mix ratios. With the increasing of W/G ratio, F/C ratio and A/G ratio, the 1 day UCS of mortar tends to be lower. The reason for decrease is that less cement content leads to less the products of cement hydration. The action of water-reducing retarder agent is to retard the hydration of cement, which is not beneficial to enhance 1-day UCS of mortar.

$$Q = 485.1 + 1489.7x_2^2 - 1581.2x_1x_2 + 359.3x_1x_4 - 519.7x_2x_4 - 2513.5x_4x_5 \quad (8)$$

4. Mix proportioning determination procedures

4.1. Calculation methods

The desired grout engineering properties were assigned as goal values, goal programming approach was used to calculate W/G ratio (x_1), G/S ratio (x_2), M/W ratio (x_3), F/C ratio (x_4), and A/G ratio (x_5) based on the obtained functional relationships of engineering properties and mix ratios.

Firstly, the mostly important engineering property was assigned as priority goal, and the other engineering properties were set as constraint conditions, and the upper and lower values of each orthogonal design levels were set as boundary conditions. Secondly, mix ratios of cemented mortar corresponding to the desired grout engineering properties were solved out by DPS Data Processing System (TANG Qi-yi, 2006).

4.2. Calculation example

The 1-day UCS was defined as optimal goal, the other engineering were set as constraint conditions (Initial fluidity was greater than 22cm and 8-hours fluidity was greater than 20cm, segregation degree was less than 2mm, bleeding rate was smaller than 3%, setting time was limited in the range of 8 and 12 hours, Shrinkage rate was smaller than 10%), and the upper and lower values of each levels were set as boundary conditions. Goal programming procedures were listed in detailed from Eq. (9) to Eq. (11).

The maximum 1-day UCS was defined as target value.

$$Z = \max(Q) \tag{9}$$

The other grout engineering properties were defined as constraint conditions.

$$\begin{cases} F_{0h} \geq 220 \\ F_{8h} \geq 200 \\ L \leq 2 \\ B \leq 3 \\ 8 \leq C \leq 12 \\ S \leq 10 \end{cases} \tag{10}$$

The upper and lower levels were defined as boundary conditions.

$$\begin{cases} 0.7 \leq x_1 \leq 1.0 \\ 0.6 \leq x_2 \leq 0.9 \\ 0.05 \leq x_3 \leq 0.20 \\ 1.0 \leq x_4 \leq 2.5 \\ 0 \leq x_5 \leq 0.02 \end{cases} \tag{11}$$

Firstly, based on boundary conditions in Eq. (11), the ranges of each mix ratio were limited which satisfied Eq. (10). Finally, based on Eq. (9), the specific mix ratios were established. The mix ratios were solved as x_1=0.7, x_2=0.897, x_3=0.065, x_4=1, x_5=0.012 respectively.

4.3. Experimental confirmation

Grout engineering properties were verified by experimental confirmation based on the goal programming results. Experimental and calculated grout engineering properties were listed in Table 6. It was shown that calculated results are in perfect agreement with experimental results. It also proved that, based on the quantitative functional relationships between engineering properties and mix ratios, we can obtain the desired grout mix proportioning which meet the demand of shield tunneling.

Table 6 Comparisons of Grout Properties Between Calculated and Experimental Results

	before injection			After injection			
	Initial fluidity cm	8-hour fluidity cm	Segregation degree mm	Bleeding rate %	Setting time h	Shrinkage rate %	1-day UCS kPa
Calculation	22.71	20.21	1.40	1.03	10.48	8.03	296.55
Experiment	23.6	19.8	2	1.5	10	8.83	285.32

5. Conclusion

The relationships of cemented mortar mix ratios and grout engineering properties were studied in combination with orthogonal experimental design, variance analysis, multivariate nonlinear regression and goal programming method. The main conclusions drawn from this study were summarized as follows:

- The qualitative and quantitative relationships of grout engineering properties and mix ratios can be obtained by orthogonal experimental design, variance analysis, and regression analysis method.
- Based on the quantitative functional relationships of engineering properties and mix ratios, mix proportioning corresponding to the desired grout engineering properties can be obtained immediately by goal programming method.
- The W/G ratio has significant influence on all of grout engineering properties, which is the mostly effective factor for cemented mortar performance. With the increase of W/G ratio, fluidity, segregation degree, bleeding rate, setting time, shrinkage rate tend to increase, and 1-day UCS tends to decrease. Therefore, the amount of water must be strictly controlled.
- The M/W ratio is an important effective factor due to the high water adsorption characteristic of bentonite. With the increase of M/W ratio, fluidity, segregation degree, bleeding rate, shrinkage rate tend to decrease.
- The hydration products of cement are dispersed by surface activity of water-reducing retarder, which retard coagulation reaction of cement. Therefore, with the increase of water-reducing retarder addition, setting time will be prolonged, and 1-day UCS is reduced.

Acknowledgements

The authors would like to acknowledge the financial support given by China Railway Shisiju Group Corporation Construction Headquarters of Yangtze River Tunnel of Nanjing (Grant No. NJCJSD20080401) and New Teacher Subjects of Special Research Fund of Colleges and Universities Doctor Disciplines (Grant No. 20070294029). The authors also wish to express gratitude to Prof. Wei ZHU of College of Environmental Science and Technology of Hohai University for his kind help and valuable guiding to this paper.

References

Loganathan, N, Poulos, H.G., (1998). Analytical prediction for tunneling-induced ground movements in clays. Journal of Geotechnical and Geoenvironmental Engineering 124, 846-856.

Rowe, R.K., Lo, K.Y., Kack, G.J., 1983. A method of estimating surface settlement above tunnels constructed in soft ground. Can. Geotech. J. 20(8), 11-12.

JIANG Xin-liang, CUI Yi, LI Yuan, et al, (2005). Measurement and simulation of ground settlements of Tianjin subway shield tunnel construction. Chinese Journal of Rock and Soil Mechanics 26, 1612-1616.

Kasper, Thomas, Meschke, Gunther, (2004). A 3D finite element simulation model for TBM tunnelling in soft ground. International Journal for Numerical and Analytical Methods in Geomechanics 28, 1441-1460.

Kasper, Thomas, Meschke, Gunther, (2006a). On the influence of face pressure, grouting pressure and TBM design in soft ground tunnelling. Tunnelling and Underground Space Technology 21, 160-171.

Kasper, Thomas, Meschke, Gunther, (2006b). A numerical study of the effect of soil and grout material properties and cover depth in shield tunnelling. Computers and Geotechnics 33, 234-247.

A. Bezuijen, A. M. Talmon, (2003). Grout, the foundation of a bored tunnel. Foundations: Innovations, observations, design and practice. Thomas Telford, London, 129-138.

S.Sakajo, T.Yoshimaru, M.Kamimura, (1996). Analytical and geotechnical consideration on ground settlement induced by tail void closure of shield tunnel construction. Geotechnical Aspects of Underground Construction in Soft Ground. Mari & Taylor(eds), 585-590.

Bezuijen, A, Talmon, A.M, Kaalberg, F.J, et al, (2004). Field measurements of grout pressures during tunnelling of the Sophia Rail Tunnel. Soils and Foundations 44, 39-48.

Bezuijen, A, Talmon, A.M, (2006). Calculation models based on monitoring during tunnel construction. Tunnelling and Underground Space Technology 21, 380-381.

JGJ70-90, (1991). Building mortar basic performances test methods, Shananxi Provincial Academy of Building Reseach. China building industry press, Beijing, 1991.

FANG Kai-tai, MA Chang-xing, (2001). Orthogonal and Uniform Experimental Design. China Science Press, Beijing, 2001.

TANG Qi-yi, FENG Ming-guang, (2006). DPS Data Processing System. China Science Press, Beijing, 2006.

Fall, M, Benzaazoua, M,Saa, E.G., (2008). Mix proportioning of underground cemented tailings backfill. Tunnelling and Underground Space Technology 23, 80-90.

Experimental test on communication cable tunnel constructed by shield tunneling method for maintenance

Linxing Guan[1], Hiroshi Irie[2], Toru Shimada[3], Atsushi Koizumi[4]

[1]Research associate, Dept. of Environment and Civil Engineering, Waseda University. 3-4-1 Ohkubo, Shinjuku-ku,Tokyo,169-8555,Japan. guanlinxing@aoni.waseda.jp
[2]Senior research engineer, Nippon Telegraph and Telephone Corporation. 1-7-1 Hanabatake, Tsukuba-city, Ibaraki, 305-805,Japan.
[3]Research engineer, Nippon Telegraph and Telephone Corporation. 1-7-1 Hanabatake, Tsukuba-city, Ibaraki, 305-805,Japan.
[4]Professor, Dept. of Environment and Civil Engineering, Waseda University. 3-4-1 Ohkubo, Shinjuku-ku,Tokyo,169-8555,Japan.

ABSTRACT: Cable tunnels for communication are mostly constructed by the primary lining of steel segment and the secondary lining of concrete. Some leakage can be found from the primary lining of the tunnels constructed before 25 years. The secondary lining is loaded by water force due to the leakage where there is a cavity between the primary lining and the secondary lining. The secondary lining is made of concrete, because it is assumed that the secondary lining is loaded only by gravity during designing some cable tunnels in Japan. It is essential to evaluate the health of secondary lining due to the leakage of primary lining. In this paper, the two prototype cable tunnel models were used and the cavities between the primary lining and the secondary lining were man-made. The models were tested by loading the water pressure in the cavities and the strains of central section of secondary lining were recorded. The models were analyzed by using three dimensional finite element method and the results were compared with the experimental values. The behavior of secondary lining and the adaptability of numerical models to the specimen were discussed.

KEY WORDS: Shield tunnel, Primary lining, Secondary lining, Maintenance

INTRODUCTION

The telegraph and telephone corporations have been trying to place the cables in the underground space for the stable service. The cable tunnel with a diameter of 2-6 meter is usually used as a storage space for the various communication cables and electric cables. The cable tunnels were mostly constructed of the primary lining and secondary lining. Fig.1 shows the typical section of the cable tunnel. It is important to appropriately evaluate the health of tunnel for the manager of tunnel, because some tunnels were constructed before 25 years. Some cavities were found in the crown of tunnel in the periodical inspection of cable tunnels. Fig.2 shows the cavity in the crown of tunnel. The cavity between the primary lining and the secondary lining is filled with water and the secondary lining is loaded by water pressure. But the gravity is only taken into account during designing some secondary linings in Japan (JSCE 2006). It is essential to evaluate the stress of secondary lining for the safety of tunnel, because the technical workers have to enter into the tunnels to maintain and repair the cables.

In this paper, the two prototype models were made and loaded by water pressure in the man-made cavity between the secondary lining and primary lining. Then the models were analyzed by using three-dimensional finite element method and the numerical results are compared with the experimental values.

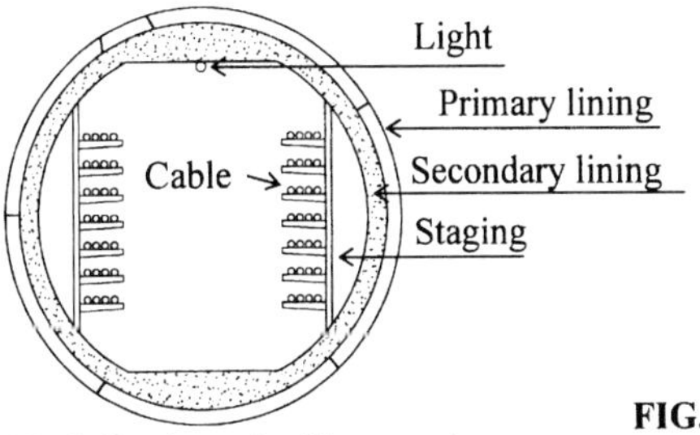

FIG. 1. Section of cable tunnel

FIG. 2. Cavity between primary lining and secondary lining

EXPERIMENTAL INVESTIGATION

Details of Test Specimens

The prototype test models were made of primary lining and secondary lining. The primary lining was assembled by steel segments and the secondary lining was cast by in-situ concrete in form. Fig.3 shows the details of test specimen and Fig.4 shows the steel segment for the primary lining. The thickness of skin plate is 3mm and the height of main girder is 125mm. Table.1 shows the dimensions of specimens. The two steel segments were assembled with bolt joints at the top of specimen. The two lateral sides of cavity were closed by steel plates and the water pressure was loaded through the hole in the skin plate. Fig.5 shows the experimental setup.

Material property

The high-early-strength Portland cement was used and the slump of concrete was 12cm. The maximum size of coarse aggregate was 20mm. The cylinder tests were conducted to find out the uniaxial compressive strength of concrete. The uniaxial compressive strength, and Young's modulus of concrete are 34.0kN/mm^2, 27.3kN/mm^2, respectively. The Young's modulus of steel is 210kN/mm^2.

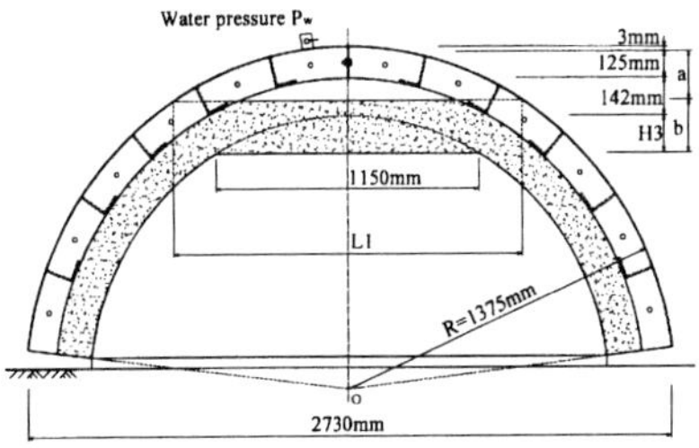

FIG. 3. Details of test specimen

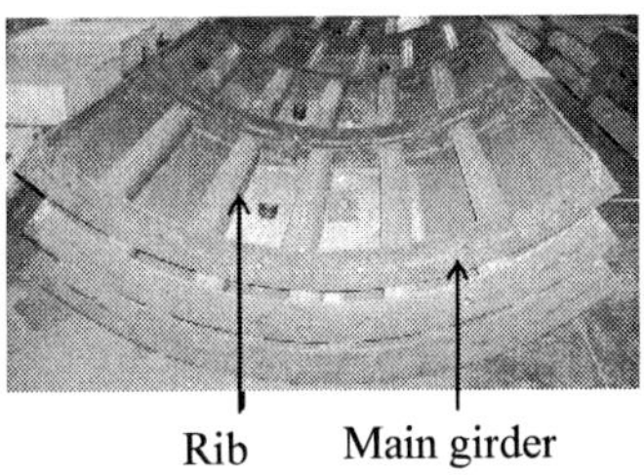

FIG. 4 Steel segment

Table 1. Dimensions of specimens

Specimen ID	Cavity height a (mm)	Concrete height b (mm)	Crown width L_1 (mm)	Crown height H_3 (mm)
No.1	278	160	1660	172
No.2	340	100	1780	170

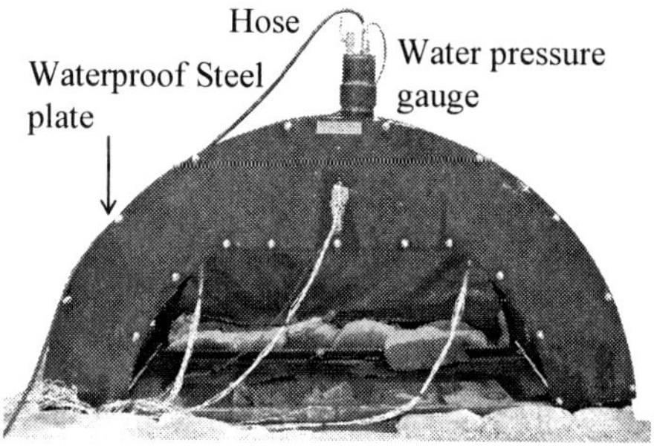

FIG. 5. Experimental setup

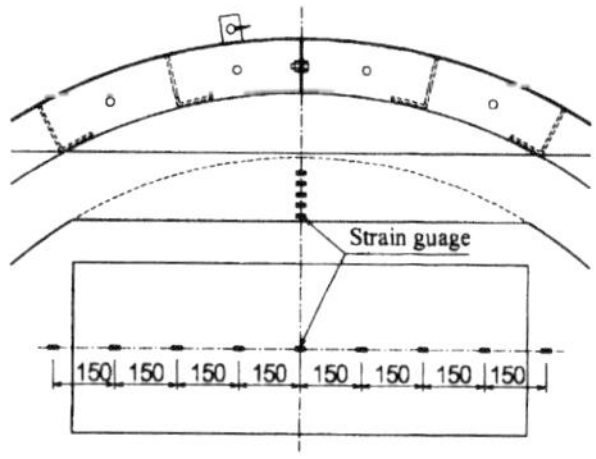

FIG. 6. Location of strain gauges

Measured Items and Instrumentation

The strains of concrete in tensile zone of central section were recorded for each load increment. The strain gauges were affixed to central section as shown in Fig. 6. The strain gauges were placed symmetrically to avoid the differences in dimension and material of the segment. Meanwhile the strain values at the symmetrical points were averaged for the experimental results.

FINITE ELEMENT ANALYSIS

FEM model

The three-dimensional finite element method was used to analyze the specimens by MSC.Marc (2005). The concrete secondary lining was modeled using solid element (Element 7 in MSC.MARC) and the primary lining made of steel segments using shell element (Element 75 in MSC.MARC). Fig.7 shows the finite-element model of specimen No.2. The interaction between primary lining and secondary lining was modeled as springs of x, y, z direction. Fig.8 shows the springs between the primary lining and the secondary lining. If a compressive force develops in the normal spring, the spring coefficient k_x is infinite and if a tensile force develops in the normal spring, k_x is equal to zero. The tangent spring coefficients on the circumferential and longitudinal direction k_y , k_z were infinite and the slip was not considered. The main girder was modeled as shell and the interaction between the main girder and the secondary lining was ignored.

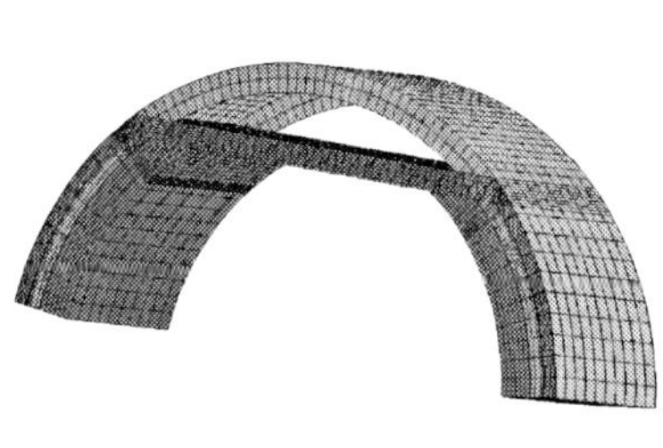

FIG. 7. Finite-element model of specimen No.2

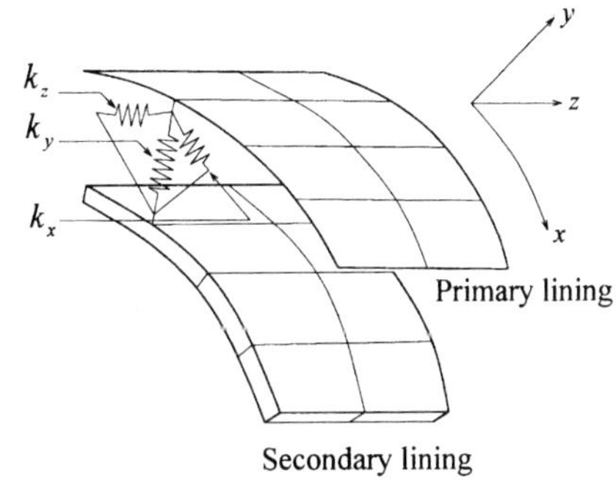

FIG. 8. Springs between primary lining and secondary lining

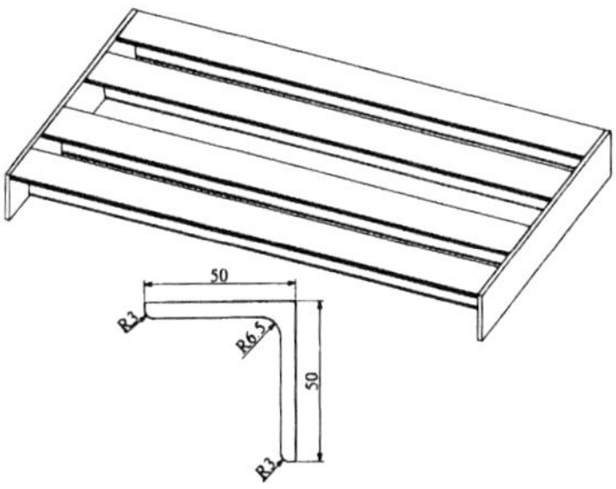

FIG. 9. Connection of equilateral angle steel

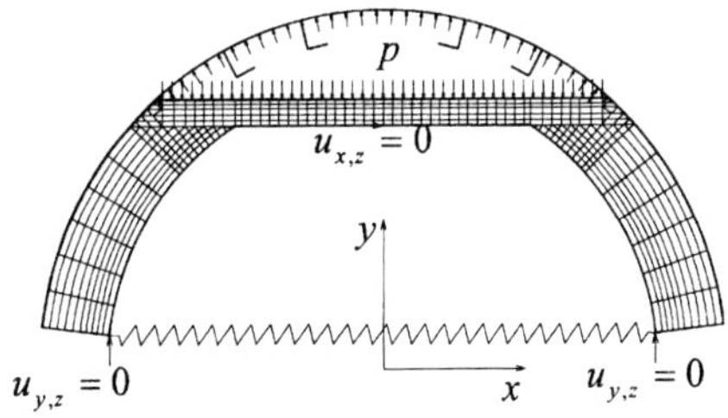

FIG. 10. Boundary Condition

Boundary condition

The ends of specimen were connected with four pieces of equilateral angle steel as shown in Fig.9. The connection is modeled as springs shown in Fig.10. The coefficient of spring is obtained from Eq. (1)

$$k = \frac{4A_s E_s}{l} \tag{1}$$

Here, k : Coefficient of spring; A_s : Section area of one piece equilateral angle steel; E_s : Elastic modulus of steel; l : Length of equilateral angle steel.

The x, z direction displacement of crown's middle point was set to zero because the dimension and boundary were symmetric with respect to the y-axis. The y, z direction displacement of ends was set to zero and the water pressure was loaded on the surface of skin plate and crown.

EXPERIMENTAL AND FEM RESULTS

Crack propagation

Fig.11 shows the crack propagation of specimen No.2. The bottom surface of crown

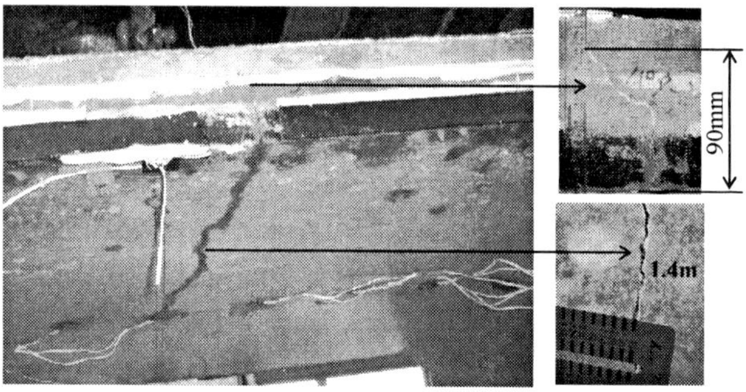

FIG. 11. Crack propagation of specimen No.2

started cracking and the cracks widened as the load increased. The crack propagated in the direction of crown thickness until 90mm from the bottom of crown. The maximum width of crack was 1.4mm. It was impossible to increase the water pressure until the collapse of specimen because of the leakage between the waterproof steel plate and specimen.

Relationship between strain and water pressure

Fig.12 shows the relationship between strain of crown and water pressure for No.1 and No.2. Here, the strain is the averaged value of symmetric points. The number in legend is the distance of test point from the center of crown. The maximum tension strain occurred at the center of crown and the compressive strain occurred at the ends of crown. The strain of specimen No.1 is elastic until 0.06MPa water pressure. The crack was not found because it was impossible to increase the water pressure because of leakage. The relation of water pressure and strain is linear at the initial phase and it becomes nonlinear thereafter for specimen No.2 with a thinner crown than specimen No.1. The crack started when the water pressure attained 0.043MPa. Cracks that developed in the crown resulted in a reduction of stiffness that caused a larger strain with a same load increment.

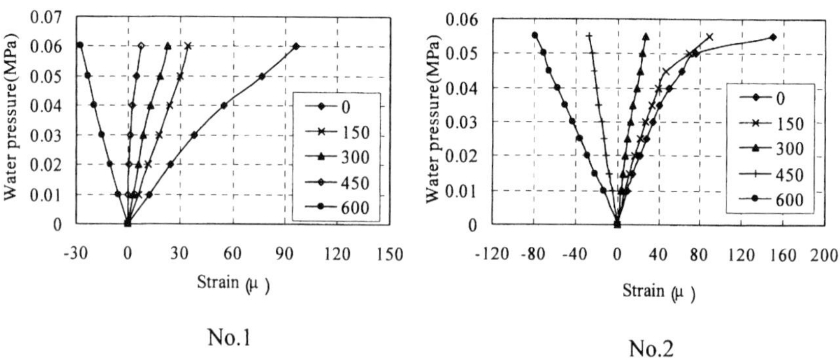

FIG. 12. Relationship between strain and water pressure

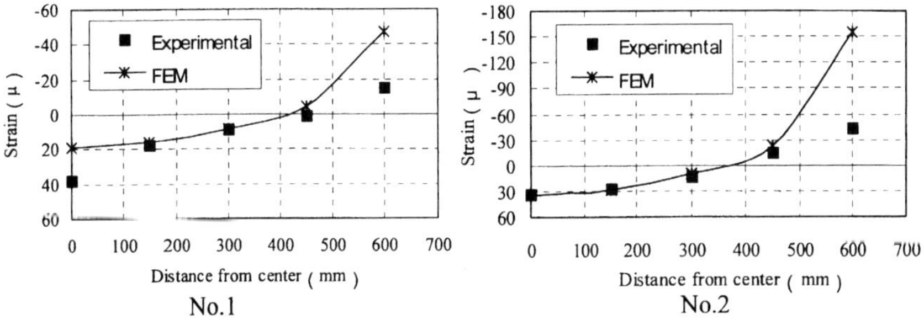

FIG. 13. Strain Distribution along the span of crown (p= 0.03MPa)

Strain distribution of crown

Fig.13 shows the strain distribution along the span of crown, when the water pressure attains 0.03MPa. From the experimental and numerical results, the strain decreases from the center and attains minus value at the end of crown. The numerical results are compatible with the experimental results at the central area. The difference between the two results becomes larger at the end of crown where the compressive strain is found. In Fem model, the point at the distance 600 is the intersection between line crown and curved secondary lining. The stress concentration was contributed to the difference between the experimental results and numeric results. In general, the numerical modeling is capable of predicting the behavior of cable tunnel.

CONCLUSIONS

(1)The maximum strain occurred at the center of crown and decreased from the center to the end of crown.

(2)The compressive strain is found at the bottom surface of crown end. It can be assumed that the minus moment occurred at the end of crown. The pattern of strain distribution resembled with the beam fixed at both ends. It becomes possible to use the simple beam theory to calculate the strain of crown.

(3)It is reasonable to model the steel primary lining with shell element, the concrete secondary lining with solid element and the interaction between the primary lining and the secondary lining with springs. The numerical modeling is capable of predicting the behavior of cable tunnel.

REFERENCES

Japan Society of Civil Engineers(JSCE), (2006). "Standard specifications for tunneling-2006, Shield Tunnels." JSCE, Tokyo (in Japanese).

Sakuma S., Ishida T., Uehara S. and Koizumi A. (1990): Effects of the axial stiffness of a shield tunnel including effects of shearing interaction between primary and secondary linings. *J. Japan. Soc. Civ. Eng.*, 424(III-14), 251-259 (in Japanese).

Takamatsu N., Murakami H. and Koizumi A. (1993): Study on the analytical model on bending behavior in longitudinal direction of shield tunnel with secondary lining. *J. Japan. Soc. Civ. Eng.*, 481(III-25), 97-106 (in Japanese).

MSC.Marc theory and user information manual, version 2005. (2005). MSC.Software corporation.

The Determination of Geometric Characteristics of Irregular Underground Bodies

Mehdi Zamani [1]

[1]*Department of Technology and Engineering, Yasouj University, Daneshjoo Avenue, Yasouj, IRAN 75914*

mahdi@mail.yu.ac.ir

ABSTRACT: The main concern is with nonuniform shapes and bodies in most of engineering problems. For example, determination of nongeometric volumes in underground structures such as karstic and salty caves to store hydrocarbon materials specially petroleum and discharge of superfluous materials. The most of aquifers consists of unsymmetric boundaries. For determination of aquifer volume and amount of water storage, creation of applied functions of boundaries is required. For determining the geometric characteristics of nonuniform and nonregular bodies, the classic models and functions cannot be used, because of their specific shapes and surfaces. Therefore the interpolation models are used. In this research, two models for one and two dimensional interpolation are presented. The advantages of models are the simplicity and the lower computation operations as compared with current classic methods such as cubic spline and B-spline. The presented model is proper to use for large amount of data and it has continuity in function and slope in the element boundaries.

KEYWORDS

Interpolation; approximation; prediction; cubic spline; prediction; simulation

INTROUDUCTION

One of the important parameters in coal mining is the determination of ore evaluation. The ore evaluation is obtaining the volume of coal seams and their unit weight to calculate the amount of available coal for extraction. These parameters are inputs for obtaining the economical evaluation and the life of coal seam. Coal seams are usually uniform layers between sedimentary formations Figure 1.

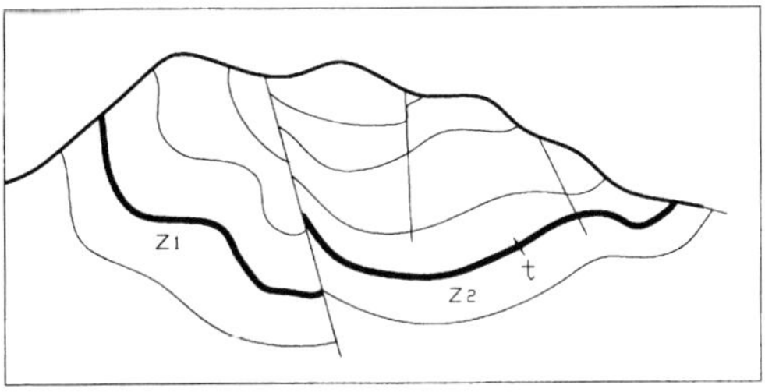

FIG. 1. Coal seam in sedimentary formations.

Its thickness regarding to its extension in two-dimension is very small usually about 30 cm for thin layer to 250 cm for thick layer. The coal seams can be considered as surfaces with uniform thickness and the other dimensions (length and width) of several kilometers. By the drilling exploration the point data of the upper and lower surfaces of coal seam can be determined. Then the functions or equations of the limiting surfaces can be developed by method of interpolation and approximation. The weight of coal for each seam can be calculated from the following equation,

$$W = \int_D \sqrt{f_x'^2 + f_y'^2 + 1}\; t(x,y)\; \gamma(x,y)\; dA \qquad (1)$$

Where

W = weight of coal seam,

f = f(x,y) = the functions of the surface of coal seam,

t = t(x,y) = the thickness of coal layer, and

$\gamma = \gamma(x,y)$ = the unit weight or the average unit weight of coal seam.

The surface function f(x,y) can be obtained by two-dimensional interpolation analysis using field data from core drilling exploration. For t and g constants or their average values the integral of Equation (1) becomes simpler to calculate. If the coal layer surface can be approximated by one-dimensional interpolation analysis then Equation (1) can be simplified as,

$$W = \int_{l} \sqrt{f_x'^2 + 1}\ t(x)\ \gamma(x)\ dx \tag{2}$$

In the analysis of subsurface groundwater flow and water supply the volume of recoverable water in unconfined and confined aquifers is very important to be determined. This is necessary for consumption uses such as drinking water, using water for agricultural and industrial purposes, especially for big cities. For the simplicity of calculations the boundaries of aquifers are usually assumed as planar or smooth surfaces. This is not true in most cases.

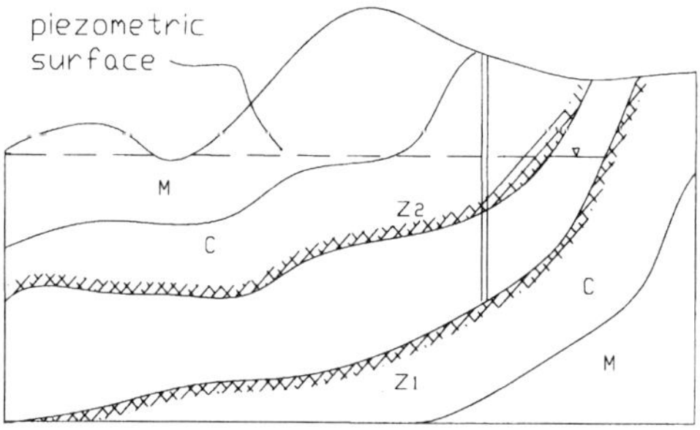

FIG. 2. Confined aquifer nonuniform boundaries

In reality these boundaries besides water table surface are not planar but curve surfaces, Figure 2. This is because of bending and folding of the geological formations having water deposits. For the correct and accurate evaluation of volume of water available in aquifers this should be done by integrating the difference between the upper surface and lower surface boundaries functions as Equation (3). For case of unconfined aquifer the upper surface boundary is not usually planar but curve surface that follows topography shape of ground surface.

$$v = \int_{D} [f(x,y) - g(x,y)]\ \Phi(x,y)\ dA \tag{3}$$

Where

v = volume of water existed in aquifer,

f (x,y) = the function of upper impermeable surface boundaries,

g (x,y) = the function of lower impermeable surface boundaries, and

Φ(x,y) = porosity of aquifer.

The functions of f(x,y) and g(x,y) can be obtained by the two-dimensional interpolation analysis from the governing field data. If the aquifer geometry be uniform in one direction or if this can be assumed, then the integration of Equation (3) can be simplified as follows,

$$v = \int_{l} [f(x) - g(x)]\, \Phi(x)\, w\, dx \,, \tag{4}$$

where w is the thickness of aquifer in direction of y coordinate.

In the same way as above procedure it is possible to determine ore deposit volumes that have nonregular and complex geometries. This is possible by obtaining the functions of upper and lower boundaries of ore deposit by using the interpolation approach of field data and applying the equations similar to Equations (3) and (4). Therefore the volumes of underground caves for gas or oil storage also for hazardous materials disposal can be determined accurately, Figure 3.

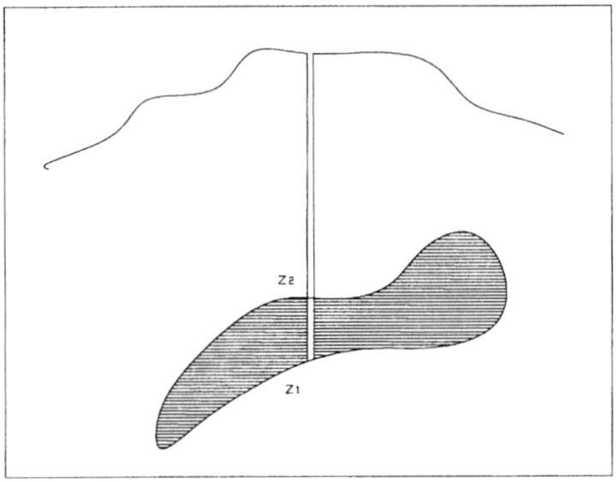

FIG. 3. Underground nonuniform ore deposit.

The most popular methods for interpolation of data are Lagrange method, Neville iteration approach, Newton divided difference methods, Cubic spline, Hermit spline, Bspline and Bezier methods. For the Neville and Newton divided difference method, the higher degree of polynomial generation is straightforward. In cubic spline approach each curve consists of a number of about n segment curves. The general way is to divide the interval to collect subintervals or segments and construct different approximating polynomial on each interval. Approximation by functions of this type is called piecewise polynomial approximation. This method has the advantage of removing the oscillatory nature of high degree polynomials. In this approach there exists the continuity of C^2 on the segments boundaries. However; the continuity order is optional to users willing. The formulation for creating cubic spline curve results in the solution of a three-diagonal system of equations. Hermit interpolation is based on two points p_1 and p_2 and two tangent values $p_1^{'}$ and $p_2^{'}$ at those points. It computes a curve segment that starts at p_1, going in direction to $p_1^{'}$ and ends at p_2 moving toward

$p_2^{'}$. Hermit interpolation has an important advantage. The Hermit curve can be modified by changing the tangent values.

The Bezier curve is a parametric curve p(t) that is a polynomial function of the parameter t. The degree of the polynomial depends on the number of points used to define the curve. The method applies control points and presents an approximating curve. The Bezier curve does not pass through internal points but the first and last points. Internal points influence the direction and position of the curve by pulling it toward themselves Conte & de Boor (1981), Saxena & Sahay (2005) and Salmon (2006).

B-spline methods were first proposed in the 1940 for curves and surfaces. The B-spline curves can be approximating or interpolating curves. The advantage of B-spline curves to Bezier curve is the obtaining the higher continuity for the individual spline segments Conte & de Boor (1981), Burden & Fairs (1989), March (2005) and Saxena & Sahay (2005). If there are a set of triple data $c = \{(x_i, y_i, y_i^{'}), i = 0,1,2,\ldots,n\}$ The cubic spline functions $S_i(x)$ can be obtained on each interval $[x_i, x_{i+1}]$. With this model the continuity of C^1 exists on the boundary of each segment. The spline function of degree 3 for each interval is,

$$s_i(x) = a_i + b_i\, x + c_i\, x^2 + d_i\, x^3 \tag{5}$$

If the Equation (2) is written based on the first point of each segment $[x_i, x_{i+1}]$ then,

$$s_i(x) = a_i + b_i\,(x - x_i) + c_i\,(x - x_i)^2 + d_i\;(x - x_i)^3 \tag{6}$$

The first derivatives of the spline function (3) is as follows,

$$s_i^{'}(x) = b_i + 2\,c_i(x - x_i) + 3\,d_i\,(x - x_i)^2 \tag{7}$$

The coefficients a_i, b_i, c_i and d_i for each interval can be obtained explicitly from the Equation (8).

$$\begin{cases} a_i = f(x_i) = f_i \\ b_i = f^{'}(x_i) = f_i^{'} \\ c_i = \dfrac{3}{h_i^2}(f_{i+1} - f_i) - \dfrac{1}{h_i}(f_{i+1}^{'} + 2f_i^{'}) \\ d_i = -\dfrac{2}{h_i^3}(f_{i+1} - f_i) + \dfrac{1}{h_i^2}(f_{i+1}^{'} + f_i^{'}) \end{cases} \tag{8}$$

The problem with this approach is that we generally don't have the tangent values at n+1 points. This can be provided by applying the following approximation equations for the first derivatives at the points and for uniform intervals.

$$\begin{cases} f_i' = \dfrac{1}{2h_i}[-\underset{\uparrow}{3}, 4, 1] \\ f_i' = \dfrac{1}{2h_i}[-1, \underset{\uparrow}{0}, 1] \\ f_i' = \dfrac{1}{2h_i}[1, -4, \underset{\uparrow}{3}] \end{cases} \tag{9}$$

The following two examples show the interpolation of two sets of data by the above approach.

Example 1

This example is in the field of groundwater engineering. The problem is about unsteady radial flow in isotropic leaky artesian aquifer with fully penetrating wells with water released from storage in aquitard. Where a well penetrating an artesian aquifer overlain by an aquitard and underlain by an aquiclude. The overlying aquitard are permeable deposits in which there is water table. The aquifer is homogeneous, isotropic, infinite in areal extent, and is of uniform thickness. The well completely penetrates the aquifer, and flow in the aquifer is radial throughout. The partial differential equation governing for this problem in polar coordinate is, Walton (1970),

$$\frac{\partial^2 s}{\partial r^2} + \frac{1}{r}\frac{\partial s}{\partial r} + \frac{P'}{T}\frac{\partial^2 s}{\partial z^2} = \frac{S}{T}\frac{\partial s}{\partial t} \quad , \tag{10}$$

With the boundary conditions,

$$\begin{cases} s(r,0) = 0 \\ s(\infty, t) = 0 \\ \lim_{r\to 0} r\dfrac{\partial s}{\partial r}(r,t) = -\dfrac{Q}{2\pi T} \end{cases} \quad . \tag{11}$$

The asymptotic solution for the above differential equation is,

$$s = \frac{Q}{4\pi T}\int_u^{\infty}\frac{e^{-y}}{y}\, erfc\left[\frac{\psi\sqrt{u}}{\sqrt{y(y-u)}}\right]dy = \frac{Q}{4\pi T}W(u,\psi), \tag{12}$$

where $u = r^2 S/Tt$, $\psi = \frac{r}{4}\sqrt{S'P'/TSm'}$, W(u) is well function, s = drawdown, r = distance from pumped well to observation point, t = time after pumping started, T = transmisivity coefficient, S = storage coefficient of aquifer, P' = hydraulic conductivity of aquitard, m' = aquitard thickness, and S' = storage coefficient of aquitard.

The data for interpolation have logarithm values of W(u, y=2) respect to 1/u Walton (1970). For the interpolation of data about four control points are considered. Figure 4 shows the comparison between the data and the model for the three spline curves. Based on the figure there is a suitable and good relationship between those.

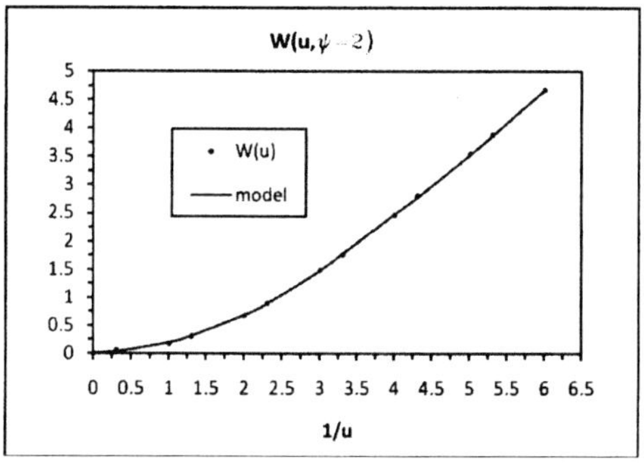

FIG. 4. The comparison between the data and the model.

Example 2

In this problem about five pairs of data (control points) are obtained from the parametric Equation (13) with non-uniform segments.

$$\begin{cases} x = t\cos\dfrac{\pi}{12} - 2\sin\sqrt{2t}\sin\dfrac{\pi}{12} \\ y = t\sin\dfrac{\pi}{12} + 2\sin\sqrt{2t}\cos\dfrac{\pi}{12} \end{cases}, \quad t \in [0,24] \tag{13}$$

About four cubic spline curves are generated according to the formulations of Equations (6), (8) and (9). Figure 5 shows the relationship between the real values and interpolated ones. As it can be seen, there is a relative expectable fitness between them.

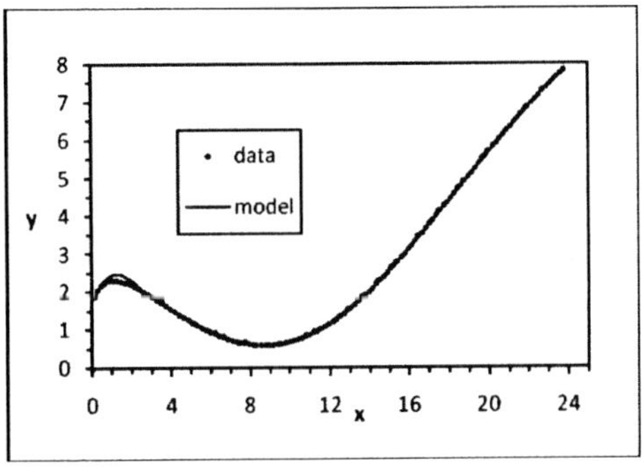

FIG. 5. The comparison between the data and the model.

RESULTS AND CONCLUSIONS

In this research two models for 1-D and 2-D interpolations were presented (the two-dimensional part is omitted from this paper because of pages limitation). The models are easy and straightforward to handle. Therefore; it can be applied for problems having huge data. It has the advantage of less computational efforts respect to the cubic spline, Bezier and B-spline methods especially for 2-D problems. The continuity for each segment is C^1 therefore; this is sufficient for many problems such as calculation the geometry of nonuniform bodies. It is recommended to compare the 2-D model with Bezier and B-spine methods respect to the fitness and accuracy.

References

Burden R., Fairs D. (1989) .Numerical Analysis. PWS-KENT publishing company, Boston, USA.

Conte S. D., de Boor C. (1981). Elementary Numerical Analysis, an Algorithm Approach. Kin Keong printing CO. PTE. LTD., Republic of Singapore.

March D. (2005). Applied Geometry for Computer Graphics and CAD. Springer-Verlag, 2nd Edition, USA.

Salmon D. (2006). Curves and Surfaces for Computer Graphics. Springer, USA.

Saxena A., Sahay B. (2005). Computer Aided Engineering Design. Springer, Anamaya publisher, New Delhi, India.

Walton W. C. (1970). Groundwater Resource Evaluation. McGraw-Hill Comp., New York, 1st Edition, USA.

STABILITY ANALYSIS OF THE FRONT SLOPE OF SMALL-DISTANCE HIGHWAY TUNNEL WITH VERY LARGE SECTION BASED ON 3D MONITORING

Dongwu XIE[1,2] and Wenqi DING[1,2]

[1]Department of Geotechnical Engineering, Tongji University, Shanghai 200092, China

[2]Key Laboratory of Geotechnical and Underground Engineering of the Ministry of Education, Tongji University, Shanghai 200092, China

Abstract: Landslide due to highway tunneling is a threat to the site safety and the stability of tunnel lining, and leads to large economic losses. As to the tunnel front slope deformation behaviors of small-distance highway tunnel, it is more complex for the tunnel of large section than that of small section. Generally, the small-distance tunnel with large section will experience many complex construction steps and many support transformations, therefore, the excavation of workface will disturb the slope many times. Longtoushan tunnel, which is located in Guangdong Province of China, is a four-lane large section highway tunnel. With consideration of weathered rock and fracture zone existing in surrounding rock of the tunnel, the both side drift method has been adopted. 3D (three-dimensional) monitoring has been introduced to monitor the movement of the front slope. The basic characteristic of 3D deformation of the slope has been analyzed to estimate the stability of the slope. The results show that: (1) the deformation of the front slope is of obvious 3D features and with no significant correlation between horizontal displacement and vertical settlement; (2) the deformation of the front slope is mainly influenced by tunneling; (3) the front slope can be divided into three regions with different characteristics of displacements; and (4) the displacement characteristics of the front slope depends with great degree on the type of slope rock. Accordingly, based on monitoring results, the construction schedule and the construction program have been often adjusted to ensure safety.

Key words: highway tunnel; front slope; 3D monitoring; very large section; small-distance

INTRODUCTION

China is a mountainous country. There are more and more highway tunnels are constructed during the development of road traffic. In order to raise the standard of highway, large section tunnel often becomes options. Three-lane tunnels have been constructed in many mountainous highways in China. Four-lane tunnel is also on the emergence.

The NATM method has been adopted in tunnel construction. Monitoring, as one of three pillars of NATM, has been given full attention by all related construction parties (XIA Caichu et al. 1999). The stability of the front slope is very important in tunnel construction with bad rock mass and shallow depth. At present, the Technical Specifications for Construction of Highway Tunnel (2009, in Chinese), a national code for road tunnel of China, does not give clear suggestions for the monitoring requirements of the front slope of tunnel.

The influence factors of the stability of highway tunnel front slope are different from natural slope and that of other types. Construction disturbance greatly influence the stability of the front slope. WANG Jianxiu et al. (2006) has made some analysis of the Xiaomansa Tunnel front slope of Yuanmo Highway, Yunnan Province of China, to give some advises for the slope treatment based on the 3D monitoring data. LI Xinxing et al. (2007) has established FEM model to analyze the influences of construction acting on the twin-tunnel front slope and concluded that it is opportune for the stability of the front slope to excavate the external line of the twin-tunnel with unsymmetric load. JIA Minghui (2007) has summarized the failure modes of the tunnel front slope of different sections but ignored the 3D characteristics of the slope displacements.

Based on the investigation of the construction of very large section tunnel (HUANG Chengzao et al. 2008), the four-lane tunnels had been constructed are almost in the hilly regions of the developed cities in the south-east China. The distance of the tunnels of two adjoining lines is comparatively small because of constrained construction site condition. At the same time, there used to be a lot of accidents about the front slope because its significance, complexity, randomicity, difficulty, and riskiness have been underestimated for its provisionality. So it is very important to make intensive research on the displacement features of the front slope of small-distance road tunnel with very large section.

The main objective of this research is to analyze the stability of the front slope of highway tunnel based the 3D monitoring data. The conclusion may be utilized as reference in similar projects.

PROJECT OVERVIEW

Longtoushan Mountain tunnel, which belongs to the Guangzhou east-ring segment of TONGSAN, JINGZHU national highway, is a four-lane large section highway tunnel. The layout of the tunnel comprises two separate lines with biggest excavation area of the tunnel of 241 m^2 and excavation width of 21.60 m. The right line has a length of 1006 m while 1010 m the left one. Maximum buried depth of the tunnel is 98 m. The smallest distance between the two separate lines (left and right line) is 23.0 m in the intake and 20.8 m in outtake. The rock mass in the tunnel entrance is completely and strongly weathered adamellite and slightly weathered adamellite at the most part of the tunnel. There are two fault zones in the tunnel site, Wenchong Fault and Shougouling Fault. The geological condition of the grade V rock based on Chinese national code at tunnel exit is poor. In this section, the both side drift method has been adopted which is shown in Fig. 1.

FIG. 1. Both side drift method of Longtoushan Tunnel

MONITORING AND ANALYSIS OF THE FRONT SLOPE'S STABILITY

1 Numerical simulation of the displacement tendency of the front slope

Finite element method (FEM) model had been established to analysis the displacement tendency of the front slope(HUANG Chengzao et al. 2008). The monitoring program had been modified based on the displacement trends.

(1) Analysis principle

Strength reduction method has been used to simulate the slip trend of the front slope in FLAC3D, which is one of the famous geotechnical analysis software.

(2) Parameters

Table 1. Parameters of FEM model

Parameters	Elastic modulus E (GPa)	Poisson's ratio ν	Specific weight γ (kN/m^3)	Cohesion C (kPa)	Friction angle Φ(°)
Residual soil	0.4	0.37	16.3	21	23.58
completely and strongly weathered adamellite	1	0.34	18	40	25

(3) The results

The safety factor of the front slope in nature state is 1.24. Fig. 2 shows the displacement vector of the front slope.

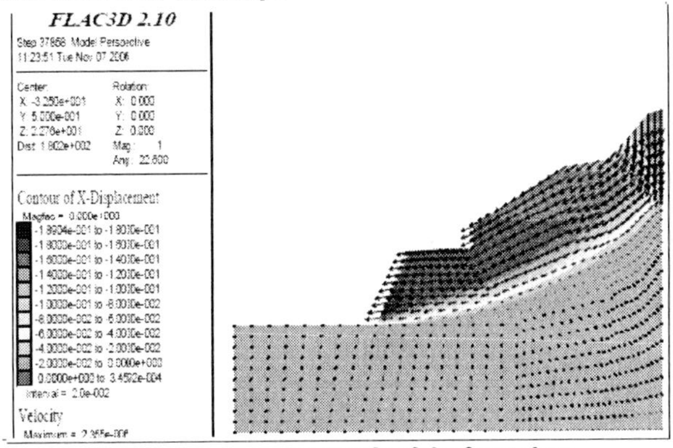

FIG. 2. Displacement trends of the front slope

Based on the FEM results, it can be found that the front slope slide in the direction to the portal when tunneling. Thus, it is obviously insufficient to only monitor the settlement of the front slope, so monitoring the 3D displacement of the front slope using Total Station has been added to the monitoring program.

2 Monitoring scheme

There consists of two group of points on the front slope of the tunnel to monitor the stability of the rock mass above the two lines, RP points and LP points. RP means the point is on the front slope of right line and LP means on the left one. Fig. 3 is the schematic layout of the monitoring points. In Fig. 3, the Y direction is along with the tunnel axis direction from the outtake to the intake; the X direction is perpendicular to the tunnel axis in the horizontal; the Z direction is perpendicular both to the X and Y direction.

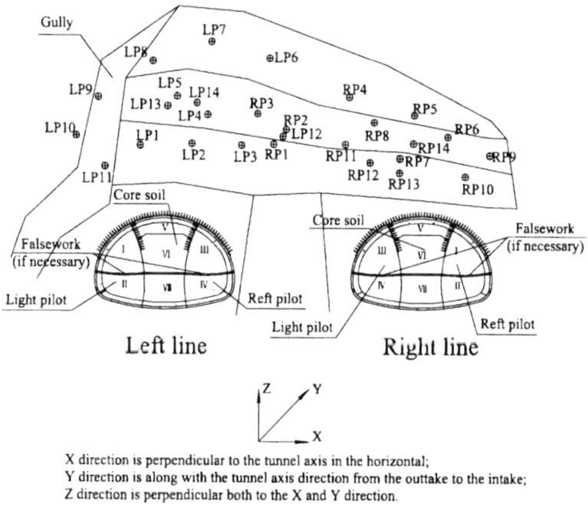

FIG. 3. Layout of the front slope monitoring points

3 Monitoring data and arrangement of construction status

The monitoring of the Longtoushan Tunnel had lasted for twenty months. There are 140 groups of data had been obtained and some of which is shown in Fig. 4 and Fig. 5.

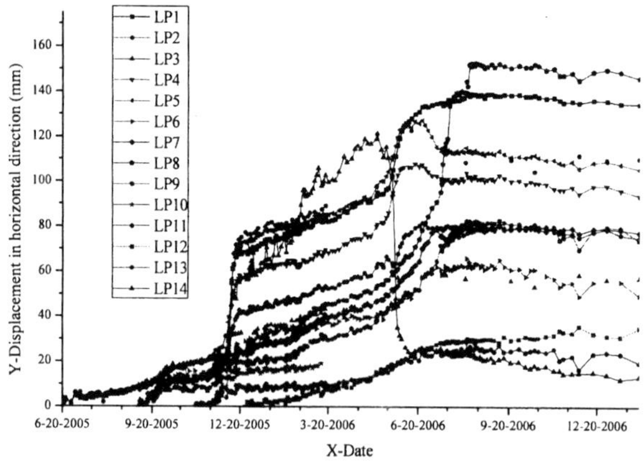

FIG. 4. Displacement-time curve in horizontal of the monitoring points on left-front-slope

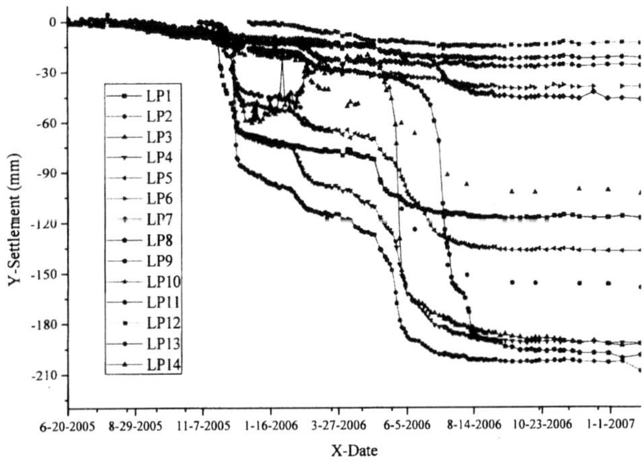

FIG. 5. Settlement-time curve of the monitoring points on left-front-slope

Construction progress curve of the left outtake has been drafted based on the construction progress which is shown in Fig. 6.

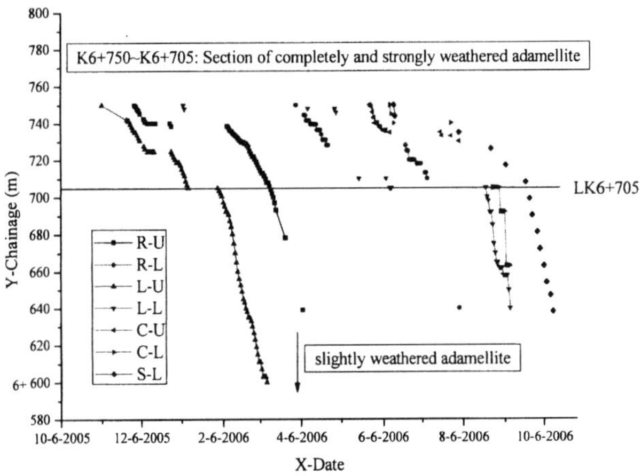

FIG. 6. Construction progress curve of the left entrance

(R-U means the upper bench of the right pilot of the left line; R-L means the lower bench of the right pilot of the left line; L-U means the upper bench of the left pilot of the left line; L-L means the lower bench of the left pilot of the left line; C-U means the upper bench of the core soil of the left line; C-L means the lower bench of the core soil of the left line; S-L means the second lining of the left line.)

4 Trend analyses of the monitoring data

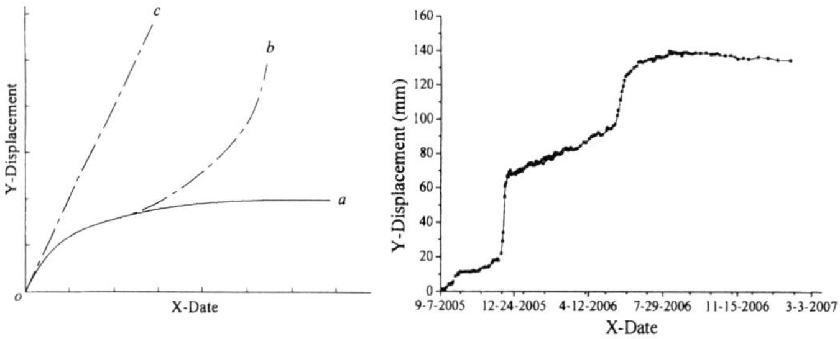

(a) (b)

FIG. 7. Shape of surrounding rock deformation time-history curve

When the method of the whole section excavation is used in tunneling, the trend of the displacement curve is shown as ***curve a*** under the good surrounding rock. The ***curve b*** in Fig. 7 (a) appears when the surrounding rock is fractured or the disturbance is strenuous in the follow-up construction. When the rock mass is completely and strongly weathered and the support is delayed after excavation, the ***curve c*** in Fig. 7 (a) can be represented. The shape of the curves in Fig. 7 (a) is very simple and it is easy to fit the curve using simple function and predict the trend.

But the trend and the shape of the curve will become very complex and hard to forecast when segment excavation method is adopted in tunneling for some inflection points in the curve. As shown in Fig. 7 (b), it is impossible to fit the displacement curve using the simple exponential function, logarithmic function, hyperbolic function, nonlinear polynomial function and so on. ZHOU Jianchun et al. (2002) had found that BP neural network algorithm is a good method in fitting and forecasting the trend of the complex curve like the Fig. 7 (b). Maybe multivariate approximation is also a suitable method because there are many factors such as time, excavation, disturbance of blasting, support rigidity, supporting opportunity, which contributing to the development of the displacement curve, but this topic is beyond this paper.

5 Spatial analysis of the displacement regularity of the monitoring points

(1) As Fig. 8 shows, the horizontal displacement of the left line is less than that of the right one. Because there are many big boulders in the rock mass of the right line, the drilling blast method had been used in excavation which had brought more disturbance than manual work, while the rock mass of the left line is in shatter, drilling blast method is not need and manual work had been used. It can be guessed that excavation method is the reason for the displacement difference.

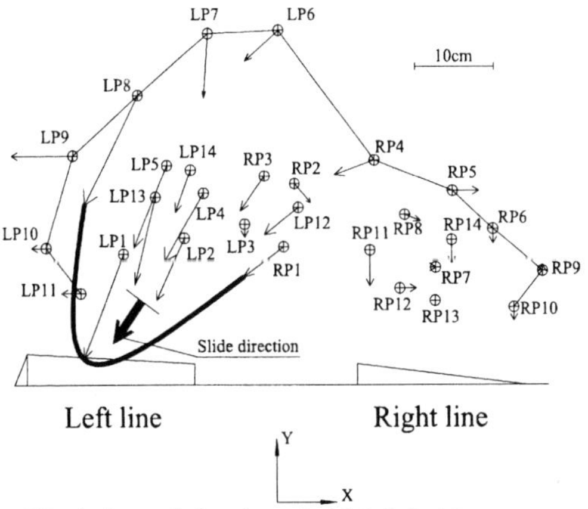

X direction is perpendicular to the tunnel axis in the horizontal;
Y direction is along with the tunnel axis direction from the outtake to the intake.

FIG. 8. Horizontal displacement trends of monitoring points (Top view)

FIG. 9. First lining instability caused by landslide

(2) The settlements of the monitoring points are showed in Fig. 10. The settlements of the points above the lines are bigger than that of the both sides' and the settlement trough had been taken shapes clearly.

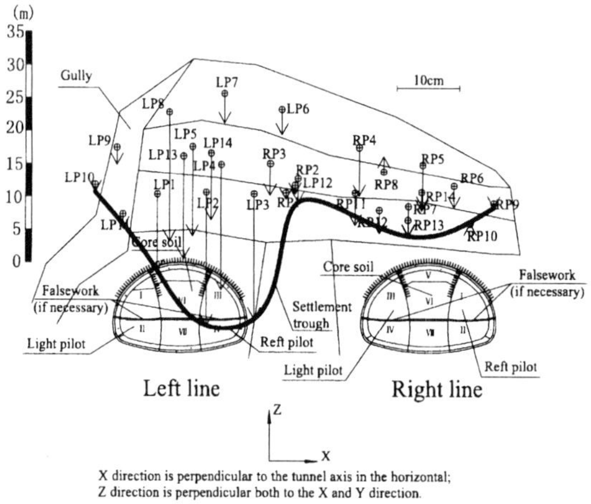

FIG. 10. Settlement trough of monitoring points (Front View)

(3) The ratio of the displacement vectors of horizontal plane and the vectors of settlement is been shown in Fig. 11. There is no regularity can be drawn from the picture. That is to say there is no correlation between the displacement vectors of horizontal plane and the vectors of settlement. It is impossible to judge the stability of the front slope based on the settlement of monitoring points. So it is essential to modify the monitoring scheme so that the horizontal displacement can be obtained.

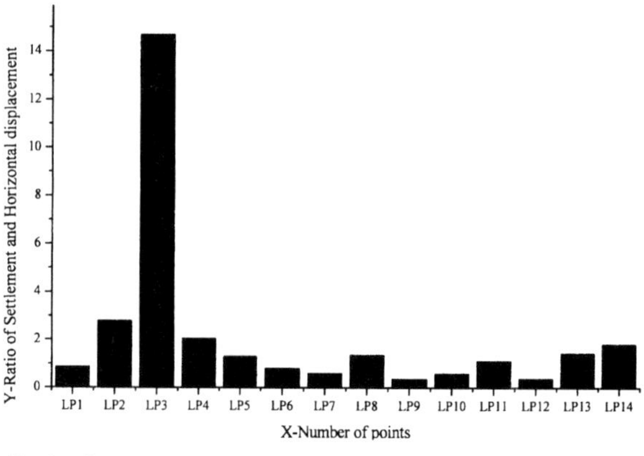

FIG. 11. Ratio of settlement and horizontal displacement of monitoring points on left-line

(4) The distance between the two adjoining lines of Longtoushan tunnel is very small. The excavation of one line can make the displacement of the front slope on the other one. The displacement trend of the front slope of the small-spacing tunnels had become more complex. The front slope can be divided into three areas based on the analysis of monitoring data.

It is shown in Fig. 12. The displacement trend of monitoring points in the Ⓐ area showed that they had been influenced by the excavation of both lines. But the excavation of right line had more influences on the monitoring points in the Ⓐ area. The © area is to the contrary. The displacement trend of monitoring points in the Ⓑ area also showed that they had been influenced by the excavation of both lines.

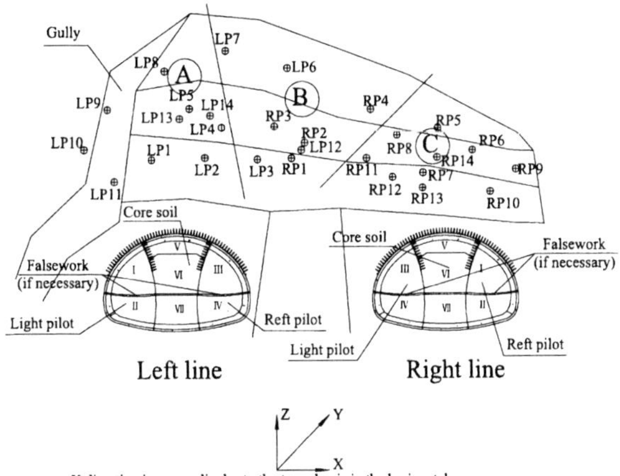

X direction is perpendicular to the tunnel axis in the horizontal;
Y direction is along with the tunnel axis direction from the outtake to the intake;
Z direction is perpendicular both to the X and Y direction.

FIG. 12. Division of Slope monitoring points (Front View)

(5) According to the recording of the excavation, the excavation of whole section of the left line finished in July 14, 2006 which has been shown in Fig. 6. After completion, the front slope tends to stabilization. Then the excavation section could be divided into two parts, sensitive to excavation part and insensitive to excavation part. In the sensitive part, the disturbance of construction should be controlled strictly in order to avoid dangerous situations.

CONCLUSION

(1) The stability of the front slope of tunnel, which is related to the safety of participants and the smooth construction of the project, should be paid sufficient attentions. The horizontal displacement trend of the front slope is the direct information to judge the possibility of landslide.

(2) The settlement of front slope has close relations with the excavation in shallow-buried section. The settlement trough had been taken shape on the surface. The surface settlements along the axis of tunnel are bigger than that of the both sides.

(3) The size of displacement is related to the construction process, excavation disturbance, and the condition of rock mass. The construction control measures should be timely adjusted in the sensitive part to ensure the stability of front slope and the safety of constructor.

(4) There is no correlation between the displacement vectors of horizontal plane and the vectors of settlement. The horizontal displacement and the settlement should be monitored separately.

(5) The displacement trend of the front slope of the small-spacing tunnels is more complex. The front slope can be divided into three areas based on the analysis of monitoring data.

ACKNOWLEDGMENTS

The authors thank Dr. LAI Yunjin (Tongji University, China) for revising the manuscript. The work described in this paper was carried out with the financial support of the "Shanghai Leading Academic Discipline Project, Project Number: B308" and "Science and Technology Project of Western Transportation Construction, Project Number: 2007-318-799-80".

REFERENCES

Huang Chengzao, Zhu Hehua, Tan Zhongsheng et al (2008). "Technical report of the research on the key technologies in the construction of tunnels with twin bore and eight lanes". HuangPu Bridge of Pearl River Construction Ltd. in Guangzhou, Guangzhou. (in Chinese).

Xia Caichu, Li Yongsheng. (1999). "Testing Theory and Monitoring Technique for Underground Engineering". *Shanghai: Tongji University Press.* (in Chinese).

The Professional Standards Compilation Group of People's Republic of China. (1995)"JTJ 042-94 Technical specifications for construction of highway tunnel". Beijing: *ChinaConununications Press.* (in Chinese).

Wang Jianxiu, Tang Yiqun, Zhu Hehua, et al. (2006). "3D Monitoring and Analysis of Landslide Deformation Caused by Twin-arch Tunnel". *Chinese Journal of Rock Mechanics and Engineering*, Vol. 25(11), 2226-2232. (in Chinese).

Li Xinxing, Zhu Hehua, Cai Yongchang et al. (2007). "Influence of Construction Method of the Twin Tunnel on Slope Stability". *Chinese Journal of Underground Space and Engineering*, Vol. 3(3): 483-487. (in Chinese).

Jia Minghui. (2007). "Research on the Slop Stability and Control Technique at the Tunnel Entrance". Shanghai: *A dissertation submitted to Tongji University in conformity with the requirements for the degree of Master*, 2007.3. (in Chinese).

Zhou Jianchun. (2002). "Study on fitting of surrounding rock deformation time-history curve by BP neural network algorithm". *Journal of Hunan University (Natural Sciences Edition)*, Vol. 29(4), 79-84. (in Chinese).

Research on 3D Numerical Model of Segment Lining for Large-section River-crossing Shield Tunnel

JIANG An-long[1,2], YANG Zhao[2].

[1]School of civil Engineering, Nanchang Hangkong University, Nanchang 330063, China, jaltjdx@163.com
[2]Department of Geotechnical Engineering, Tognji University, Shanghai 200092, China, yangzhaolp@126.com

ABSTRACT: The design of tunnels requires a proper of surface settlements and lining forces. In engineering practice, different design methods tend to be used, varying from simple empirical and analytical formulations to advanced finite element analyses. This paper begins with a review of empirical and analyses for settlements and lining forces. Then, 3D Numerical model of continuous soil-shell-spring is presented so as to analyze the 3D member forces of the segmental linings of shield tunnel. With the application of FEM, the model employs several types of elements, such as thick shell, GOODMAN, contact and solid elements, and concerns the extrusion between segments in the joints, the stiffness contribution of the joints, the interaction between segments and surrounding soil. Finally, based on 3D numerical model of continuous soil-shell-spring, the member forces and displacements of stagger-jointed segmental lining are computed separately for Shanghai Yangtze River Tunnel. The result of this dissertation will be provided to the staking company for working reference.

INTRODUCTION

The design of any tunnel structure has to fulfill some basic requirements concerning structure stability, durability, water tightness and reliability, low costs and low-risk profile. To achieve these goals it is necessary that engineer have ability to predict unwanted effects. Although it is difficult to translate the reality of tunnel lining behavior into a mechanical mathematical engineering model, it is absolutely essential that this be done.

Today, most design of shield tunnel is based on the guideline which is published on Tunneling and Underground Space Technology at 2000 by International Tunneling Association. Conventional structural models assume plane strain conditions for lining and ground. The stiffness of lining is considered as a constant. Active soil pressures on the lining are assumed to be equal to the primary stresses in undisturbed ground. In order to consider the interaction between soil and segment, there exists a bond between the lining and ground, both for radial and tangential deformations. Furthermore the material properties of soil and lining are generally assumed to be elastic. Grout pressure and grout material are not taken into account when modeling. This conventional model ignores the influences of segment and assembling imperfections, the construction imperfections, the construction method, the nonlinear property of joint material, jack forces, nonlinear effects, ground freezing, grouting, etc. That model can not be applicable the in stagger-jointed segmental lining.As a result, the three-dimension member force distribution in concrete segments using thick shell element develops with corresponding unwanted deformations .

Development of 3D-FEM model for shield driven tunnels

In order to consider the extrude action between segments, the fastening action of bolt, the nonlinear property of bending stiffness, as well as the interaction between soil and lining segment, a 3D numerical model of continuous soil-shell-spring is presented in this paper. In this model, a routine is coded on ANSYS secondary development platform which is used to consider the coupling effect on stress field and seepage field. The lateral and longitude structure characteristic is important to the large diameter shield tunnel , which can be reflected in this model exactly.

1. The model of shield tunnel

Shield tunnel is composed of segments. The segments are jointed together by longitude and circumferential bolt. There are two different methods to assemble the segment lining. The first method is continuous-jointed assemble. In this method, the longitude slots of adjacent lining ring are on a line. The second method is stagger-jointed assemble. In this method, the longitude slots of adjacent lining rings are not on a line. Because of there are a lot of joints and slots in the shield tunnel structure and the lining ring can be calculated as a thick shell, each lining segment can be considered as a thick shell.

1.1 Shell element

In element library of ANSYS, there are two shell elements, thick shell element and thin shell element. The thin shell element is established based on classical Kirchhoff-Love theory. The principal assumption of Kirchhoff-Love theory is that:

a. The deflection of the shell relative to the shell size is very small.

b. The normal of middle section maintains as a straight line and to be perpendicular to the middle section after deformation.

c. The value of the direction stress which perpendicular the middle section is too small to be negligible.

In this theory, the deflection is continuous and the differential of deflection is discontinuous. In order to solve this problem, Mindin-Reissner advanced a new theory. In this theory, deflection and rotation angle are interpolated separately; besides, the second hypothesis is modified as the normal of middle section which maintains as a straight line and can not be perpendicular to the middle section after deformation. The theory of Mindin shell is the thick shell theory which can consider the lateral shear deformation.

Because of the ratio of thick and width of lining segment is more than 1/5, the laterl shear deformation should be considered, so the thick shell element is adopted to simulate the segment in this paper. .

1.2 circumferential joint element

To simulate this discontinuous characteristic of shield tunnel, joint element is adopted.We can see from the predecessors' research that to exactly simulate the circumferential joint or not may have an important impact on the results. Zhu-hehua advanced the idea of using spring element model to simulate the joint. This model is based on discontinuous medium theory which introduces the Goodman element and thought the joint element can be tensioned to simulate the link action of bolt. In this paper, we also adopt spring element to simulate the circumferential joint and extend the spring element to 3D. Each circumferential joint should use four springs to simulate, using the first spring to simulate the axial compression in local coordinate system, the second to simulate the shear force transfer along longitude direction, the third to simulate the radial shear force transfer, and the last one to simulate the rotation around the longitude axis.

From the empirical data, the first, the second and the third springs may be considered as a constant coefficient spring. The stiffness of the last spring is not a constant, which vary with the member force of the joint. Through the joint experiments, we can generalize the stiffness of rotation as a function of rotation angle as follows:

$$k_{\theta_z} = \begin{cases} -4.21e3/2\sqrt{-\theta} & \theta < 0 \\ 2 \times 1.945e7\theta & 0 \le \theta < 0.0034 \\ 5.6e3/2\sqrt{\theta - 0.0034} & 0.003 < \theta \end{cases}$$

Under external loads, the relationship between internal force and deformation of each joint element node can be defined as follow:

$$\begin{Bmatrix} U \\ V \\ W \\ 0 \\ 0 \\ M_{\theta_Z} \end{Bmatrix} = \begin{bmatrix} K_x & & & & & \\ & K_y & & & 0 & \\ & & K_Z & & & \\ & & & 0 & & \\ & 0 & & & 0 & \\ & & & & & K_{\theta_Z} \end{bmatrix} \begin{Bmatrix} u \\ v \\ w \\ 0 \\ 0 \\ \theta_Z \end{Bmatrix}$$

Where U、V、W are the internal force of joint element in local coordinate system, $M_{\theta z}$ is the moment.

1.3 Longitude joint element

In the beam-spring model, there are two kinds of shear springs used to simulate the shear transfer between the adjacent rings, the first being radial spring and the second longitude spring in the local coordinate. In this paper, the longitude joint element is extended to 3D, so a kind of spring is added to simulate the axial extension and compression.

1.4 The contact element between soil and lining

In the convention model, bonds are used to simulate the interaction between the lining and soil. This model can not exactly simulate the interaction action between the lining and soil. In this paper, the contact element is used to simulate such interaction action.

According to the different relative stiffness of interaction materials, there are two types of contact element. The first type is used to simulate the interaction between rigid and deformable body, the second to simulate the interaction between deformable body-deformable bodies. In this paper, eight-node hexahedron solid element is selected to simulate the ground. A contact element is established between the shell element and the solid element which is used to simulate the ground. The slave surface of the contact element is the lining surface, the master surface of the contact element being the ground surface. This can effectively consider the compatibility of deformation and load transferring.

a. normal contact relationship

In theory, in the process of contact, the two elements are not allowed to invade each other, but in the numerical calculation, invasion is not inevitable. If invasion take place on the last iteration, there will be normal force on the penetration area.

b. tangent contact relationship

Coulomb friction is widely used to describe the tangent mechanical behavior in rigid-deformable body. In this relationship, tangent friction force is not allowed to excess the normal pressure multiply frictional coefficient. In the elastic contact analysis, the relationship of coulomb friction can be defined as follows:

$$|\tau| \leq \mu p + c$$

Where, τ is the equivalent shear stress, μ is the friction coefficient, P is the normal pressure and c is the cohesive force between lining and ground.

2 Construction process

Shield tunnel construction may be divided into four typical construction stages.

2.1 Excavation stage

At this stage, soil pressure is in balance with support pressure and with the friction force between the soil and the shield. Along the length of shield machine, the soil pressure around the shield machine can not be relaxed for the shield. The perturbation of shield driving can not be negligible to the ground.

2.2 Grouting stage

There are seven gap centimeters between the inner diameter and the external diameter. In order to reduce the soil deformation due to the gap, before the lining departure from the shield with the shield machine driving forward, cement slurry should be injected to the interspaces.

2.3 Departure from shield

Because of the departure from the shield, the gap between the ground and lining will be closed. There will occur unexpected settlement occur. The value of the settlement depends on the grouting technology.

2.4 Consolidate stage

Grouting material with time solidification and soil consolidation will cause the interaction between lining and ground. On the one hand, with the grouting material solidification, ground deformation of will be restrained by the lining structure; on the other, such deformation will increase with the time.

3 The simulation of shield driving process

By simulating shield driving process, the impact of shield driving on ground settlement can be obtained. In order to get the impact exactly, in simulation process we should comply with such assumption:

a. The soil's constitutive relationship is D-P plastic model.

b. Use pressure on driving face to simulate the motion of soil.

c. Apply grouting pressure to simulate the backfill grouting.

d. Change the parameters of grouting material is used to simulate the backfill effect of shield interspaces.

e. Use the fluid-solid coupling theory to calculate the seepage force which is used to describe the seepage in the tunnel and the partial seepage in the driving stage.

In this paper, the impact of shield driving process on soil disturbance is not simulated for simple calculation. We assume injecting grout immediately after shield tunnel excavation. In the simulation process, we just keep the excavation, backfill grouting and change the material's property to accomplish the simulation of shield driving process.

Engineering application

The length of Shanghai Yangtze River tunnel is 8.9km. The diameter of shield machine is 15.43m, which is the largest diameter in the world at this stage. Each ring is assembled by ten segments. In order to increase the whole stiffness of the tunnel and reduce the feasibility of creep, the stagger-jointed method is selected to assemble the segments.

Table 1. parameters of the ground

The number of ground	Bulk density γ(kN/m^3)	Passion's ratio μ	Deformation modulus E (MPa)	Internal cohesion c (kPa)	Internal friction angle φ(°)
①$_2$	17.3		6.78	11	28
①$_3$	18.6	0.26	7.18	9	27
④	16.8	0.35	1.41	12	10.5
⑤$_{3t}$	17.6	0.33	2.21	16	14.5
⑤$_3$	18.0	0.32	3.00	18	18
⑦$_{1-2}$	18.6	0.26	9.33	6	30
⑦$_2$	18.0	0.25	20	1	34.5
⑨	18.0	0.24	18	4	33.5

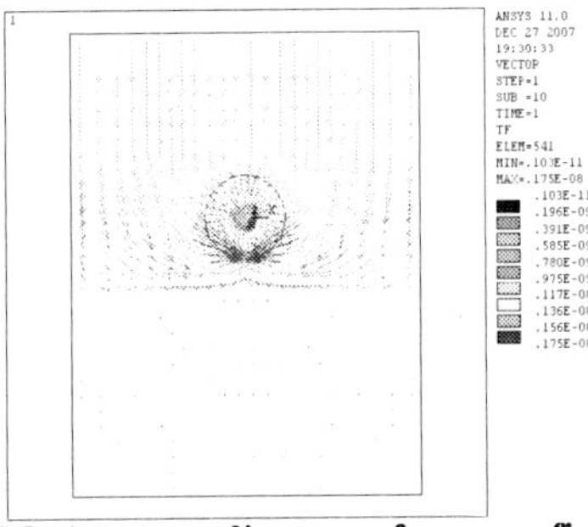

FIG. 1. vector diagram of seepage field.

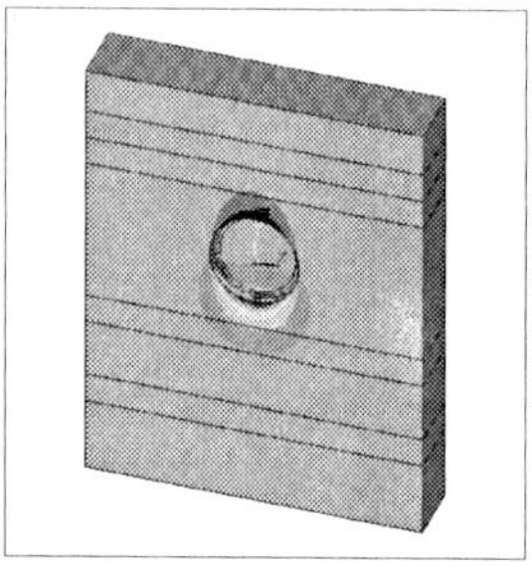

FIG. 2. Ground displacement after the first excavation

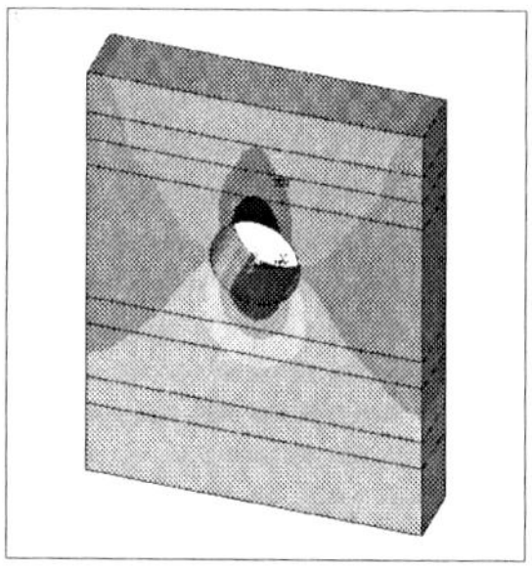

FIG. 3. Ground displacement after the sixth excavation

The maximum of settlement on crown is 27.9 centimeters, and the maximum of uplift on invert arch segment is 21.8 centimeters.

FIG. 4. The ground settlement after the first excavation

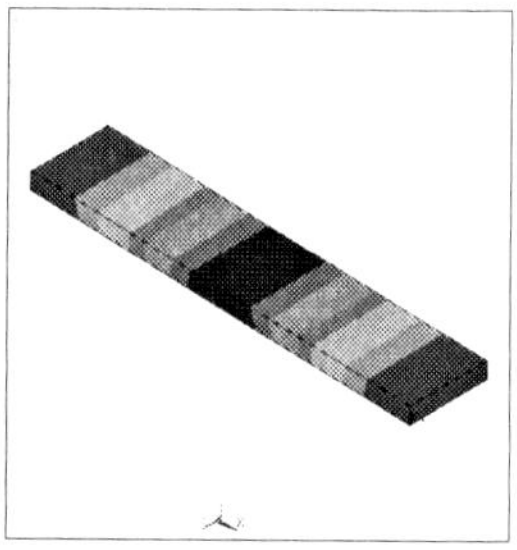

FIG. 5. The ground settlement after the sixth excavation

From fig.4, we can see that under the propellant force of shield machine, the surrounding ground will fall behind the shield machine and bulge before the shield machine. With the shield driving, the maximum settlement increases. At the end of the excavation, the value of maximum settlement is 15.99 centimeters.

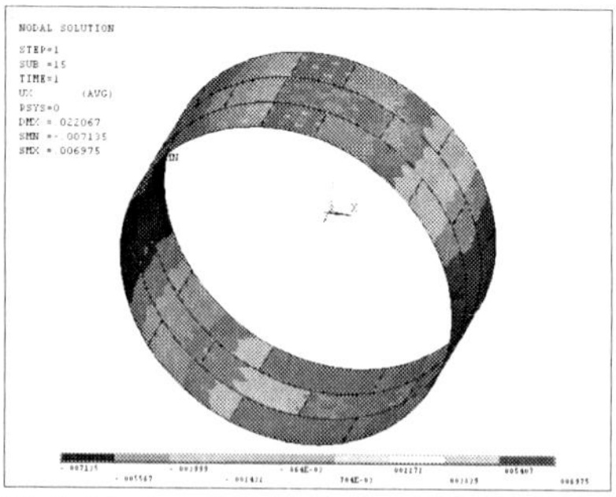

FIG. 6. Horizontal displacement of segment linings

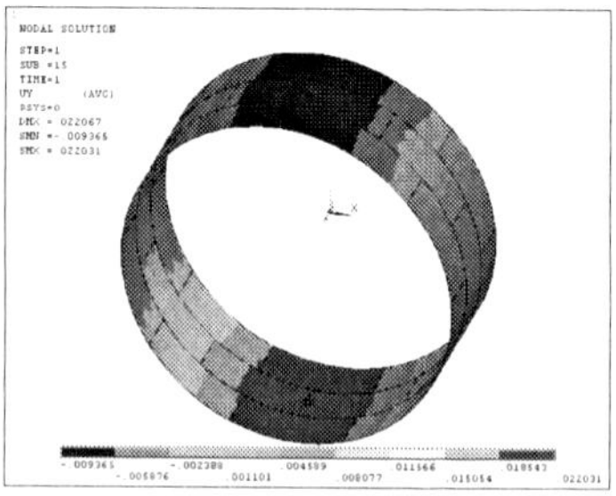

FIG.7. Vertical displacement of segments linings

The maximum horizontal displacement is in the direction of the horizontal diameter, with the value being 7mm. The maximum vertical displacement is located on the bottom of tunnel with the value being 2.20cm. The maximum displacement of the segment lining can not exceed the allowable deformation limits.

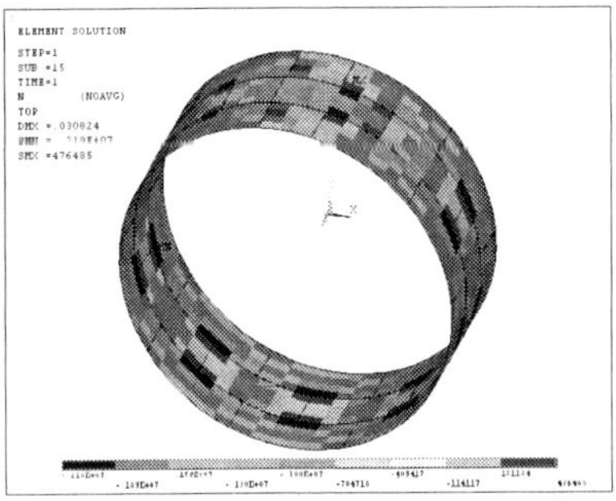

FIG. 8. Axial force of segment section

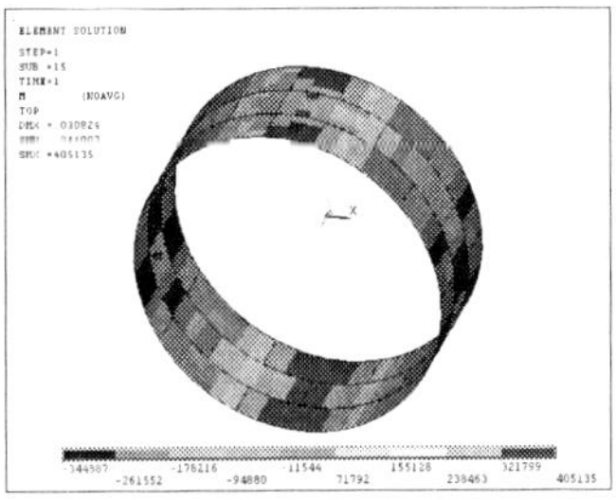

FIG. 9. Moment of segment section

The value of axial compression is negative. The maximum axial compression is located on the circumferential joint. Its -2180kN. The value of moment is positive when the inner segment is in tension state and the value of moment is negative when the outside segment is in tension state. The maximum positive moment is located at the top and bottom of the tunnel; the value is 405.13kN.m/m. The maximum negative moment is located on the horizontal diameter, the value being 344.89kN.m/m.

CONCLUSIONS

In cross-river calculation, we need to consider the interaction between soil and linings, the impact of stagger-jointed, the nonlinear of joint material, the grouting pressure, the fluid-solid coupling etc. In order to consider the impact of those, we employ several types of elements, such as thick shell, GOODMAN, spring element to simulate joint, solid element. By using such elements and the theory of fluid-soil coupling, 3D numerical model of continuous soil-shell-spring is presented in this paper to analyze the 3D member forces of the segmental linings of shield tunnel. The result of this dissertation will be provided to the staking company for working reference.

ACKNOWLEDGMENTS

The authors appreciate the support of the Shanghai Urban Construction Group.

REFERENCES

Adams J.(1996). "The engineering behavior of a Canadian Muskeg. Proceedings. "*sixth international Conference on soil mechanics and foundation engineering*: 1-37.

C.B.M.Blom, E.J.Van der Horst, P.S.Jovanovic.(1999). "Three-dimensional Sturctural Analyses of the Shield-Driven "Green Heart" Tunnel of the High-Speed Line South. " *J.Tunnelling and Udergound Space Tchnology*, Vol. 14(2): 217-224

C.B.M.Blom, van Oosterhout GPC.(2001). "Full-scale laboratory tests on a segmented lining. "*Summary Report*, Delft University.

Hellmich C.(1999). "Shotcrete as part of the New Austrian tunneling method: from thermochemomechanical material modeling to structural analysis and safety assessment of tunnels. "*Doctor Paper*, Technical University Vienna: 1999.

J P His, J C Small.(1992). "Analysis of Excavation in an Elasto-plastic Soil Involving Drawdown of the Water Table. " *J.Comp. & Geotech*, Vol.13: 1-19.

J P His, J C Small. (1992). "Simulation of Excavation in a Poro-Elastic Material. " *J.Int J Num. Anal. Meth Geomech*, Vol. 16: 25-43.

Kasper T, Meschke G.(2004). "A 3D finite element simulation model for TBM tunnelling in soft ground. " *J.Int J Numer Anal Meth Geomech*, Vol.28:1441-1460.

Komiya K, Soga K, Akagi H, Hagiwara T, Bolton MD.(1999). "Finite element modelling of excavation and advancement processes of a shield tunnelling machine. " *J.Soils Foundations*, Vol.39(3):37-52.

Lee KM, Ji HW, Shen CK, Liu JH, Bai TH.(1999). "Ground response to the construction of Shanghai metro tunnel-line 2. " *J.Soils Foundations*, Vol.39(3):113-134.

Melis M, Medina L, Rodriguez JM. (2002). "Prediction and analysis of subsidence induced by shield tunnelling in the Madrid Metro extension. " *J.Can Geotech*, Vol.39:1273-1287.

Mesri G, Choi Y K.(1985). "The uniqueness of end of primary(EOP) void ratio effective stress relationship. " *11th International Conference on soil mechanics*: 587-590.

Mesri G, Godlewski P M.(1995). "Time-and stress-compressibility interrelationship", *Geotechnical Engineering*(GT5), ASCE 103: 417-430.

Ronaldo I Borja.(1992). "Free Boundary, Fluid Flow, and Seepage Force in Excavations. " *J. of Geot. Eng*, Vol. 118(1):125-146.

Working Group No.2, International Tunneling Association.(2000). "Guidelines for the design of shield tunnel lining." *J.Tunneling and Underground Space Technology*, Vol.15(3):303-331.

Zhu Hehua, T. Hashimoto el al.(1996). "Back-analysis for shield tunnel using beam-joint model. " *Geotechnical Aspect of Underground Construction in Soft Ground*: 15-17.

Upper Bound Solutions for the Face Stability of Shallow Circular Tunnels Subjected to Nonlinear Failure Criterion

Fu Huang[1], Xiao-Li Yang[2]

[1]Doctor candidate, School of Civil and Architectural Engineering, Central South University, Changsha 410075, China, Email: hfzndx2002@yahoo.com.cn
[2]Professor, School of Civil and Architectural Engineering, Central South University, Changsha 410075, China, Email: csuyang@yahoo.com.cn

ABSTRACT: Based on the previous failure mechanism, the face stability of shallow circular tunnels is investigated using the upper bound theorem of limit analysis in conjunction with the nonlinear failure criterion. By using generalized tangential technique, the nonlinear failure criterion is introduced to optimize the retaining pressure of tunnel face. The optimum value of retaining pressure of tunnel face is obtained by employing nonlinear programming which is characterized by two variables with constraint. In order to evaluate the validity of the numerical results, the optimum solutions of this paper are compared with the value of retaining pressure calculated by linear failure criterion. The agreements indicate the proposed method is effective. The numerical results show that the nonlinear failure criterion has an important influence on the retaining pressure of tunnel face.

INTRODUCTION

Owing to little influence on environment, and unrestricted driving for ground environment, shield tunneling method has taken the dominant place among various construction methods of subway tunnel. However, how to keep the stability of tunnel face is a key factor in shield tunnel construction. Consequently, the stability analysis of the front of a shield tunnel driven in shallow stratum is a hot issue. Several authors (Soubra et al., 2008; Soubra, 2002; Leca and Dormieux, 1990; Lee and Nam, 2001) investigated the face stability of shallow tunnel by limit analysis method since Davis et al. (1980) studied the face stability of shield tunnel which was based on upper and lower bound theorem of limit analysis, and derived some useful conclusions. According to an upper bound approach in limit analysis and three-dimensional numerical modeling Li et al. (2009) studied the face stability of large slurry shield-driven tunnels, and concluded that the global blow-out of the partial upper part of the tunnel face occurs when the slurry pressure is too great while the global collapse of the whole tunnel face occurs when the slurry pressure is too small. Taking into account of seepage force, the face stability with steel pipe-reinforced multi-step grouting in underwater tunnels was evaluated by Lee et al. (2004) using limit analysis methods.

These above mentioned studies of stability analysis for tunnel are mainly focused on the linear Mohr-Coulomb criterion. However, numerous experiments and researches (Hoek and Brown, 1997; Agar et al., 1985) have demonstrated that the strength envelopes of geomaterials are nonlinear, and the linear relationship is just a special case. Therefore, how to analyze the face stability of shallow tunnel in the framework of limit analysis based on nonlinear criterion is a key issue confronted by many researchers (Park et al., 2008; Lee and Pietruszczak, 2008). By proposing a generalized tangential technique, Yang and Yin (2006) estimated the seismic passive earth pressures on the rigid walls in seismic conditions using the upper bound theorem of limit analysis in conjunction with the nonlinear yield criterion.

In this paper, based on the upper bound theorem, the stability of the front for a shallow tunnel driven in cohesive material with a nonlinear yield criterion is investigated by applying the generalized tangential technique. For the purpose of evaluating the validity, the retaining pressure σ_T of shallow tunnel face which is calculated with linear MC yield criterion and the results of σ_T in this paper with nonlinear MC yield criterion are compared.

NONLINEAR YIELD CRITERION AND TANGENTIAL TECHNIQUE

Based on different lateral confinement conditions for two types of rock, triaxial test is carried out by Agar et al. (1985), the results suggest that the relationship between maximum and minimum principal stress is nonlinear, and the nonlinear relationship can be expressed as

$$\tau = c_0(1+\sigma_n/\sigma_t)^{1/m} \tag{1}$$

where σ_n and τ are the normal and shear stress on the yield surface respectively, c_0 is the initial cohesion, σ_t is axial tension stress, and m is the nonlinear coefficient. For the purpose of introducing the nonlinear yield criterion to limit analysis, a generalized tangential technique is applied by Yang and Yin (2006). The tangential line to the nonlinear yield criterion in Eq.(1) can be described as

$$\tau = c_t + \sigma_n \tan\varphi_t \tag{2}$$

where c_t and $\tan\varphi_t$ are the intercept and slope of the tangent respectively. As a result, c_t can be expressed by the following equation.

$$c_t = \frac{m-1}{m} c_0 \left(\frac{m\sigma_t \tan\varphi_t}{c_0}\right)^{\frac{1}{1-m}} + \sigma_t \tan\varphi_t \tag{3}$$

FAILURE MECHANISM

In order to maintain the stability of tunnel face, it is necessary to provide a proper supporting pressure which is applied by the pressurized slurry in slurry shield-driven tunnel Therefore, it is significant to establish the minimum retaining pressure for tunnel face. Leca and Dormieux (1990) proposed two failure mechanisms, collapse and blow-out failure mechanisms, and optimized the weighting and surcharge coefficients by using the upper-bound method of the limit analysis theory. In the present paper, upper bound solution for the face stability of a tunnel subject to nonlinear yield criterion is derived based on the failure mechanism which proposed by Leca and Dormieux (1990). Furthermore, failure mechanism is assumed as blow-out failure mechanism which is shown in Fig.1.

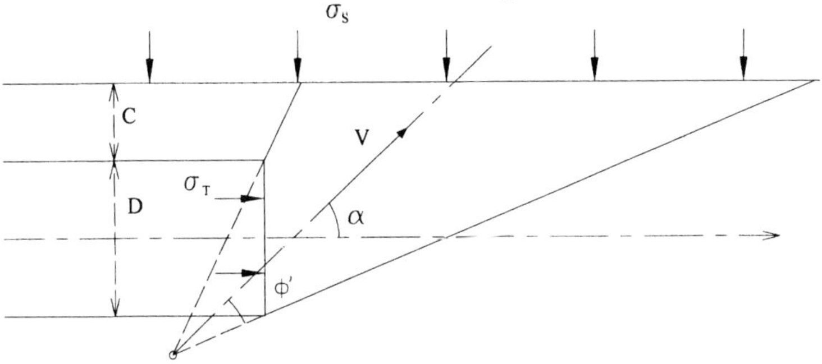

FIG.1. Failure mechanism of tunnel face (Leca and Dormieux, 1990)

UPPER BOUND SOLUTIONS FOR SHALLOW CIRCULAR TUNNEL

The upper bound theorem illustrates that if plastic admissible deformation mechanism satisfied displacement boundary condition, the load which is determined by the power of external loads equal to the dissipation power is larger than practical ultimate load. According to this theorem and failure mechanism of tunnel face, the following relationship can be obtained

$$p_e \leq p_v \tag{4}$$

where, p_e is the power of external loads, and p_v is the dissipation power. Since the failure mechanism is assumed as blow-out failure mechanism, the power of external loads is composed of three components, while the power of dissipation is the energy dissipation which occurs along velocity discontinuity surfaces of the failure mechanism. The concrete computation process is given by Leca and Dormieux (1990). Substituting the above mentioned expressions into formula (4), the retaining pressure of tunnel face in a cohesionless material can be expressed as

$$\sigma_T = \gamma D N_\gamma + \sigma_s N_s \tag{5}$$

where N_γ is soil weight coefficient and N_s is surcharge loading coefficient, γ is unit weight of the soil, σ_s is surcharge loading and D is the tunnel diameter. N_γ and N_s are function of the variable α when ϕ' is given, and can be optimized when α is chosen such that N_γ and N_s are at minimum. Besides, α is the angle between the axis of the cone adjacent to the tunnel and the horizontal, and ϕ' is soil friction angle. By regarding this issue as a nonlinear programming which take into account of constraint, the sequential quadratic programming is employed to obtain the minimum value of N_γ and N_s. Based on linear MC criterion, Leca and Dormieux (1990) calculated the optimum value of N_γ and N_s for common values of the friction angle ϕ' (20° ~ 40°), and concluded that if the assumed mechanisms are close to the actual ones, the surcharge loading has little influence on face stability.

In order to calculated σ_T in a cohesion material, base on upper bound theorem a modified formula is proposed by Soubra et al. (2008), which can be expressed as

$$\sigma_T = \gamma D N_\gamma + c N_c + \sigma_s N_s \tag{6}$$

where N_c is cohesion coefficient, c is soil cohesion, γ is unit weight of the soil, σ_s is surcharge loading and D is the tunnel diameter. N_c may be written as

$$N_c \tan\phi' + 1 - N_s = 0 \tag{7}$$

Therefore, substituting the optimal value of N_γ and N_s into (6) and (7), the optimum value of retaining pressure of shallow tunnel face σ_T is obtained. As shown in Fig. 2, the retaining pressure of tunnel face σ_T is plotted as function of the depth ratio C/D (C is depth of tunnel) for given values of the friction angle ϕ' (20° ~35°), when $c = 4kpa, D = 5m, \gamma = 18KN/m^3, \sigma_s = 10kpa$. From Fig.2, it can be seen that σ_T increase with the increasing of C/D, and when the C/D is the same, σ_T increase with the increasing of ϕ'. Besides, this variety is the same as the optimal value for N_γ and N_s calculated by Leca and Dormieux (1990).

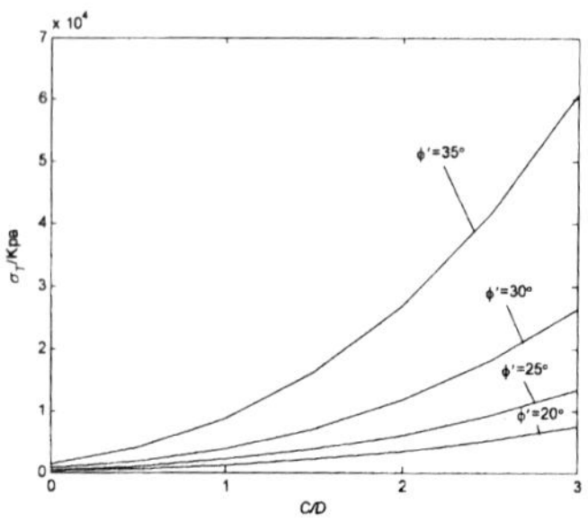

FIG.2. Upper bound values of σ_T by linear MC criterion

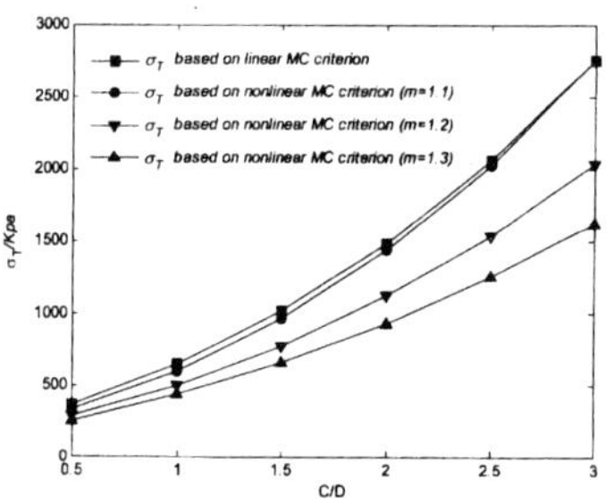

FIG.3. The comparison of retaining pressure for tunnel face by linear and non linear MC criterion

UPPER BOUND SOLUTIONS WITH NONLINEAR YIELD CRITERION

Based on the upper bound theorem, the generalized tangential technique is used to develop retaining pressure of tunnel face σ_T with the nonlinear yield criterion in this paper. In the upper bound solution with linear yield criterion, the soil parameters, cohesion c and friction angle ϕ', are constants which are given before the calculation of upper bound solution. However, in the upper bound solution subjected to nonlinear yield criterion the cohesion c and friction angle ϕ' are no longer constants, but c_t and φ_t which are determined by generalized tangential technique in formula (2). Since c_t is the function of φ_t and determined by formula (3), the optimization of retaining pressure of tunnel face can be regarded as a nonlinear program which is characterized by the two variables of φ_t and α with constraint. Once the above formulas are substituted into the nonlinear optimization program, the program will search for the optimum value of c_t and α automatically. Finally, the optimum value of c_t φ_t, and σ_T will be obtained.

NUMERICAL RESULTS

In order to evaluate the validity of the method which employed in this paper, the optimum value of retaining pressure of tunnel face σ_T in conjunction with nonlinear criterion is calculated when $c = 4kpa$, $\sigma_t = 15kpa$, $\gamma = 18KN/m^3$, $D = 5m$, $\sigma_s = 10kpa$, and m =1.1, 1.2 and 1.3 based on the blow-out failure mechanism. In Fig.3, optimal values of σ_T are plotted as function of the C/D with linear and nonlinear MC yield criterion respectively, and their results are compared. It was found that the σ_T using the linear MC yield criterion are almost equal to those using the nonlinear MC yield criterion when m =1.1, with the maximum difference being less than 5%. However, according to formula (1), we can see that when m=1, the nonlinear MC yield criterion decreases to the linear MC yield criterion. As a result, the comparison shows that the technique used in this paper is an effective method for optimizing the retaining pressure of tunnel face σ_T with the nonlinear MC yield criterion. Furthermore, the figure suggests that σ_T increases with the increasing of m when C/D is the same and increases with the increasing of C/D when m is the same.

CONCLUSIONS

Based on the failure mechanism proposed by Leca and Dormieux (1990), this paper employs the generalized tangential technique to analyze the retaining pressure of tunnel face with nonlinear failure criterion in the framework of the upper bound theorem of limit analysis. The results are optimized by nonlinear programming, and are compared with the optimum value of σ_T which calculated by linear failure criterion. The comparison illustrates that the retaining pressure of tunnel face which is computed in this paper is reasonable and valid. According to the discussion above, the following conclusions can be drawn:

(1) By using generalized tangential technique, the nonlinear MC failure criterion is introduced to calculate the retaining pressure of tunnel face σ_T, and the optimization process can be regarded as a nonlinear programming which characterized by the two variables of ϕ' and α with constraints.

(2) According to the research results by Leca and Dormieux (1990), the upper solution of retaining pressure of tunnel face σ_T is calculated, and the results

show that σ_T increases with the increasing of depth ratio. When the depth ratio is the same, σ_T increases with the increasing of ϕ'.

(3) For the different depth ratios, using generalized tangential technique, the effects of the coefficient m on retaining pressure of tunnel face have been investigated. It is found that the parameter m has a large influence on the σ_T.

REFERENCES

Davis, E.H., Dunn,M.J., Mair,R.J. and Seneviratne,H.N. (1980). "The Stability of Shallow Tunnels and Underground Openings in Cohesive Material." *Geotechnique*, 30 (4): 397–416.

Soubra, A.H., Dias, D., Emeriault, F. and Kastner, R. (2008). "Three-dimensional face stability analysis of circular tunnels by a kinematical approach." *Proceedings of the GeoCongress, Characterization, Monitoring, and Modelling of Geosystems,* USA New Orleans. 9-12.

Soubra, A.H. (2002). "Kinematical approach to the face stability analysis of shallow circular tunnels." *Proceedings of the Eight International Symposium on Plasticity,* Canada, British Columbia. 443–445.

Leca, E. and Dormieux, L. (1990). "Upper and lower bound solutions for the face stability of shallow circular tunnels in frictional material." *Geotechnique*, 40 (4): 581–606.

Lee, I.M. and Nam, S.W. (2001). "The study of seepage forces acting on the tunnel lining and tunnel face in shallow tunnels." *Tunnelling and Underground Space Technology*, 16 (1): 31-40.

Li,Y., Emeriault, F., Kastner, R. and Zhang, Z.X.. (2009). "Stability analysis of large slurry shield-driven tunnel in soft clay." *Tunnelling and Underground Space Technology*, 24 (4): 472-481.

Lee, I.M., Lee, J.S. and Nam, S.W. (2004). "Effect of seepage force on tunnel face stability reinforced with multi-step pipe grouting." *Tunnelling and Underground Space Technology*, 19 (6): 551-565.

Hoek, E. and Brown, ET. (1997). "Pratical estimate the rock mass strength." *International J. of Rock Mechanics and Mining Sciences*, 34 (8): 1165-1186.

Agar, J.G., Morgenstern, N.R. and Scott, J. (1985). "Shear strength and stress-strain behavior of Athabasca oil sand at elevated temperatures and pressure." *Can. Geotech. J*, 24 (1): 1-10.

Park, K.H., Tontavanich, B. and Lee, J.G..(2008). "A simple procedure for ground response curve of circular tunnel in elastic-strain softening rock masses." *Tunnelling and Underground Space Technology*, 23(2): 151-159.

Lee, Y.K. and Pietruszczak S. (2008). "A new numerical procedure for elasto-plastic analysis of a circular opening excavated in a strain-softening rock mass." *Tunnelling and Underground Space Technology*, 23(5): 588-599.

Yang, X.L. and Yin, J.H. (2006). "Estimation of seismic passive earth pressures with nonlinear failure criterion." *Engineering structures*, 28 (3): 342-348.

Analytical to the issuc of spherical cavities expansion with the non-linear Mohr-Coulomb failure rule

ZOU Jin-feng[1,2] PENG Jian-guo[2] ZHANG Jin-hua[2] LUO Heng[1] AN Ai-jun[1]

(1. School of Civil Engineering & Architecture, Central South University, Changsha 410075, China; 2. Hunan Provincial Communicaitons Planning,Purvy and Design Institute, Changsha 410008, China)

Abstract: In order to gain the analytical solutions for spherical expansion considering large deformation, the spherical cavity expansion in saturated soils was investigated. The nonliear **Mohr-Coulomb** failure criterion was adopted. The deformation was small in elastic region and large in plastic region. According to the equilibrium equations and compatibility condition, the stress distribution in two regions, the plastic radius, the excess pore pressure and the limit expansion pressure were derived. The numerical results are compared with tests. From the comparisons, it is found that the limit pressure and excess pore water pressure of the present solutions almost equal the test results. The limit pressure increases and the excess pore pressure decreases with the parameter of m.

Key words: large deformation, nonlinear **Mohr-Coulomb**failure criterion, limit expansion pressure, excess pore water pressure, radius in plastic region, undrianed

1 Introduction

The cavity expansion theory is widely studied in engineering mechanics. At present, these studies are almost based on Mohr-Coulomb (MC) and Cam-Clay models' for drained cavity expansion. The undrained cavity expansion seldom takes the nonlinear failure criterion into account. YU and HOULSBY [1] and COLLINS and YU [2] presented the analytical solutions to the cavity expansion under dilation and critical state by introducing the large deformation theory, using MC failure criterion. However, they didn't find an effective calculation method for the effective stress and the excess pore water pressure. It hypothesized that the small deformation in elastic region and the large-deformation and non-associated flow flue in plastic region [3-4]. They estimated the pore water pressure and the effective stress in cavity expansion for undrained strain-hardening soils. AU et al [5] obtained the approximate solutions of cavity expansion in elastic-plastic region for undrained soils. GALLIPOLI D et al [6-8] derived the approximate solutions for drained cavity expansion in cohesionless soil. However, those works are focused on the linear MC failure criterion. The above solutions seldom considered the joined effects of large deformation, undrained condition and nonlinear failure criterion.

In engineering practice, the soils not only took place serious pull crack and shear, but also had plastic flow property. The large displacement and large deformation often occur in engineering of pile driving, cone penetrating, and grouting compaction. To obtain the distributions of effective stress, strain, the plastic radius, and excess pore water pressure in elastic and plastic region in cavity expansion, the analytical solutions are derived, with the nonlinear failure strength criterion, undrained conditions and large strain being considered. In order to see the validity of the proposed solutions, numerical results are presented and comprised test results.

Foundation item: Project (50408020) supported by the National Natural Science Foundation of China
Corresponding author: ZOU Jin-feng, PhD; Tel: +86-731-2656248; E-mail: zoujinfeng_0@163.com

2 Nonlinear Mohr-Coulomb failure criterion

According to the literature[9], nonlinear Mohr-Coulomb failure criterion can be written as

$$\sigma_1 = N_p + M_p\left(\frac{\sigma_3}{N_p}\right)^{\frac{1}{m}} \tag{1}$$

Where σ_1 and σ_3 are the maximum and minimum principal stress when rock and soil mass failured. σ_c is the uniaxial compressive strength. m is the parameters determined by triaxial test. M_p and N_p are the constant which gained by $M_p = (1+\sin\phi)/(1-\sin\phi)$ and $N_p = 2c\cos\varphi/(1-\sin\varphi)$. So M_p and N_p relate with shear strength index c and φ.

For spherical cavity expansion, $\sigma_1 = \sigma_r, \sigma_3 = \sigma_\theta$. So equation (1) may be rewritten as

$$\sigma_\theta = N_p\left(\frac{\sigma_r - N_p}{M_p}\right)^m \tag{2}$$

3 Definition of problem

Figure 1 shows a cavity with an initial radius a_0 and an element at an initial distance r_0 from the center of the cavity wall moves to a new radial position r from the center. The property of the soil in the elastic region is defined by Young's modulus E, and Poisson's ratio ν.

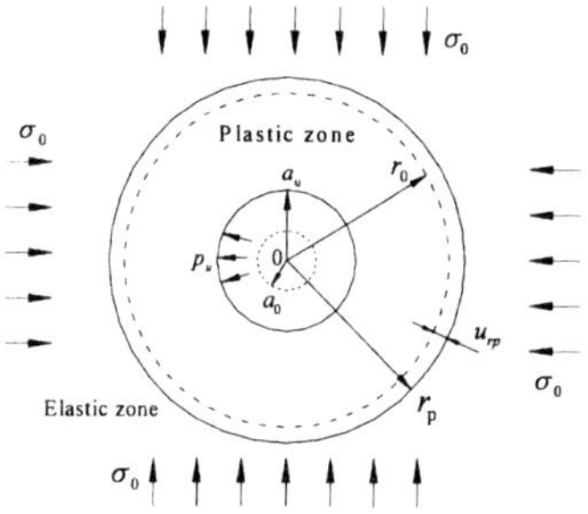

Fig.1 Sketch of cavity expansion

In Figure 1, a_u is the ultimate expansion radius, r_p is the plastic region radius, p_u means the ultimate expansion pressure on the circular cavity wall, and σ_0 is the applied static soil pressure at an infinite distance. The soil around the cavity is regarded as uniform, isotropic elasto-plastic material. There is an initial stress in the soils before loading.

In order to account for the large strain effect in the plastic region, logarithmic strains are adopted. These strains are defined as

$$\varepsilon_r = -\ln\left(\frac{dr}{dr_0}\right), \quad \varepsilon_\theta = -\ln\left(\frac{r}{r_0}\right)$$

(3)

where ε_r and ε_θ are the radial strain and the tangential strain respectively. Therefore, the volumetric strain can be written as

$$\varepsilon_v = \varepsilon_r + 2\varepsilon_\theta \tag{4}$$

For the undrained condition, the volumetric strain, ε_v, is zero in the plastic region. And the radial strain is

$$\varepsilon_r = -2\varepsilon_\theta \tag{5}$$

Substituting equation (3) into equation (5), the following solution can be obtained by considering the boundary conditions at the cavity wall

$$r^3 - r_0^3 = a^3 - a_0^3 \tag{6}$$

4 Analytical solutions in elastic region

4.1 Equilibrium equation

For the cavity expansion problem, the differential equation of equilibrium for the axisymmetric property, without body forces, can be written as

$$\frac{d\sigma_r}{dr} + 2\frac{\sigma_r - \sigma_\theta}{r} = 0 \tag{7}$$

The two boundary conditions for the problem are $\sigma_r \big|_{r=a_u} = p_u$ and $\lim_{r \to \infty} \sigma_r = \sigma_0$.

Where a_u is current radius for the cavity wall, σ_0 is initial mean effective stress and p_u is the cavity pressure.

4.2 Elastic stress and displacement

As a cavity is expanded, according to reference [9], the solution for stress and displacement can be easily shown to be

$$\left.\begin{aligned} \sigma_r &= \sigma_0 + \left(\sigma_{rp} - \sigma_0\right)\left(\frac{r_p}{r}\right)^3 \\ \sigma_\theta &= \sigma_0 - \frac{1}{2}\left(\sigma_{rp} - \sigma_0\right)\left(\frac{r_p}{r}\right)^3 \end{aligned}\right\} \tag{8}$$

$$u = \frac{1+\nu}{E}\left(\sigma_{rp} - \sigma_0\right)\frac{r_p}{2}\left(\frac{r_p}{r}\right)^2 \tag{9}$$

where σ_{rp} is radius stress between the elastic and plastic region. By substituting $r = r_p$ into Equation (9), the strain at the interface between the elastic region and the plastic region may be obtained as

$$\varepsilon_r^e = -2\varepsilon_\theta^e = \frac{(1+v)}{E}\left(\sigma_{rp} - \sigma_0\right) = B_0 \tag{10}$$

where ε_r^e and ε_θ^e are elastic strain of radius and tangential, respectively. The radial stress σ_{rp}, and tangential stress, $\sigma_{\theta p}$, at the interface between the elastic region and the plastic region may be obtained according to stress boundary condition corresponding to $r = r_p$.

$$\sigma_{rp} + 2q_p\left(\frac{\sigma_{rp} - q_p}{M_p}\right)^m = 3p_0 \tag{11}$$

where p_0 is the hydrostatic in situ stress at infinity.

At the interface between the elastic zone and the plastic zone, $r = r_p$, equation (6) becomes

$$r_p^3 - r_0^3 = (r_p - u_p)^3 - a_0^3 \tag{12}$$

For cavity expansion problem, the octahedral deviator stress q_p at the interface between elastic region and plastic region is

$$q_p = \sigma_{rp} - \sigma_{\theta p} \tag{13}$$

Substituting equation (9) into equation (12) and ignoring higher-order terms of q/G. Equation (12) becomes

$$\left(\frac{a_0}{a}\right)^3 + \frac{q_p}{2G}\left(\frac{r_p}{a}\right)^3 = 1 \tag{14}$$

As cavity continues to expand indefinitely (i.e. $a_0/a \to \infty$), $a \to a_u$,then $r \to r_p$. So equation (14) can be rewritten as:

$$\left(\frac{r_p}{a_u}\right)^3 = \frac{2G}{q_p} \tag{15}$$

5 Analytical solutions in plastic region

5.2 Stress fields

According to the conditions, $r = r_p$ and $\sigma_r = \sigma_{rp}$, $r = a_u$ and $\sigma_r = p_u$, Simultaneous equations (2) and (7), the stress fields in plastic region is gained by integral equation (7):

$$\int_{\sigma_r}^{\sigma_{rp}} \frac{d\sigma_r}{\sigma_r - N_p\left(\dfrac{\sigma_r - N_p}{M_p}\right)^m} + 2\ln\frac{r_p}{r} = 0 \tag{16}$$

Simultaneous equations (4) and (16), tangential stress in plastic region can be gained.

5.3 Limit cavity expansion pressure

Simultaneous equations (2) and (7), integral equation (7) with the conditions of $r = r_p$ and $\sigma_r = \sigma_{rp}$, $r = a_u$ and $\sigma_r = p_u$, the expression for the limit expansion pressure may be gained

$$\int_{p_u}^{\sigma_{rp}} \frac{1}{\sigma_r - N_p\left(\dfrac{\sigma_r - N_p}{M_p}\right)^m} d\sigma_r + 2\ln\frac{r_p}{a_u} = 0 \tag{17}$$

To calculate the numerical integral of equation (17), we can have the expression for the limit expansion pressure.

5.4 Pore water pressure

According to Henkel's theory of excess pore water pressure, the initial excess pore water pressure was gained [9]. The excess pore water pressure for spherical expansion expressed as

$$\Delta u = c_u\left[\frac{4}{3}\ln\frac{r_p}{r} + \left(2.0A_f - 0.67\right)\right] \tag{18}$$

where A_f is the coefficient of excess pore water pressure, and c_u is the undrianed strength. The factors influenced the excess pore water pressure relates only to the factors of r_p/r, the coefficient of excess pore water pressure, undrianed strength, and distance from the cavity center.

The expressions of excess pore water pressures are obtained when $r = a_u$ at the cavity wall for spherical cavity expansion expressed as equations (19).

$$\Delta u = c_u\left[\frac{4}{3}\ln\left(\frac{2G}{q_p}\right) + \left(2A_f - 0.67\right)\right] \tag{19}$$

5.5 Effective stress field

According to the stress field and excess pore water pressure formula in plastic region, the effective stress field in plastic region is obtained. Corresponding to the total stress, the effective stress distribution in plastic region expressed as equations (20) for spherical expansion.

$$\left.\begin{aligned}\sigma_r' &= \sigma_r - c_u\left[\frac{4}{3}\ln\frac{r_p}{r}+\left(2A_f-0.67\right)\right]\\ \sigma_\theta' &= \sigma_\theta - c_u\left[\frac{4}{3}\ln\frac{r_p}{r}+\left(2A_f-0.67\right)\right]\end{aligned}\right\} \tag{20}$$

For Vesic's theory of spherical cavity expansion, the limited cavity expansion pressure is shown as:

$$p_u = \frac{3\left(p_0 + c\cot\varphi\right)\left(1+\sin\varphi\right)}{3-\sin\varphi}\left(I_r\right)^{\frac{4\sin\varphi}{3(1+\sin\varphi)}} - c\cot\varphi \tag{21}$$

6 Analysis and discussion

6.1 Parameter studies

In order to investigate the effects of nonlinear failure criterion on the proposed theory, an example of saturated soil is considered. The values of limited expansion pressure are shown in Fig.2 for φ=11^0, p_0=120.0 kPa, c=10 kPa, m=1.235, a_u=0.25 m, E=3000 kPa, v=0.35 with parameter r_p/a_u varying from 1.0 to 10.0. Fig.3 presents values of limit expansion pressure obtained by test results for φ=20, c=6 kPa, v=0.5, E=5600 kPa, γ=17.0 kN/m^3, h=14 m and the proposed method with the m varying from 1.8 to 3.0.

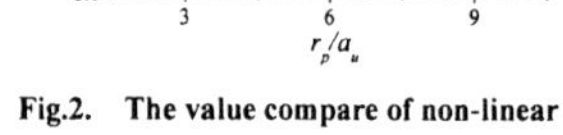

Fig.2. The value compare of non-linear failure rule and Vesic theory

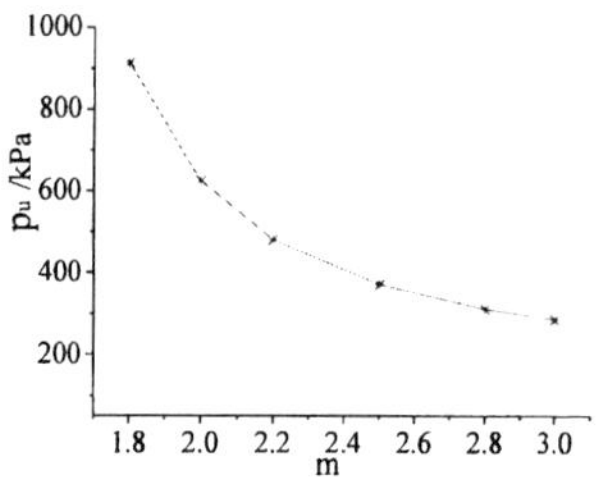

Fig.3. Curve of limited expansion pressue with parameter of m

It is found from Fig.2 that Vesic's theory is one of special cases for proposed theory. Whit the parameter of m increasing, expansion pressure increas nonlinearly with parameter of r_p/a_u increasing.

It is found from **Fig.3.** that expansion pressure decreases nonlinearly with m increasing. The differences is almost to 3 times at m=1.8 than m=3.0.

6.2 Analysis of examples

In order to investigate the effects of parameter m on the limit cavity expansion pressure, excess pore water pressure, the local test results are considered. The values of limited expansion pressure and excess pore water pressure are shown in Fig.4 and Fig.5 corresponding to φ=22, c=4 kPa, v=0.5, E=5900 kPa, γ=18.8 kN/m^3, h=13.3 m, p_u=800 kPa, Δu_a=66 kPa with parameters of m varying from 1.1 to 2.0.

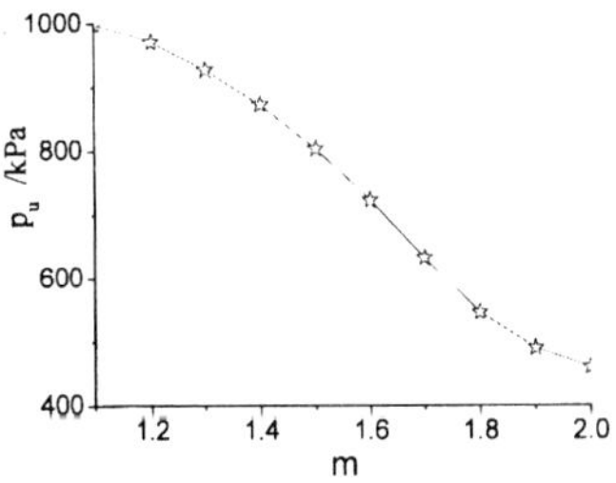

Fig.4. Comparison between test and the calculated values of the limited pressure value

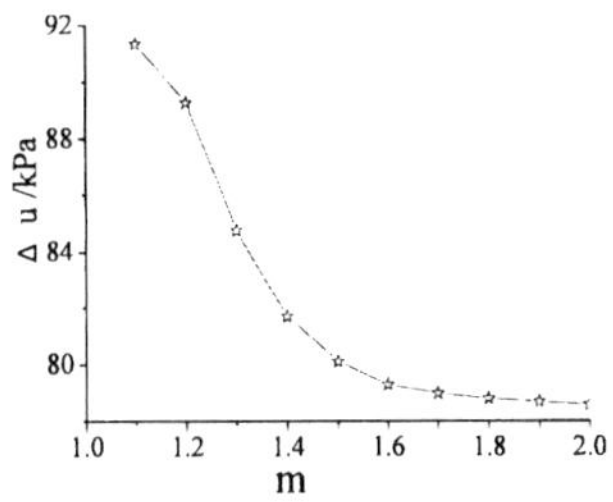

Table 5. Comparison between test and calculated values of the excess pore pressure in cylindrical expansion

It is found from Fig.4 that the proposed theory solutions of limited expansion pressure almost equaled the test results, 800 kPa when parameter of m varied from 1.45 to 1.65. limited expansion pressure increase nonlinearly with m increasing.

7 Conclusions

(1) Analytical solutions about limit expansion pressure, excess pore water pressure and effective stress field are proposed considering the large strain and nonlinear **Mohr-Coulomb** failure criterion in saturated soils. The numerical results are compared with test results. From the comparisons, it was found that the limit pressure and excess pore pressure of the present solutions almost equaled the test results as parameter of m varied from 1.4 to 1.6.

(2) The conclusions show that the cavity expansion pressure is nonlinear increasing and the excess pore water pressure is decreasing with the values of m increasing. The increase trend is more obvious for spherical expansion.

Reference

[1] YU H S, HOULSBY G T. A large strain analytical solution for cavity contraction in dilatants soil[J]. International Journal for Numerical and Analytical Methods in Geomechanics, 1995, 19 (7): 793-811.

[2] COLLINS I F, YU H S. Undrained cavity expansions in critical state soils[J]. International Journal for Numerical and Analytical Methods in Geomechanics, 1996, 20: 489-516.

[3] CAO, LF, TEH CI, CHANG M F. Undrained cavity expansion in modified cam clay theoretical analysis[J]. Geotechnique, 2001, 51(4): 323-334.

[4] CAO L F, TEH C I, CHANG M F. Analysis of undrained cavity expansion in elastic-plastic soils with nonlinear elasticity[J]. Journal for Numerical and Analytical Methods in Geomechanics, 2002, 26(1):25-52.

[5] AU A K, YEUNG A T, SOGA K., Cheng Y.M. Effects of subsurface cavity expansion in clays[J]. Geotechnique, 2007, 57(10): 821-830.

[6] GALLIPOLI D, GENS A, SHARMA R. An elastoplastic model for unsaturated soil incorporating the effects of suction and degree of saturation on mechanical behavior[J]. Geotechnique, 2003, 53(1): 123-135.

[7] GALANOV B A, KARTUZOV V V, IVANOV S M. New analytical model of expansion of spherical cavity in brittle material based on the concepts of mechanics of compressible porous and powder materials[J]. International Journal of Impact

Engineering, 2008, 35(12): 1522-1528.

[8] ROSENBERG Z., DEKEL E. Analytical solution of the spherical cavity expansion process[J]. International Journal of Impact Engineering, 2009, 36(2): 193-198.

[9] ZOU Jinfeng. Cavity expansion based on the linear and nonlinear failure criteria and it's engineering applications[D]. Central South University, 2007,11(in Chinese).

Study on maintenance technology of shield tunnel in soft ground

Yi Rui[1,2], Hehua Zhu[1,2], Mci Yin[1,2], Xiaojun Li[1,2]

[1]Department of Geotechnical Engineering, Tongji University

[2]Key Laboratory of Geotechnical and Underground Engineering of Ministry of Education, Tongji University, Shanghai, 200092, China;

ABSTRACT: Shield tunnel disease, such as lining cracks, deformation and leakage, would appear in different degree under the long-term impact of nature and operation. These diseases not only endanger the security of driving, but also affect the performance of shield tunnel. Without prompt treatment, they would inevitably cause further damage, and even failure of the tunnel. For these reasons, making research on maintenance technology for shield tunnel is necessary. In this paper, the fault-tree method was applied to perform a preliminary analysis of the causes of tunnel diseases, and then, by introducing rough set theory, the causes could get further excavated. Besides, calculation models for various diseases are erected, which can calculate the impact of these diseased in a quantitative way. All of these methods above provide a basis for applying reasonable maintenance method with a clear objective, and, combined with digital technology, they are applied in Shanghai Yangtze River Tunnel.

KEYWORDS:

1. INTRODUCTION

Shield tunnel disease, such as lining cracks, deformation and leakage, would appear in different degree under the long-term impact of nature and operation. At present, the maintenance method for these diseases is worked out by experience of engineers primarily, maybe sometimes considering the geology, climate and hydrology. Generally, after a simple analysis of the appearance, the disease then gets treated.

Theory and practice show that relying on experience alone is not enough. Obviously, the lack of systematic and scientific analysis of impact and cause of tunnel diseases would lead to unreasonable maintenance measures and programs, causing some of waste. Consequently, it's necessary to make further research on shield tunnel diseases.

2. COMMON DISEASES OF SHIELD TUNNEL

Cracks in concrete structures are very common. Many concrete structures in use are accompanied with cracks. The impact of cracks on shield-driven tunnels is reflected in the safety and durability of structures first, then in other performance (such as water-proof and moisture-proof function). If handled improperly, it will bring terrible harm.

Leakage water is the most common disease of shield tunnel in soft ground. It will lead to effects in many aspects, such as stability, facilities, traffic safety, structure performance, etc. Besides, Serious seepage would induce the differential settlement and destruction of some adjacent underground structure.

The deterioration of concrete primarily refers to carbonization of concrete in shield tunnel, which results from the $Ca(OH)_2$ in concrete reacting with CO_2 in air. The main problem caused by carbonation lies in the damaged surface of structure, which would reduce the strength and durability of tunnel.

Segment dislocation means the inner surface of one tunnel segment and its adjacent segment not aligned, usually caused by nonuniform of load effect. When a concentrated load which exceeds the load limit is applied on the segment, dislocation appears. It will result in the failure of waterproof function of shield tunnel.

Corrosion of reinforcement is also a common disease. It would reduce the bearing capacity of reinforcement to a large extent. Furthermore, under serious condition, it would make the tunnel liner discrete, reducing the strength and bearing capacity.

3. ANALYSIS OF CAUSES OF TUNNEL DISEASES

One sort of disease can be caused by various reasons. Scientific maintenance technique does not focus on the apparent causes, but dig out the real causes after logic diagnosis, according to the for the logical link between diseases and reasons. In this paper, the fault tree method is applied.

Fault tree analysis is a kind of graphical deduction, which convert various effecting factors, related to the diseases, in to a sort of tree structure, including direct and indirect factors, environmental affect, and human factors. We can see which factors are related to the disease--the top event in the fault tree, from top downward, and we can also analyze the impact on the disease from the bottom upward. The figure below shows the fault tree of cracks in shield tunnel.

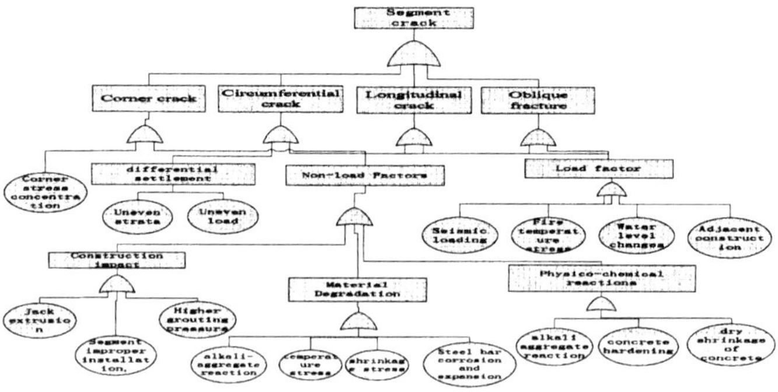

FIG.1. Fault tree of cracks

The causes deduced by the fault tree are only possible causes, containing empirical factors and uncertainties, and perhaps redundant information. Thus, the conclusions need to get further handled by introducing rough set theory. In this process, an information table of causes is established, which dig out the real and primary cause when removing the redundant ones. In the meantime, and the relevant laws of tunnel diseases are also concluded. The steps of application of rough set theory are shown as follows:

1) Collect and analyze relevant information, covering disease data, geology, design, construction, monitoring, management, etc. Based on that, a fault tree is established, and then a condition attribute set (possible causes of tunnel disease) and a decision attribute set (whether the disease appears) are established. After that, an information table of original data is created.

2) When applying rough set theory, data in attribute set should be discrete. So if continuous, the data should be converted into discrete. In this way, the initial decision table of analysis is established.

3) The next step is attribute reduction. In a system of tunnel disease information $S = < U, A, V, f >$, $U = (x_1, x_2, ..., x_n)$ refers to the set of all processing objects. $A = C \cup D$ is the set of all attributes. C is the set of condition attribute, that is the set of a various possible disease causes. D is the set of decision attribute, that is, whether the disease happens or not. Then, the steps of reducing condition attribute are shown as follows: First, remove the condition attribute c_i, and then recalculate $\gamma_{c-|c_i|}(D)$, which means the extent to which the decision attribute depends on the condition attribute. If $\gamma_{c-|c_i|}(D) = \gamma_c(D)$, it means the condition attribute c_i is redundant, and could get removed. If $\gamma_{c-|c_i|}(D) \neq \gamma_c(D)$, c_i is necessary. Through applying similar measures, the whole set of condition attribute is reduced

4) Deduce the real and primary causes of tunnel diseases, and build decision-making rules.

4. CALCULATION MODELS FOR TUNNEL DISEASES

In this paper, the beam-joint model is adopted. The joint is described by axial stiffness(k_n), tangential stiffness(k_s) and rotational stiffness stiffness(k_θ) respectively. A simplified beam-joint model is shown in the figure below.

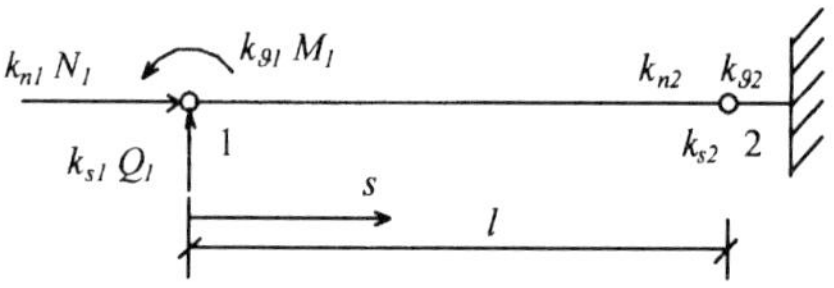

FIG.2. Beam-joint model

The figure shows a simplified beam-joint model, on the assumption that the node 2 is fixed. There are two nodes 1 and 2, simulating joints between segments, described by 6 parameters k_{ni}, k_{si}, $k_{\vartheta i}$. (i=1,2). Impose a axial force N1, a shear force Q1, and a moment M1 on node 1. According to the Castigliano's second theory, the relationship matrix between displacement $\{\delta_1\} = \{u_1 \square v_1 \square \theta_1\}^T$ and force $\{F_1\} = \{N_1 \quad Q_1 \square M_1\}^T$ can be established as below:

$$\{\delta_1\} = \begin{bmatrix} a_1 & 0 & 0 \\ 0 & b_1 & d_1 \\ 0 & d_1 & c_1 \end{bmatrix} \{F_1\} \tag{1}$$

Where a1, b1, c1, d1 mean the flexibility matrix elements, $a_1 = \frac{l}{EA} + \frac{1}{k_{n1}} + \frac{1}{k_{n2}}$, $b_1 = \frac{l^3}{3EI} + \frac{1}{k_{s1}} + \frac{1}{k_{s2}} + \frac{l^2}{k_{\vartheta 2}}$, $c_1 = \frac{l}{EI} + \frac{1}{k_{\vartheta 1}} + \frac{1}{k_{\vartheta 2}}$, $d_1 = -\frac{l^2}{2EI} - \frac{l}{k_{\vartheta 2}}$. Reverse the formula above and we can get $\{F_1\} = [k_{11}]\{\delta_1\}$.

For node 2, the expression of nodal force could be acquired by principle of static balance. In the same way, the relation between force and displacement of node 2 could be acquired. Eventually, the relation between force and displacement on nodes could be deduced as follows:

$$\begin{Bmatrix} F_1 \\ F_2 \end{Bmatrix} = \begin{bmatrix} k_{11} & k_{12} \\ k_{21} & k_{22} \end{bmatrix} \begin{Bmatrix} \delta_1 \\ \delta_2 \end{Bmatrix} = \begin{bmatrix} k_{11} & Ak_{11} \\ Ak_{11} & k_{22} \end{bmatrix} \begin{Bmatrix} \delta_1 \\ \delta_2 \end{Bmatrix} = [K] \begin{Bmatrix} \delta_1 \\ \delta_2 \end{Bmatrix} \tag{2}$$

In addition, the role of stratum resistance should be considered. In this paper, the stratum resistance is described by peripheral spring, and elastic resistance of stratum is given the formula below:

$$F_n = K_n \cdot U_n, \quad F_s = K_s \cdot U_s \tag{3}$$

Where, F_n, F_s is the axial and tangential resistance, K_n, K_s is the corresponding ground spring coefficient. U_n ,U_s is the corresponding node displacement.

From the beam-joint model, we can see that the mechanical state of shield tunnel is determined by the following parameters: material and geometric parameters E ,I ,A , parameters of joint stiffness k_n, k_s, k_θ, parameters of ground spring K_n, K_s, loads imposed on shield tunnel. Therefore, to erect the calculation models for tunnel diseases is to modify these parameters, which render the calculation outcome matching with actual mechanical state of shield tunnel. This method seems simple and clear, with a considerable accuracy, and could get easily handled by digital technology mentioned in the following chapter.

Calculation model for the deterioration of concrete lining is shown as the following picture:

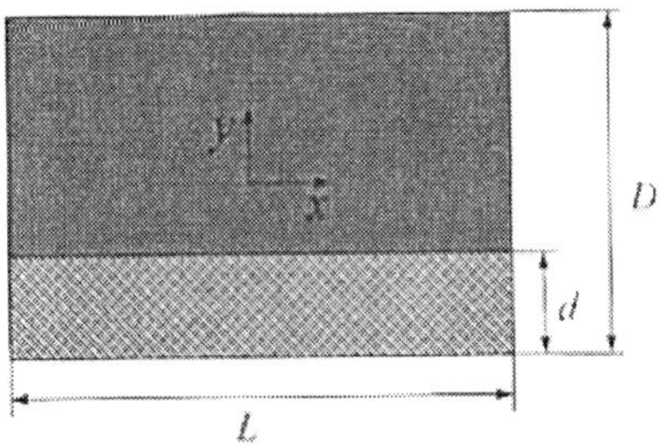

FIG.3. Calculation model for the deterioration of concrete lining

In the figure above, the initial thickness of tunnel lining is D. The elastic modulus is E and the shear modulus is G . The thickness of deterioration is d, and the elastic modulus and shear modulus change to E_1 、 G_1 . The values of these time-dependent parameters could be acquired by related research and experiment. Eventually these modified parameters are applied in the beam-joint model to simulate the impact of deterioration of concrete.

$$A_n^{'} = \frac{E_1}{E} d + (D - d) \quad (4)$$

$$A_v^{'} = \frac{G_1}{G} d_1 + (D - d) \quad (5)$$

In order for simplifying the calculation, the elastic module and shear module are assumed to remain unchanged, The effective area and shear area got to be modified accordingly, as shown above.

Lining cracks will make the tunnel lining not smooth in geometry. To simulate the change of axis curve, a special bolt spring is introduced, as shown below .

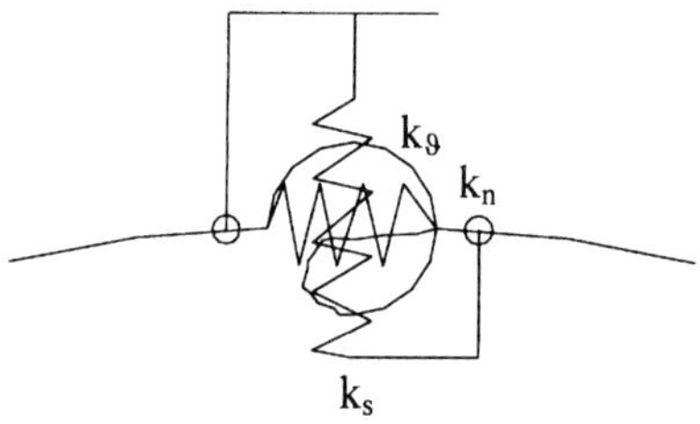

FIG.4. Calculation model for cracks

In the figure, K_N, K_θ, K_S means the axial, tangential and rotational stiffness. After measuring the crack width, depth, and orientation, these 3 parameters could get determined by back analysis.. When they are designated as ∞, it means there is no cracks in the structure. If the crack is imposed on by a tensile load, the crack would tend to open, and the axial, tangential rotational stiffness would get reduced accordingly. When the crack is under pressure, the crack will gradually close up and the stiffness would remain unchanging basically.

5. Digital System for Maintenance of Shield Tunnel

In the long-term maintenance of shield tunnel, by applying those methods mentioned above, the cause of these diseases could get deduced, and the mechanical state can be determined. Based on that, a reasonable and effective maintenance measure can then be chosen and applied. However, in practical maintenance, it need to deal with massive information, covering surveying, design, construction, monitoring and management. Besides, for each disease, various fault trees and numerical calculation models need to be established accordingly, which, obviously seems impossible for manual work. Luckily, the digital technology could solve these problems effectively. In this paper, a digital shield tunnel is established to carry out the management of maintenance, and the digital platform is shown in the following figure.

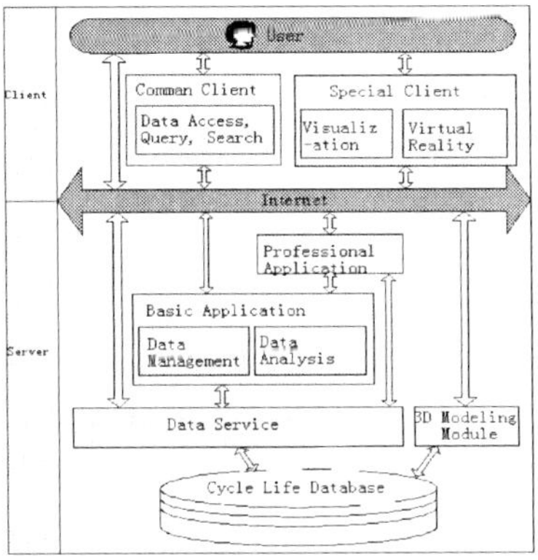

FIG.6. Structure of digital tunnel platform

In order to satisfy the needs of practical application, this digital platforms applied client - server architecture based on Internet. On the server side, a life-cycle database is established, actualizing the centralized storage and management of project data, which applies the current commercial databases to store data with advantage in data maintenance and indexing. Based on this, the server realizes the functions of data service, model service, application service and professional application. The data service is to provide data support, such as data access, deletion and search. Model service is a visual model generated by project data according to requirements of various users, facilitating a direct and visualized browse. The basic application service consists of two areas: data processing including data quality control, query, database maintenance, security and backup; data analysis including spatial-temporal data interpolation, prediction and spatial analysis. Professional application service is based on those services above, by the means of space analysis, professional computing and data digging. The professional analysis and application are realized such as carrying out the mechanical analysis about tunnel diseases, and eventually digital management and decision-making are realized.

6. Application in Yangtze River Tunnel

The Yangtze River tunnel is located at the South Channel waterway of Yangtze River mouth in the northeast of Shanghai, which is a significant part of national expressway. It starts from Wuhaogou in Pudong New Area, connected with lands on Changxing Island. This shield tunnel is designed as dual 6 lanes expressway, with the internal diameter about 13.7m. The relief of onshore area of the project is 'river mouth, sand mouth, sand island which is within the major four relief units in

Shanghai.

Facing such a huge and complex shield tunnel, several basic principles should be established first, such as "prevention first", “detection early" "timely maintenance" and “specific solution for different problem". Combined with logic diagnose and numerical calculation model, a digital information system is established, and relevant maintenance measures are enacted. Based on integration of digital technology and numerical calculation, the damage degree could get determined in a quantitative way. In the end, specific maintenance method is chosen and applied according to professional analysis by the system. The Shanghai Yangtze River Tunnel project will finish construction in near future, and the digital system for maintenance has already been erected, as shown in following figures.

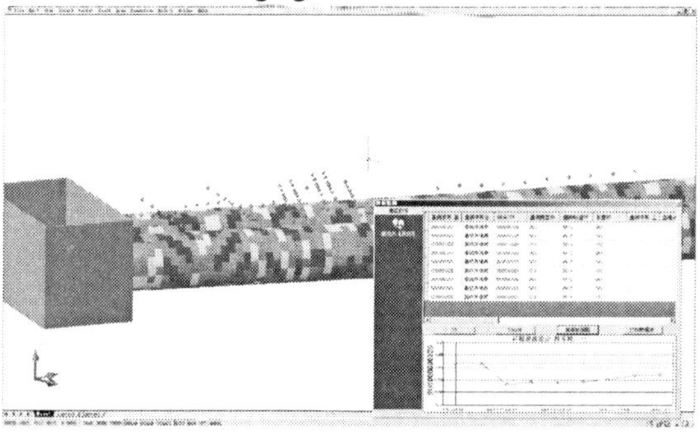

FIG.6. Visual query of monitering data

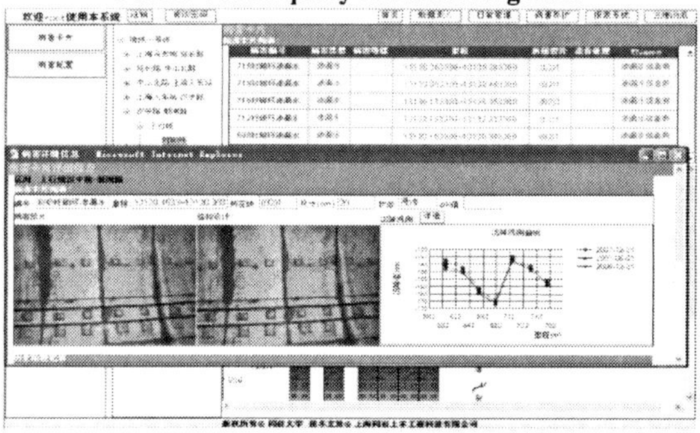

FIG.7. Quiry of seepage water information

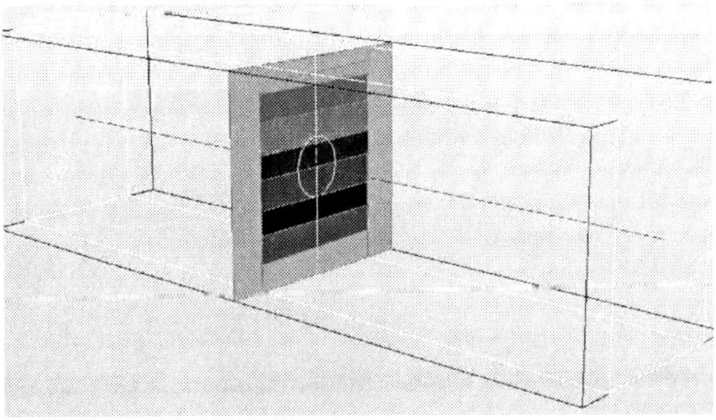

FIG.8. Slice of 3D strata in integration of digital technology and numerical calculation

7. Conclusion

(1)It is possible to carry out logic diagnose on the causes of tunnel diseases and determine the mechanical state by numerical calculation. Combined with digital system, these methods could be applied in maintenance of shield tunnel, providing reliable basis for implementation of effective and reasonable measure.

(2)Information technology such as database, visualization and network can be used as an effective way of organizing and managing maintenance information, thereby enhancing the efficiency to a great extent.

(3)Numerical analysis is effective in determining the mechanical state of shield tunnel with various diseases. To establish an integration between digital system and numerical analyses could provide valid professional information for decision-making.

REFERENCES

[1] ZHU Hehua, LI Xiaojun. Digital Underground Space and Engineering. Chinese Journal of Rock Mechanics and Engineering. 2007,26(11):2277—2288.(in Chinese)

[2] Elkadi, A.S. & Huisman, M.2002. 3D-GSIS geotechnical modelling of tunnel intersection in soft ground: the Second Heinenoord Tunnel, Netherlands. Tunnelling and Underground Space Technology 17(4):363-369

Seismic Response of Large Span Shallow Tunnels in Dilative Rocks

Xiao-Li Yang[1], and Bo Huang[2]

[1]Doctor, School of Civil Engineering and Architecture, Central South University, Changsha 410075, China; yxnc@yahoo.com.cn
[2]Master candidate, School of Civil Engineering and Architecture, Central South University, Changsha 410075, China

ABSTRACT: Seismic response induced by earthquakes is investigated using finite differential code for large span shallow tunnels. The mechanical behavior of weak rock is studied considering the influences of stress dilatancy on the failure mechanisms. Quasi-static representation of earthquake using a seismic coefficient concept is adopted for seismic estimations. The numerical results are presented. The results show that the dilatancy angle has great influences on the rock displacement and the failure face shape for the shallow tunnels, and that break angle deduced by earthquake force becomes larger. The failure face forms and extends to the ground surface with nonassocated flow rule. With the dilatancy angle increasing, the loose region decreases gradually. With the seismic coefficient increasing, loose region is distributed in a lager range. Therefore, the influence of nonassociated flow rule should be considered for determining the failure mechanisms of shallow tunnels subjected to earthquake force.

INTRDUCTION

Tunnels and underground structures have better seismic performance than ground building, because of the restriction effect of surrounding rock and soil on both tunnels and buried structures. Therefore, the seismic design of tunnels and buried structures do not cause attentions, which makes the research on the mechanical properties of surrounding rocks to tunnels considering the influence of horizontal seismic force not to be developed rapidly. The Osaka-Kobe earthquake, occurred in Japan in 1995, brought extraordinary destruction of many large underground structures including subway stations and interval tunnels, so people began to recognize the structure seismic resistance safety (Hashash et al. 2001). The Niigata earthquake, which happened in Niigata of Japan in 2004, destroyed twenty four tunnels seriously, and many of them needed to be repaired and to be reinforced. Tunnel collapses show that earthquake has important influences on the seismic stability of shallow tunnels. After the Wen-Chuan earthquake in Sichuan of China, large scale infrastructure was reconstructed gradually in disaster area. Large numbers of shallow tunnels were reinforced. The mechanical behavior of shallow tunnels have not been studied sufficiently considering the influence of horizontal seismic force, because of the complex mechanical behavior of weak rock, the uncertainty of surrounding rock parameters and complex geological conditions.

This paper attempts to develop the influence of earthquake on the mechanical behavior of shallow tunnel subjected to unsymmetrical force, considering dilatancy property of weak rock. Due to lack of seismic analysis in dilative rock in the current design codes of tunnels, the results benefit to the seismic design and constructions.

DILATANCY PROPERTY

Large quantity researches of shear dilation on the mechanical behavior of the rock and soil mass have been done both at home and abroad. These results show that the effects of shear dilation on the rock mechanical behavior are significant, and that the effects are different for different rocks. Wang and Xie (1998) studied the break angle of the shallow tunnel by the model tests. They concluded that it was unsafe to calculate the lateral pressure coefficient just according to the pseudo-friction angle, while the disappearance of cohesion was not considered. In order to study crack and break angle of the shallow tunnel, Qian et al.(1999) made model tests, which are based on the result of the two above scholars. Considering the influence of stress dilatancy on stability factor of shallow tunnel, Yang and Sui (2008) drew the conclusion that the stability of the shallow tunnel increased as the increase of dilatancy angle. By modeling a large underground cavern, Kudoh et al.(1999) proposed that the plastic property of rock mass should be evaluated by the dilatancy of cock mass. Alejano et al.(1999) presented that the dilatancy of rock masses should be assessed completely before the mechanical behavior of the rock were understood. However, some researchers also investigated the dilatancy in rock with the non-associative flow rules (Detournay 1986; Farmer 1993; Sterpi 1999). Obviously, these researches do not distinguish the performance of the rock masses based on different rock conditions. Furthermore, Vermeer and Borst (1984) proposed that the dilatancy angle was significantly smaller than the friction angle, and derived formula to assess the dilatancy angle (ψ) in granular soils. Hoek and Brown (1997) presented that the values of dilatancy angle $\psi = \varphi / 4$ (φ is friction angle) for and good quality rock, $\psi = \varphi / 8$ for average-quality rock, and $\psi = 0$ for poor-quality rock.

Therefore, the influence of shear dilation on the mechanical behavior of the rock masses cannot be neglected, when the horizontal seismic force is calculated. In this paper, using the finite differential software, the mechanical behavior of the shallow tunnels subjected to earthquake are simulated, and the influence of shear dilation on the shape of the failure face are studied. The conclusions could provide some useful references to the seismic design and constructions in the seismic areas.

NUMERICAL SIMULATION

For the shallow tunnels with four-lane highway, numerical simulation is conducted to investigate the influences of horizontal seismic force and shear dilation on the shape of the failure faces. The portal section of the Longtoushan tunnel is chosen as the four-lane road tunnel section. The buried depth is 10m, the horizontal length of the tunnel modeling is 120m long, the maximum vertical length is 94m, and the slope angle of the tunnel is 25°. The left and right boundary of the model is fixed for the horizontal direction, the bottom boundary of the model is fixed for the vertical direction, and the top boundary of the model is free. Moreover, Mohr-coulomb criterion and elastic-plastic constitutive model are employed in this numerical simulation. The explicit finite difference code is used to calculate the stability of shallow tunnels. Quasi-static representation of earthquake effects using a seismic coefficient concept is adopted for seismic calculations. Using the seismic coefficient, only horizontal seismic force is considered in the calculation, and the direction of the horizontal seismic force is chosen by the worst load orientation. According to this principle, the horizontal seismic force varies from 0.05g to 0.2g (g is the acceleration of gravity) when the dilatancy angles are 0, $\varphi/3$, and $2\varphi/3$, respectively. The rock parameter are shown in Table 1, where γ, E, μ, C, φ represents the gravity, elastic modulus, poisson ratio, cohesion and friction angle, respectively.

Table 1. Material Parameters

parameters	γ	E	μ	C	φ
four lane road tunnel	19 $kN{\cdot}m^{-3}$	1GPa	0.35	110 KPa	25°

Analysis of Displacement Vector

When the dilatancy angle $\psi = 0$, $\varphi/3$ and $2\varphi/3$, as illustrated in Fig.1, Fig.2 and Fig.3 respectively, the top surrounding rock moves from the deep-buried side to the shallow-buried side. The maximum displacement vector of tunnel occurs in the arch shoulder of the tunnel, and the displacement vector of the shallow-buried side of tunnel is large than the one in the deep-buried side. Furthermore, the surrounding rock, which closes to the surface, moves along the slope. On the vertical plane above the shallow-buried side of tunnel, the results show that the displacement vector direction of the deep-buried side of tunnel is downward, and that the displacement vector direction of the shallow-buried side of tunnel is upward.

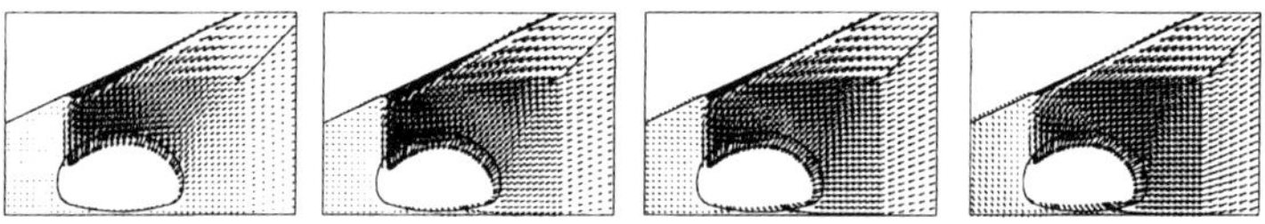

(a): W=0.05g; (b): W=0.10g; (c): W=0.15g; (d):W=0.20g

FIG.1. Displacement vector of surrounding rock with dilatancy angle $\psi = 0$

As a result, the shear failure zone happens at the shallow-buried side of tunnel, and failure occurs in the shallow-buried side of the tunnel is earlier than the deep-side of the tunnel. With the horizontal seismic force increasing from 0.05g to 0.15g, the maximum displacement vector of tunnel occurs in the arch shoulder of tunnel. When the horizontal seismic force increases up to the 0.20g, the maximum displacement vector is transferred to the crown of the tunnel. With the horizontal seismic force increasing, a large displacement vector occurs in the bottom arch of tunnel in the deep-buried side. The bottom arch of tunnel is extruded and damaged, according to the great change of displacement vectors.

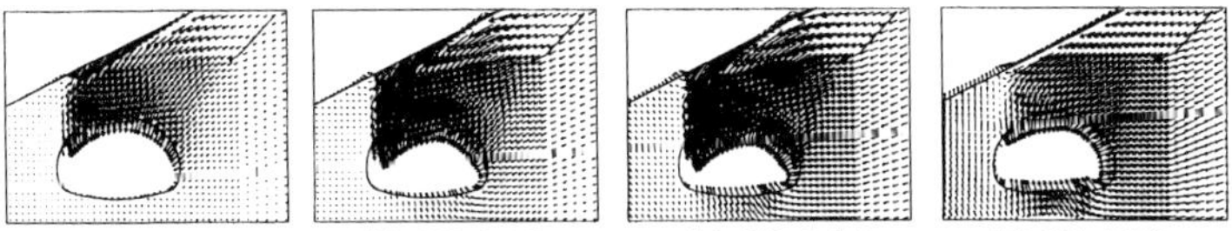

(a): W=0.05g; (b): W=0.10g; (c): W=0.15g; (d):W=0.20g

FIG.2. Displacement vector of surrounding rock with dilatancy angle $\psi = \varphi/3$

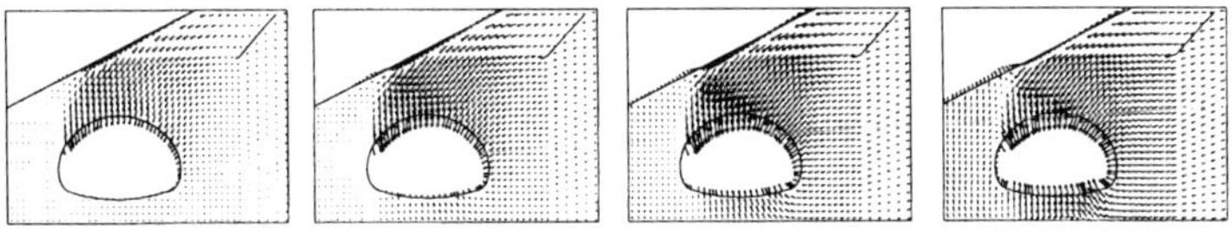

(a): W=0.05g; (b): W=0.10g; (c): W=0.15g; (d):W=0.20g

FIG.3. Displacement vector of surrounding rock with dilatancy angle $\psi = 2\varphi/3$

Failure Plane Location

As illustrated in Fig.4, the break angle of shallow-buried side of tunnel increases from 84.20° to 86.32°, when the horizontal seismic force increases form 0.05g to 0.20g with the dilatancy angle $\psi = 0$, and the failure plane of the shallow-buried side of the tunnel extends to the ground. The failure plane is nearly vertical, and the location of the failure plane anastomoses the plane of relative displacement in the displacement vector figure. Both of them are occurred in the arch shoulder of the shallow-buried side of tunnel. Surrounding rocks occur relative displacement, so shear failure zone happens at the shallow-buried side of tunnels, and the failure plane exists in the arch shoulder of the deep-buried side of tunnels. But it never extends to the ground, it only occurs relative displacement in a certain range.

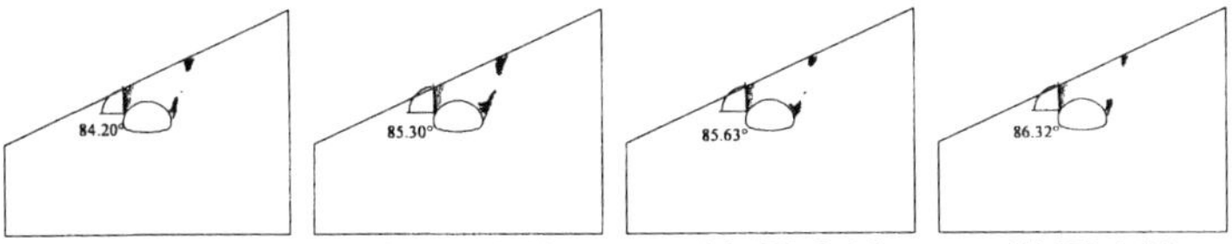

(a): W=0.05g; (b): W=0.10g; (c): W=0.15g; (d):W=0.20g

FIG.4. Maximum shear strain rate of rock with dilatancy angle $\psi = 0$

As shown in Fig.5, the break angle of the shallow-buried side of tunnel increases from 73.48° to 81.20°, when the horizontal seismic force increases form 0.05g to 0.20g with the dilatancy angle $\psi = \varphi/3$, and the failure plane of the shallow-buried side of the tunnel also extends to the ground. The failure plane is nearly vertical, the location of the failure plane anastomoses the plane of relative displacement in the displacement vector figure. Both of them occur in the arch shoulder of the shallow-buried side of tunnel, and failure occurs in the shallow-buried side of the tunnel earlier than the deep-side of the tunnel. The range of failure plane on the arch shoulder of the deep-buried side tunnel reduces as the horizontal seismic force increases form 0.05g to 0.20g.

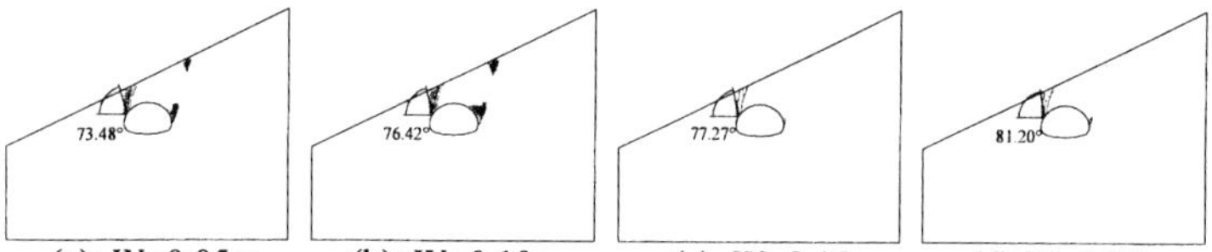

(a): W=0.05g; (b): W=0.10g; (c): W=0.15g; (d):W=0.20g

FIG.5. Maximum shear strain rate of rock with dilatancy angle $\psi = \varphi/3$

As illustrated in Fig.6, the break angle of the shallow-buried side increases from 73.08° to 86.81°, when the horizontal seismic force increases form 0.05g to 0.20g with the dilatancy angle $\psi = 2\varphi/3$. The failure plane of the shallow-buried side of the tunnel extends to the ground. The failure plane is nearly vertical. It occurs in the arch shoulder of the shallow-buried side of the tunnel. Furthermore, failure occurs in the shallow-buried side of the tunnel is earlier than the deep-side of the tunnel. The range of failure plane of the arch shoulder of the deep-buried side of the tunnel reduces and disappears at last as the horizontal seismic force increases form 0.05g to 0.15g. The range of failure plane of the arch shoulder of the deep-buried side of the tunnel occurs again, when the horizontal seismic force increases form the gravity multiple of 0.20. The maximum displacement vector of tunnel transfers to the crown of the tunnel, and the displacement vector of tunnel occurs in the arch shoulder of deep-buried side of tunnel according to the stress redistribution. Therefore, the failure plane is limited a certain range.

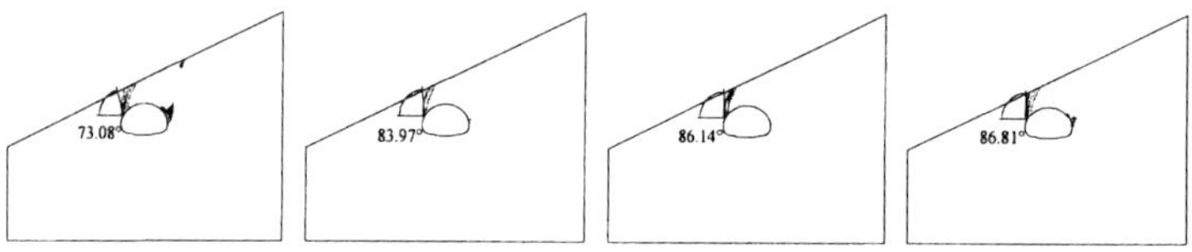

(a): W=0.05g; (b): W=0.10g; (c): W=0.15g; (d):W=0.20g

FIG.6. Maximum shear strain rate of rock with dilatancy angle $\psi = 2\varphi/3$

With associated flow rule, the failure plane occurs in the arch foot of the tunnel deep-buried side with the horizontal seismic force increases from 0.05g to 0.10g. The failure plane occurs in the arch foot of shallow-buried side of tunnel when the horizontal seismic force increases from 0.15g to 0.20g. The range of failure plane of tunnel reduces as the dilatancy angle increases.

CONCLUSIONS

Considering the influence of horizontal seismic force and shear dilation, this paper investigates the break angle and surrounding rock displacement of shallow tunnels. Numerical results show that:

(1) The failure plane range of the arch shoulder for the deep-buried side of the tunnel decreases with the increasing of horizontal seismic force. The break angle of the shallow-buried side of the tunnel increases with the increasing of horizontal seismic force. Failure plane which occurs in the shallow-buried side of the tunnel is earlier than the deep-side of the tunnel, and the failure plane is nearly vertical.

(2) Due to the effects of horizontal seismic force, the maximum displacement vector for shallow tunnel is variable. The maximum displacement vector is transferred from the arch shoulder of the shallow-buried side of tunnel to the crown of the tunnel. The bottom arch of tunnel is extruded and damaged with the horizontal seismic force increases.

(3) The range of failure plane of tunnel decreases with the increasing of dilatancy angle. The arch foot of tunnel is extruded and damaged with associated flow rule.

REFERENCES

Hashash, Y., Hook, J., Schmidt, B., and Yao, J.J. (2001). "Seismic design and analysis of underground structures." *Tunneling and Underground Space Technology*, Vol.16: 247–293.

Wang, B., and Xie, J.C. (1998). "Model test and reliability analysis for bias tunnel." *Engineering Mechanics*, Vol.15(1): 85–93. (in Chinese)

Qian, D.F., Wang, B., and Xie, J.C. (1999). "Experimental study for the break angle and the breaking up of the shallow tunnel in soil and sand interbred." *Journal of Lanzhou Railway University*, Vol.18(3): 25–29. (in Chinese)

Yang, X.L., and Sui, Z.R. (2008). "Influences of stress dilatancy on stability factors of shallow tunnels." *Journal of Central South University*, Vol.39(1): 190–195.

Kudoh, K., Koyama, T., and Nambo, S. (1999). "Support design of a large underground cavern considering strain-softening of the rock." *9th ISRM Congress*, Paris, Balkema: Rotterda. 407–411.

Alejano, L.R., Garca, B.F., Alonso, E., and Taboada, J. (1999). "Back analysis of a rock-burst in a shallow room and pillar gypsum exploitation." *9th ISRM Congress*, Paris, Balkema: Rotterda. 1077–1080.

Detournay, E. (1986). "Elasto-plastic model of a deep tunnel for a rock with variable dilatancy." *Rock Mechanics and Rock Engineering,* Vol.19: 99–108.

Farmer, I.W. (1993). "*Engineering behavior of rocks*." London: Chapman & Hall.

Sterpi, D. (1999). "An analysis of geotechnical problems involving strain softening effects." *International J. Numer. Anal. Meth. Geomech.*, Vol.23: 1427–1454.

Vermeer, P.A., and Borst, R. (1984). "Nonassociated plasticity for soils." C*oncrete and Rock*, Vol.29(3): 3–64.

Hoek, E., and Brown, E.T. (1997). "Practical estimates of rock mass strength." *International J. of Rock Mechanics* and *Mining Sciences*, Vol.34(8): 1165–1187.

Experimental Study on Anchoring Effect of Rock Bolts to Fractured Rock Mass

Wei-min Yang[1], Nuan-dong Wen[2], Shu-cai Li[1], Xiao-jing Li[3]

[1]Geotechnical and Structural Engineering Research Center of Shandong University, Jinan 250061, China; weimin-yang@163.com
[2]Fangshan Firefighting Detachment of Beijing Fire Station, Beijing 102400, China;
[3]Shandong Jianzhu University, School of Civil Engineering, Jinan 250101, China.

ABSTRACT: Fractured rock mass is very prevalent in the rock engineering. With excavation perturbation, the joints and cracks will propagate, coalesce and lead to rock mass failure finally. It is well known that the rock bolt can achieve perfect anchoring effect in practice, especially in fractured rock mass. However, so far, the design of bolting is still more or less empirical. The present study aims to determine the effect of rock bolt on the behavior of fractured rock mass through physical experiments. In this research, a series of tests are performed considering different anchoring positions, anchoring angles and anchoring density. The results were contrastively analyzed both in quality and quantity. Several conclusions were drawn on the basis of analysis.

0 INTRODUCTION

Fractured rock masses are very common in rock engineering practice (Hoek and Brown, 1980; Hudson and Priest, 1979). With the excavation, the original cracks propagate, coalesce and loosen the rock mass. Therefore, the surrounding rock mass may fail and threaten the safety of engineering structures (Hoek and Bieniawski, 1965; Chen et al, 2003; Ashby and Hallam, 1986). Fortunately, rock bolts can mitigate this problem efficiently. Rock bolts bind the rock mass into an integral unit which interacts through the so called suspension function, combination beam function and dowel function (Zhu et al, 2002; Cheng et al, 2003). Besides, the normal stress and shear stress near the cracks provided by rock bolts, they also prevent cracks from propagating (Bobet and Einstein, 1998; Dyskin and Germanovich, 1993). So, the bearing capacity of the rock mass is promoted. Comparing to the wide use and efficient anchoring effect observed in practice, mechanistic studies of rock bolts are meager. So far, the anchoring mechanism can not be interpreted very well. The design of bolting is still more or less empirical. To utilize the rock bolts better, it is necessary to make the mechanism clear. In the present study, physical tests were performed to study the characteristic of anchored samples with different anchoring positions, anchoring angles and anchoring density.

1 EXPERIMENT SETUP

1.1 Choice of analogous material

To simulate the rock material, several principles should be considered.

- Brittleness
- Density
- Ratio of tensile strength to compressive strength

With such considerations, a kind of mixture was chosen, which was developed by Shandong University of China. This kind of material was composed of barite powder, ferrous powder, quartz sand, rosin, alcohol and plaster powder. With different mixture ratio, the material performed different characteristic. After trial testing, a mixture weight ratio was chosen: F:B:S=1:1.5:0.28, G%=2.5%, β=25%, RA%= 5%. Where F indicated ferrous powder, B indicated barite powder, S indicated quartz sand, G% indicated plaster power proportion by weight, β indicated the molarity of rosin-alcohol solution, RA% indicated the rosin-alcohol solution proportion by weight. The mechanical parameters of the mixture were listed in Table 1.

Table 1. Mechanical Parameters of Rocky Material

Young's Modulus (GPa)	Cohesion (MPa)	Internal Friction Angle	Compressive Strength (MPa)	Poisson's Ratio
1.5	0.083-0.108	35-40	5.0	0.22-0.25

The choice of rock bolt material was always a key point, because it was hard to meet all the analogy theory for the mechanical parameters. With reference to literature and trial testing (Benmokrane et al, 1995; Kilic et al, 2000; Zhu and Ren, 2001; Wen, 2007), a kind of bamboo was selected to simulate the rock bolt. The mechanical parameters of the bamboo were E=10. 10GPa, σ_t=132.05MPa. The diameter of bolt and hole were 2.5mm and 3.5 mm respectively.

The grout material was mixture of barite powder and rosin alcohol solution. The weight ratio was barite: rosin: alcohol=1:0.1:0.2.

1.2 Preparation for samples

The sample was in parallelepiped shape. The crack ran through the sample and with an angle of 45 degree to the loading axes. The preparation procedure was listed below. For simplification, only one crack was set in the center of sample.

- Weigh the material according to the ratio and mix them;
- Cast the mixture into the mould and tamp layer by layer;

- Compress the material to parallelepiped shape using indenter of press;
- Demould;
- Do curing until the alcohol volatilize totally;
- Install bolts and grout

FIG. 1 was a picture of a sample without bolt.

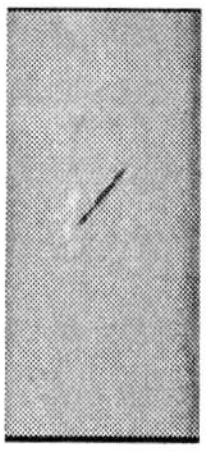

FIG. 1 Picture of unbolted fractured sample.

2 EXPERIMENTAL STUDY ON THE REINFORCEMENT EFFECT

2.1 Test program

The uniaxial tests were carried out considering different anchoring angles, anchoring positions and anchoring density. The bolts were installed perpendicular to loading direction (marked as L) and crack strike direction (marked as S) respectively. For each direction, the bolts were placed in the middle of crack (marked as M) and at the ends of crack (marked as E). The arrangement of bolts was illustrated as FIG. 2. So, altogether there were four groups of tests, which were LM, LE, SM and SE. In each group, the bolts were performed in different quantity.

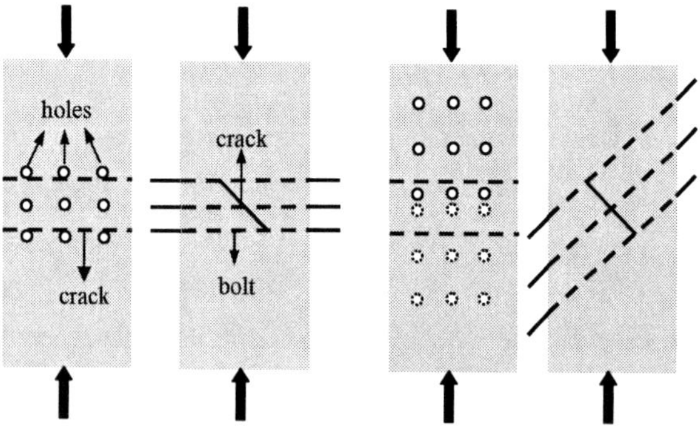

FIG. 2 Sketch of bolts installation.

2.2 Test results

(A) LM series

One, two and three bolts were installed in the middle of crack respectively in this series. FIG. 3 showed the configuration of bolts.

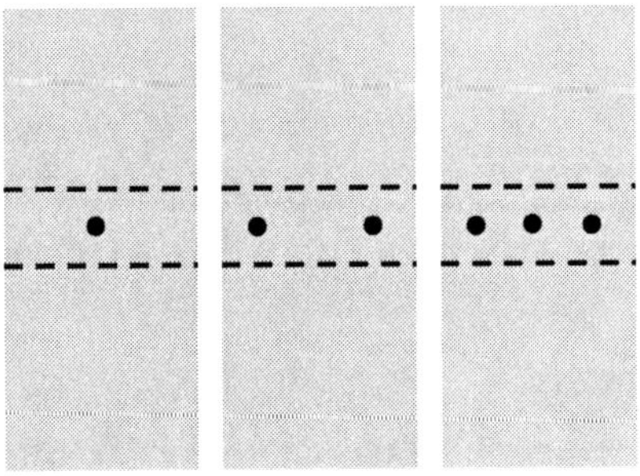

FIG. 3 Different anchoring density in the middle of crack.

The stress-strain curves of these tests were obtained, together with that of the intact sample and jointed sample without bolts, as illustrated in FIG.4.

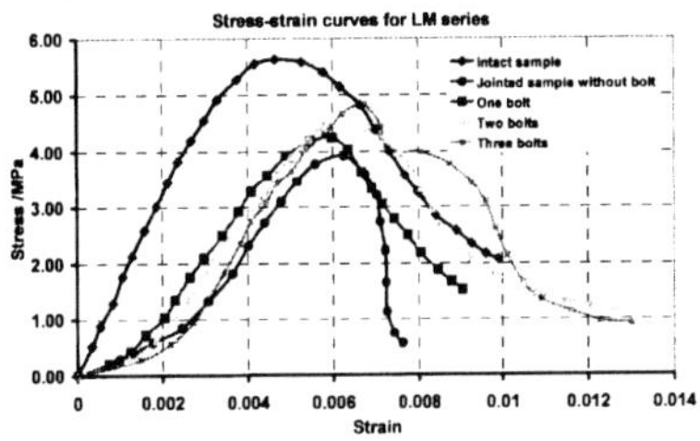

FIG. 4 Stress-strain curves for LM series.

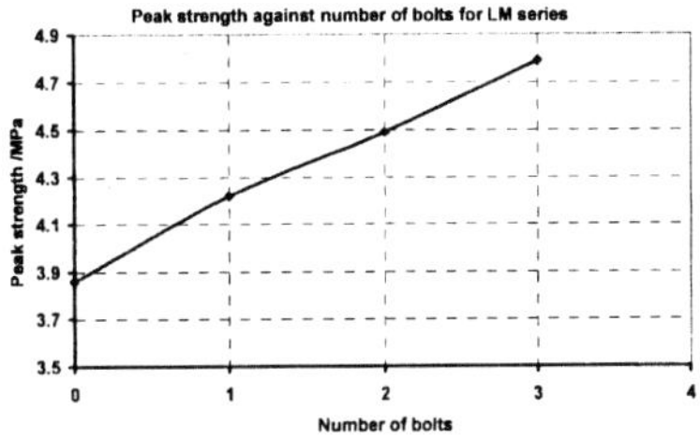

FIG. 5 Curve of peak strength against number of bolts for LM series.

From the stress-strain curve, some information could be derived. Comparing to the jointed sample without bolts, both the peak strength and the ultimate strain of the bolted samples increased. The more bolts were installed, the more increment strength and strain achieved.

The bolts provided shear resistance and normal force to the crack, which could prevent or delay the crack opening and propagation. The sample could endure more loading and more deformation. The failure mode was converted from brittle failure to ductile failure with bolt installation. Therefore, the strength and strain was improved.

The improvement of peak strength for LM series was shown by FIG.5 in quantity. It could be clearly noticed that the peak strength of samples increased linearly against the number of bolts. The slope of curve fitting was 0.306. So, good anchoring effect was obtained in this series.

(B) LE series

Two, three, four and six bolts were installed at the ends of crack respectively in this series. Particularly, for the three bolts case, the bolts were located at one end, as shown in FIG. 6.

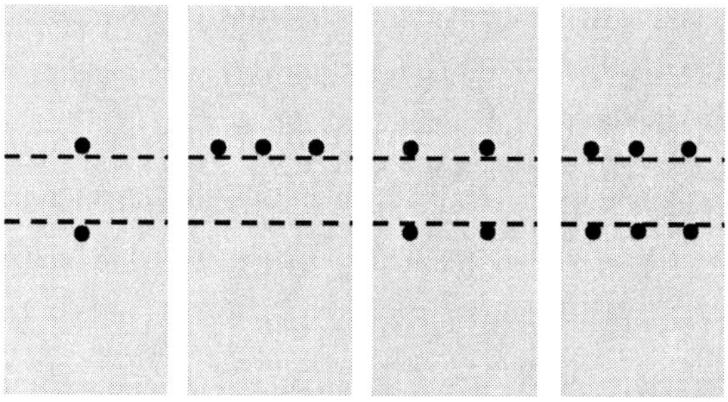

FIG. 6 Different anchoring density at the ends of crack.

In LE series, it was same to LM series that the peak strength and peak stain increased with increment of bolts quantity, except for three bolts case. Anchoring at one end of the crack led to the crack propagation at the other unbolted end. So, this was the reason that the strength of three bolts case was even lower than that of two bolts case.

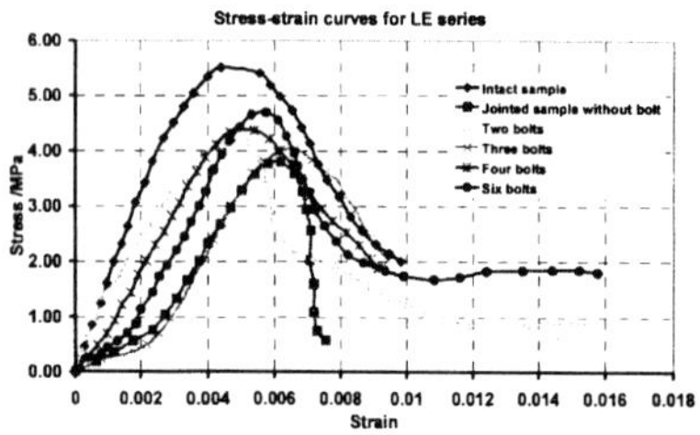

FIG.7 Stress-strain curves for LE series.

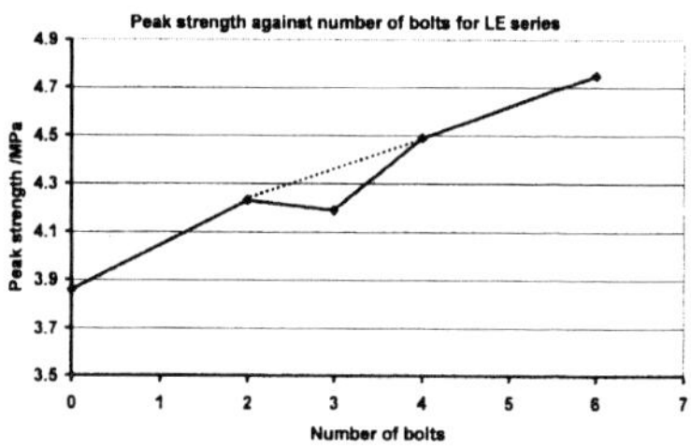

FIG.8 Curve of peak strength against number of bolts for LE series.

As shown by FIG. 8, if the data point of three bolts case was neglected, the peak strength increased almost linearly against the increment of bolts quantity. However, the slope of the line trended towards decreasing. The curve fitting slope of the line was 0.146.

(C) SM series

In this series, the bolts were installed in the middle of crack and perpendicular to the crack strike direction.

There was the same law that the peak strength increased while the number of bolts raising, as illustrated by FIG.9. However, not like "L" series (i.e., LM series and LE series), the ultimate strain didn't follow this law. That was to say, in this anchoring mode, the bolts did not contribute too much to the deformation ability of samples.

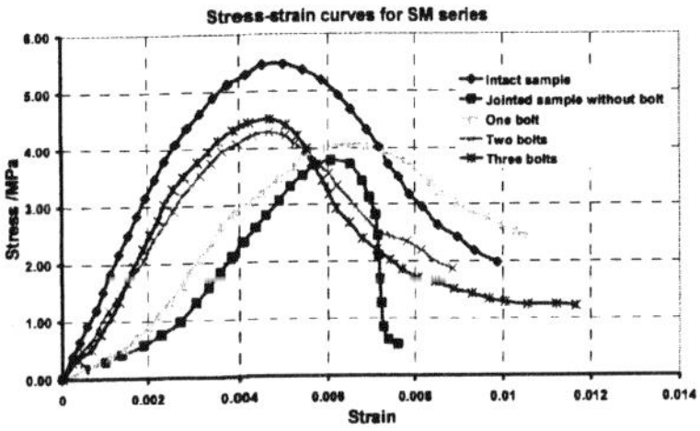

FIG. 9 Stress-strain curves for SM series.

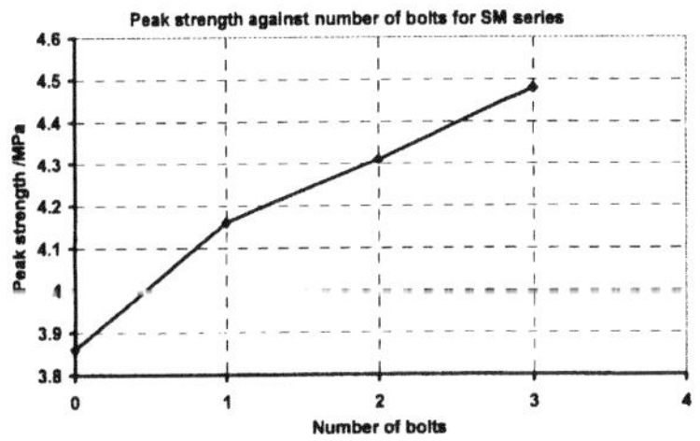

FIG.10 Curve of peak strength against number of bolts for SM series.

The curve of peak strength against number of bolts was shown by Fig. 10. There was an obvious increment between the unbolted case and one bolt case, while the increments for the two bolts case and three bolts case were smaller. The slope of curve fitting was 0.201.

(D) SE series

Two, three, four and six bolts were installed at the ends of crack respectively, just like LE case, while the bolt was perpendicular to the crack strike direction.

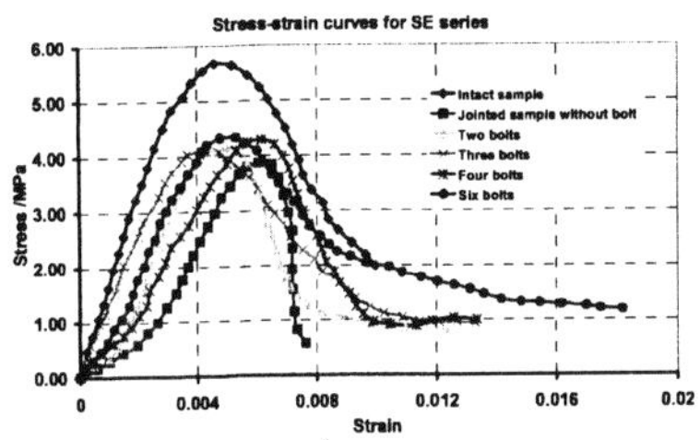

FIG. 11 Stress-strain curves for SE series.

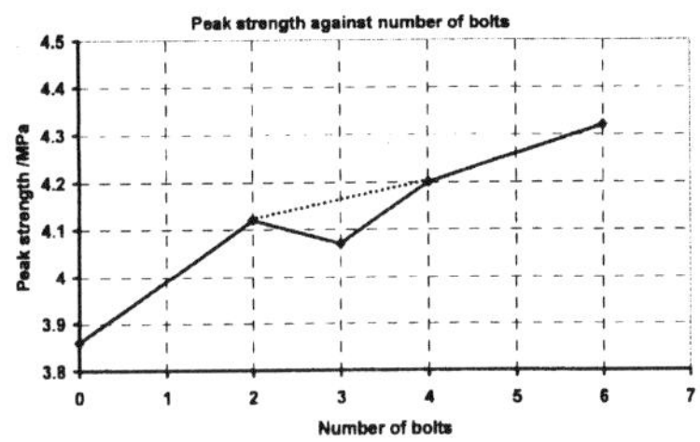

FIG.12 Curve of peak strength against number of bolts for SE series.

It could be noticed that more bolts led to more peak strength generally, as shown in FIG. 11. Similarly to LE series, three bolts case was an exception. Its strength was even smaller than that of two bolts case, illustrated as FIG. 12. Like SM series, there was no clear law about improvement to the ultimate strain with bolts quantity enhancement.

If the point of strength for three bolts case was neglected, the curve about peak strength against number of bolts was an increasing curve. The slope of this curve was about 0.073.

3 ANALYSIS

If the results about the above tests were summarized and contrastively analyzed, several laws could be obtained.

(1) No matter which kind of anchoring mode adopted, the peak strength was evidently improved after bolting. When loaded, the bolt could provide shear resistance and normal force to the crack, which could prevent or retard the crack from opening and propagation. That was so called toughing function and the sample was reinforced by the bolts. Therefore, the peak strength of bolted samples was enhanced. In addition, more bolts installed, more strength increment were achieved.

(2) If the "L" series (LM and LE) and "S" Series (SM and SE) were compared, it could be noticed that the ultimate strain increased with number of bolts increasing for L series, while there was no such a law for S series. That meant that bolting in the "L" mode could improve the ductility of samples more effectively. While the bolt was perpendicular to the crack strike, it contribute the least shear resistance to the crack (Pellet, 1994).

(3) The slopes of the curve fitting about peak strength against number of bolts were summarized in Table. 2. It could be clearly found that the slopes of "L" series were bigger than that of "S" series correspondingly and the slopes of "M" series were bigger than that of "E" series. That was to say, bolting perpendicular to the loading direction was more efficient than the to the crack strike direction. Moreover, better effects could be achieved if the bolts were installed in the middle of the crack than at the ends of crack. So, the best anchoring mode was LM, i.e., installing the bolts perpendicular to the loading direction and in the middle of crack.

(4) Anchoring at only one end of crack was not very effective, because it was easy for the crack to propagate through the other end. It could be recognized from the results of LE and SE series.

Table 2 Slopes of Curves about Peak Strength against Bolts Quantity

	LM series	LE series	SM series	SE series
Slope value	0.306	0.146	0.201	0.073

4 CONCLUSIONS

In the present study, four groups of physical tests were carried out to investigate the anchoring effect of different anchoring mode. It could be concluded that the bolts could effectively improve the mechanical characteristic of fractured samples, especially for the peak strength. More bolts could provide more strength increment to the sample. But, it could be forecast that there must be a threshold value for the number of bolts, beyond which the strength of the sample would not increase anymore. For the anchoring mode, bolting perpendicular to the loading direction was more efficient than perpendicular to the crack strike. Moreover, installing the bolt in the middle of crack could achieve better effect than at the ends of crack. So, the best way was to install the bolts perpendicular to the loading direction and in the middle of crack.

ACKNOWLEDGMENTS

This research was supported by the National Science Fund for Distinguished Young Scholars of China (Grant No 50625927), the Key Project for Science and Technology Research of Ministry of Education (Grant No 108158) and China Scholarship Council.

REFERENCES

Ashby M.F. and Hallam, S.D. (1986). "The failure of brittle solids containing small cracks under compressive stress states." *Acta Metallurgica*, 34: 497-510.

Benmokrane, B., Chennouf, A. and Mitr, H.S. (1995). "Laboratory evaluation of cement-based grouts and grouted rock anchor." *Int. J. Rock Mech. Min. Sci. & Geomech. Abstr.*, Vol. 32(7):633-642.

Bobet, A. and Einstein, H.H. (1998). "Fracture coalescence in rock-type material under uniaxial and biaxial compression.International." *Int. J. Rock Mech. Min. Sci.*, Vol. 35(7): 863-888.

Chen, W.Z., Li, S.C. and Zhu, W.S. (2003). "Experimental and numerical research on crack propagation in rock under compression."(in Chinese) *Chinese Journal of Rock Mechanics and Engineering*, Vol. 22(1):18-23.

Cheng, L.K., Fan, J.L. and Han, J. (2003). Rock and Soil Anchoring.(in Chinese) Beijing: China Architecture &Building Press.

Dyskin, A.V. and Germanovich, L. N. (1993). "Model of crack growth in microcracked rock." *Int. J. Rock Mech. Min. Sci. & Geomech. Abstr.*, Vol. 30(7): 813-820.

Hoek, E. and Bieniawski, Z.T. (1965). "Brittle fracture propagation in rock under compression. " *Int. J. Fracture.*, 1: 137-155.

Hoek, E. and Brown, E.T. (1980). Underground Excavations in Rock. London: Institution of Mining and Metallurgy.

Hudson, J.A. and Priest, S.D. (1979). "Discontinuities and Rock Mass Geometry." *Int. J. Rock Mech. Min. Sci. & Geomech. Abstr.*, Vol. 16(2): 339-362.

Kilic, A., Yasar, E.and Celik, A.G. (2000). "Effect of grout properties on the pun-out load capacity of fully grouted rock bolt." *Tunnelling and Underground Space Technology,* Vol. 17: 355-362.

Pellet, F. (1994). Strength and deformability of jointed rock mass reinforced by rock bolts (Ph.D. thesis). Lausanne: Ecole Polytechnique Fédérale de Lausanne, Switzerland.

Wen, N.D. (2007). The Study of Anchorage Styles Optimization of Fractured Rock Mass on Experiment and Numerical Simulation (Master thesis). Jinan: Shandong University, China.

Zhu, W.S. and Ren, W.Z. (2001). "Research on reinforcing effect for jointed rock masses of shiplock slope by physical modeling."(in Chines) *Chinese Journal of Rock Mechanics and Engineering*, Vol. 20(5): 720-725.

Zhu, W.S., Li, S.C. and Chen, W.Z (2002). Failure Mechanism and Anchoring Effect of Jointed Rock Mass and Engineering Application.(in Chinese) Beijing: Science Press.

Study on evaluation method of fire safety of tunnel lining structure

Zhiguo YAN[1], Tongji University and Hehua ZHU[2], Tongji University

[1] Key Laboratory of Geotechnical and Underground Engineering of Ministry of Education, Tongji University ; Department of Geotechnical Engineering, Tongji University, Shanghai 200092, China; yanzguo@tongji.edu.cn

[2] Key Laboratory of Geotechnical and Underground Engineering of Ministry of Education, Tongji University ; Department of Geotechnical Engineering, Tongji University, Shanghai 200092, China; zhuhehua@tongji.edu.cn

ABSTRACT: High temperature will results in severe damage to tunnel lining structure. To evaluate safety status of tunnel lining structure in fire accidents, an evaluation method based on Fuzzy Comprehensive Evaluation is established. A series of evaluation indicators which include basic parameters of tunnel, characteristic of lining concrete, surrounding conditions of tunnel, fire scenarios and fire-prevention methods are employed in the method. The weight and quantitative score of these evaluation indicators are determined through the Analytic Hierarchy Process (AHP) based on fire experiments and investigation. Furthermore, evaluation process and safety standard are present. The method is utilized successfully in the Shanghai Yangtze River Tunnel, and according to the evaluation results, the suggestions how to improve fire-protection ability of lining structure are obtained as follows: a)mixing polypropylene fiber into lining concrete to prevent concrete from spalling, 2) to fill handhole of segment with concrete or cement mortar to protect linking bolt from fire.

INTRODUCTION

There are a number of fire accidents with high peak temperature, rapid heating velocity and long duration in traffic tunnels. For example, Nihonzaka tunnel fire in 1979, the Channel Tunnel fire in 1996, the Mont Blanc Tunnel fire in 1999 and St. Gotthard Tunnel fire in 2001, etc (Haack, 2002). All of these fire accidents show that fire results in severe damage to tunnel lining structure in terms of concrete spalling, deterioration of impermeability durability of concrete, reduction of physical-mechanical characteristics and load bearing capacity. And these fire damage to tunnel lining may lead to collapse of tunnel structure. As shown in FIG. 1, the Channel Tunnel fire causes severe damage to tunnel lining. Furthermore, lining structures are also severely damaged resulting from rapid heating velocity and high peak temperature in the full-scaled experiments carried out by Tongji University (FIG.2).

To ensure the safety of tunnel lining structure in fire accidents, how to evaluate the ability of fire prevention of structure is a very important issue in the process of tunnel design, construction and operation.

This paper briefly introduces the evaluation method of fire safety of tunnel lining structure and presents a case study to verify the method.

FIG.1. Fire damage to tunnel segments in Channel Tunnel fire (Kirkland, 2002).

FIG. 2. Concrete spalling under high temperature.

EVALUATION INDICATORS AND SAFETY STANDARD

There are a lot of indicators that contribute to safety level of tunnel lining structure in fire accidents. Based on the theoretical analysis, experiments and investigation, the key indicators are selected and divided into five classes:

1) Basic parameters of tunnel (S1, W1)
2) Characteristic of lining concrete (S2, W2)
3) Surrounding conditions of tunnel (S3, W3)
4) Fire scenarios (S4, W4)
5) Fire-prevention methods (S5, W5)

And as shown in FIG.3, each class also include numerous sub-classes, which can be evaluated quantificationally (namely give quantitative score) through Single Factor

Fuzzy Comprehensive Evaluation or Experts' Scoring Method. For example, for class 4, its sub-class of maximum temperature (S41, W41) can be defined as Table 1 through Experts' Scoring Method.

Furthermore, based on Experts' Scoring Method, the safety standard is established, as shown in Table2.

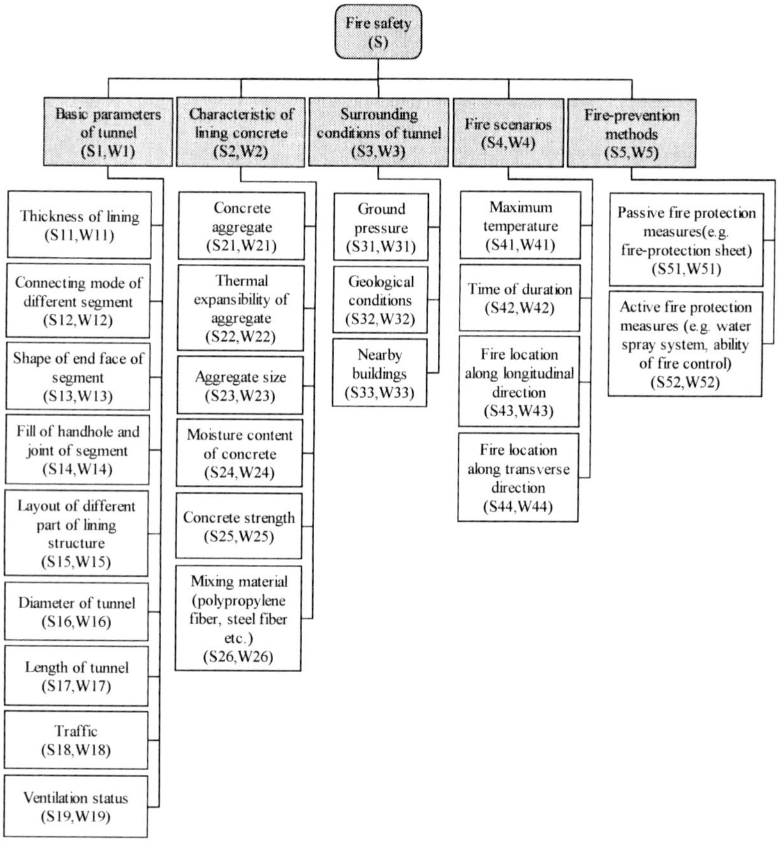

FIG. 3. Evaluation indicators in the method.

Table 1. Score of Maximum temperature (S41, W41)

Temperature(°C)	500-600	800-900	1200	1300-1400
Description	Car accident	Carriage(Bus) accident	Heavy truck accident	Tanker accident
Score	4	3	1	0

Table 2. Safety standard of tunnel lining structure

Safety standard	Value of fuzzy comprehensive evaluation (S)	Description
A	3.5-4	Lining structure has ability to prevent fire, structure is safe in fire accidents
B	2.5-3.5	Lining structure has limited ability to prevent fire, structure is safe in a certain extent in fire accidents
C	1.5-2.5	Lining structure has no enough ability to prevent fire, structure is dangerous in fire accidents
D	<1.5	Lining structure has no ability to prevent fire, structure is severe dangerous in fire accidents

EVALUATION PROCESS

Considering fuzzy relationship between safety standard and evaluation indicators and randomicity of evaluation indicators, Fuzzy Comprehensive Evaluation is employed in this method.

Cauchy distributions are selected as membership function for quantitative sub-class indicators in the process of single factor fuzzy comprehensive evaluation (FIG.4).

Furthermore, to determine the weight of evaluation indicators, Exponential Scale Method is used in the method (Hou and Shen, 1995). As shown in Table 3, the difference between indicator A and B can be described quantificationally by different weight.

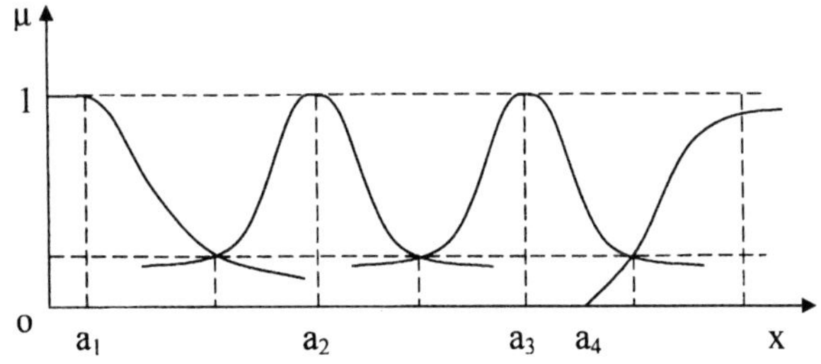

FIG. 4. Cauchy distribution in the model

Table 3. Weight of evaluation indicators based on Exponential Scale Method

Description	A is equal to B	A is little bigger than B	A is bigger than B obviously	A is bigger than B significantly
Weight	1	1.276	2.080	4.327

In briefly, the evaluation process can be summarized as follows:

$$S_{i,j} = R_{C_{i,j}} \cdot V^T = [r_{C_{i,j}A}, r_{C_{i,j}B}, r_{C_{i,j}C}, r_{C_{i,j}D}][v_1, v_2, v_3, v_4]^T \tag{1}$$

$$r_{C_{i,j}m} = \mu_m(C_{i,j}), \qquad m = A, B, C, D \tag{2}$$

$$S_i = W_{i,j} \cdot S_{i,j}^{\ T} \tag{3}$$

$$S = W_i \cdot S_i^T \tag{4}$$

Where, $S_{i,j}$ is evaluation value of sub-class indicator $C_{i,j}$, it also can be assigned a value through Experts' Scoring Method. $r_{C_{i,j}A}, r_{C_{i,j}B}, r_{C_{i,j}C}, r_{C_{i,j}D}$ is degree of membership of sub-class indicator $C_{i,j}$ to safety standard A, B, C and D, respectively. v_1, v_2, v_3, v_4 is equal to 4, 3, 2, and 1, respectively. $\mu_m(C_{i,j})$ is member ship of sub-class indicator $C_{i,j}$ to safety standard A, B, C or D. $W_{i,j}$, W_i is weight of sub-class indicator $C_{i,j}$ and class indicator C_i, respectively. S_i, S is evaluation value of class indicator C_i and final Value of fuzzy comprehensive evaluation of lining structure, respectively.

CASE STUDY

The Shanghai Yangtze River Tunnel is 7.5 km long with outer diameter 15m and inner diameter 13.7m (FIG. 5). It is the largest diameter shield tunnel in the world. The lining structure is composed by ten reinforced concrete segments per ring. The segments are connected each other by 3 bending M36 bolts in circumferential direction, and segment rings are connected each other by 38 sloping straight M30 bolts in longitudinal direction.

The traffic flow includes cars with restricted speed 80km/h and metro trains in lower-story of the tunnel.

To ensure the safety of tunnel in the long operation stage, evaluation of fire safety of lining structure is carried out based on the forementioned method. The results show that the fire safety level of the lining structure is A, which means that *Lining structure has ability to prevent fire; structure is safe in fire accidents.*

Based on the evaluating results, the suggestions how to further improve fire-protection ability of lining structure are present as follows: a) mixing polypropylene fiber into lining concrete to prevent concrete from spalling, 2) to fill handhole of segment with concrete or cement mortar to protect linking bolt from fire.

(a) (b)

FIG. 5. The Shanghai Yangtze River Tunnel.

CONCLUSIONS

Fire accidents will result in severe damage to tunnel lining structure. To evaluate fire safety of lining structure, a method is present in this paper and is used to evaluate the safety status of the Shanghai Yangtze River Tunnel in fire accident. Furthermore, based on the evaluating results, suggests how to further improve fire-protection ability of lining structure are present.

The case study show that the evaluation method is feasible to describe the fire safety level of tunnel lining structure, and evaluation indicators and safety standard are suitable.

However, considering the complexity of fire accidents, the further works includes: 1) probability nature of evaluation indicators should be considered perfectly in the method. 2) more experiments and analysis are needed to contribute to the definition of quantitative value of evaluation indicators. 3) Optimization on evaluation model to satisfy the nature of fire safety of lining structure in fire accidents.

ACKNOWLEDGMENTS

The authors appreciate the support of the Project(50808137)supported by National Natural Science Foundation of China and Key project (2006BAJ27B04 、 2006BAJ27B05) supported by National Science & Technology Pillar Program.

REFERENCES

Haack, A. (2002). "Current safety issues in traffic tunnels". *Tunnelling and Underground Space Technology*. 17,117-127.

Hou Y.H.and Shen D.J. (1995). " Index number scale and comparison with other scales." *Systems Engineering-theory & Practice*. (10), 43-46.

Kirkland, C.J. (2002). "The fire in the Channel Tunnel." *Tunnelling and Underground Space Technology*. 17(2), 129-132.

Investigation into Artificial Ground Freezing Technique for a Cross Passage in Metro

Li Dayong[1], Wang Hui[2]

[1]Professor, College of civil engineering, Shandong University of Science and Technology. Qingdao, China ,266510; ldy@sdust.edu.cn
[2]Senior engineer, Guangzhou Subway Design and Research Institute, Guangzhou, China, 510010; wanghui@dtsjy.com

ABSTRACT: This paper studied a practical application of ground freezing in a cross passage construction of the Nanjing metro in the southeastern China. During the process of the ground freezing, temperature changes in brine, and soil; deformation changes of the ground surface and of a tunnel were monitored. Based on the field measurements, this paper revealed laws of the changes of the above factors and proposed the guidance on ground freezing method for the cross passage construction.

INTRODUCTION

Ground freezing first used in the 1860s in coal mines of Wales has been practiced for over a century [1]. So far it has been performed for groundwater cutoffs, earth supports for access shafts, tunnel excavations, for creating frozen earth cofferdams for excavations, and for temporary underpinning [2]. Nowadays, it is also widely implemented in the construction of metros, for example, for cross passages in some cities of China such as Nanjing, Shanghai, and Guangzhou.

A cross-passage acting as connecting two tunnels is commonly located in the middle of tunnels where a lowest position is. It meets two objectives: One is to act as a structure against fire disaster so that passengers can escape through it when a fire occurring; the other is to drain waste water from the subway station, and from the leakage from cracks of the lining, and both gathering in a sump (See Fig.1 (a)).

Cross passages are frequently built between parallel twin tunnels with artificial ground freezing techniques in many cities of China because freezing can add significant strength to weak soils. Furthermore, frozen soil is impermeable, and the freezing process does not pollute the surrounding groundwater and soil. Therefore, the ground freezing approach is extensively attractive to engineers. However, the volume changes of the soft soil during freezing and thawing may cause adverse effects on the nearby infrastructures such as roads, buildings, and buried pipelines. To reduce this negative effect, some factors mentioned listed in the abstract of this paper should be explored in the field. Consequently, on the basis of a case history the authors were involved in, some research in this paper on a cross passage using ground freezing method was carried out in order to guarantee the safety of the construction.

DIAOYUTAI-SANSHANJIE (D-S) CROSS PASSAGE PROJECT

D-S cross passage project is affiliated to Nanjing line No.1 metro. The line No.1 is 21.7km long among which 14.5km is below ground, and it was started to construct in the end of year 2000 and completed in 2004. Since the year 2005, the line No.1 of Nanjing Metro has been in smooth operation.

The tunnels were built with earth pressure balanced (EPB) shield machine and rock bored method according to different ground conditions. One ring of the tunnel lining consists of 6 precast concrete segments with 1.2m wide and 0.35m thick. The tunnels are 5.5m in inner diameter, thus 6.2m in outer diameter. The cross passage located in the middle of the tunnels consists of the two main parts: a passage, and a sump. Ground freezing is readily performed to form frozen soil wall curtains resisting pressures from soil and water outside the frozen zone. Soil inside the frozen soil zone as retaining structure is excavated using compacted air picks. Hence it is important for us to determine the proper time to meet the needs of excavation soil. To achieve this objective, some field measurements such as monitoring brine temperatures, soil frozen temperatures, ground surface settlements, and the existing tunnels' deformations should be conducted. Based on the analyses of the monitoring results, we can determine the appropriate time for excavating the soil within the frozen zone.

ENGINEERING GEOLOGY FOR THE CROSS PASSAGE

Physical parameters for the soils surrounding the cross passage are shown in Table 1. Moreover, the water table varies yearly from 1 m to 2m below the ground surface. The cross passage situates in soil layers both silt□-2 and silty sand with abundant water and low permeability. This situation is suitable for adopting ground freezing technology.

Table 1. Physical Parameters for the Soils

Soils	Thickness(m)	Water content ω(%)	Permeability k(×10^{-6}cm/s)
Fill □	3.5	34.7	12.7
Silt□-1	2.1	33.5	4.23~7.02
Silt□-2	6.4	37.8	1.03~6.67
Silty sand □-2	3.8	32.1	7.31~256
Silty clay③-2	4.5	28	2.6

DESIGN OF GROUND FREEZING APPROACH

Setup of freezing tubes

The patterns of installation of the freezing tubes consist of three manners according to the slopes of freezing tubes with respect to the horizontal, i.e. up inclined, horizontal, and down inclined, shown in Fig.1. In a circular pattern, 58 freezing tubes' holes are bored with a spacing varying from 0.6m to 0.7m.

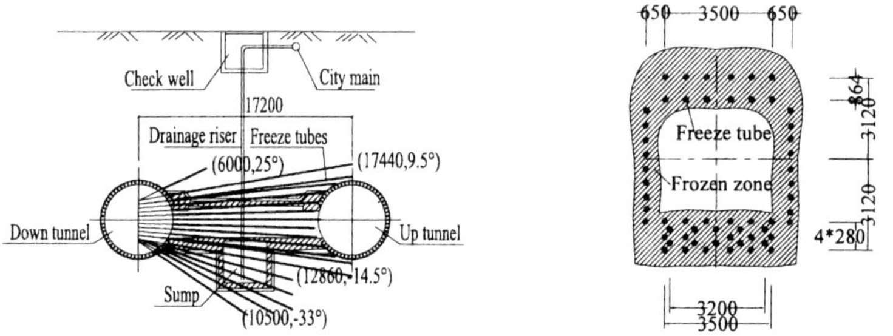

(a) Elevation view (b) Cross-section view

FIG.1 Installation of freeze pipes in soil. (units:mm)

Ground freezing equipment and design parameters for freezing

Ground freezing is carried out using brine glaciations (NaCl) since liquid nitrogen in the tunnel is excluded for safety reasons. A freezing unit is developed for soil freezing at temperatures of the freezing tubes at -28^{0}C to -30^{0}C. This unit has a content of 87500 kCal/h with a power of 110kW. The flow flux of each freezing tube is no less than 4m^3/h. The desired duration of overlapped frozen soil circles is 20 days. The duration when the frozen soil reaches the design thickness is 30 days. In addition, the sustaining duration of soil freezing is 35days. The auxiliary equipment for the ground freezing includes two sets of brine cycling pump with a flow flux of 200m^3/h and a power of 45kW, and one cooling tower providing fresh water of 15m^3/h.

MONITORING RESULTS

The in-situ monitoring in this study contains: (1) in-situ soil monitoring during the period of artificial soil freezing (2) influences of soil freezing on the existing tunnels (3) monitoring surface settlements due to soil freezing. Based on the results of the in-situ monitoring, the assessment of existing tunnels is investigated, and the strength of frozen soil satisfying with the demanded should be determined, i.e. what time is appropriate when soil inside the cross passage will be excavated.

The duration of construction for the cross passage was divided into three stages. In the first stage, it took 12 days to bore 58 holes. Ground freezing before soil excavation inside the frozen zone lasted 39days. Ground freezing during soil excavation and the construction of supporting structures lasted 34 days. Therefore, the total freezing duration was 73days. In the second stage, the duration of compensation grouting in case of ground settlements was 2 days. Finally, when the construction was completed, detaching accessories used 2 days. During the whole process, temperatures of brine, temperatures of the freezing soil, deformations of the ground surface, and movements of the existing tunnel were monitored.

Temperatures of brine in closed circuits

The system for this project consists of two independent closed circuits. One of these is comprised of a pump and freeze pipes with their accessorial hoses, valves, and fittings, and the other includes a pipe configuration that fits into a heat exchanger.

Taking # 2 freezing tube as an example, the temperature diagrams are shown in Fig.2. Results of other freezing tubes are similar to those of # 2. The freezing started on 29 Nov. 2002. It shows that at the beginning of the freezing, the difference of temperatures of the brine between supply and return pipes decreases remarkably. Until 37th day (on 7 January 2003) of the freezing, the differential temperature is no more than 0.5^0C, which shows the heat in the soil transfers little to the air through the brine. Therefore, that means the overlapped soil frozen wall curtains are in a good shape. At the least, after this moment, we may begin to take soil excavation into account.

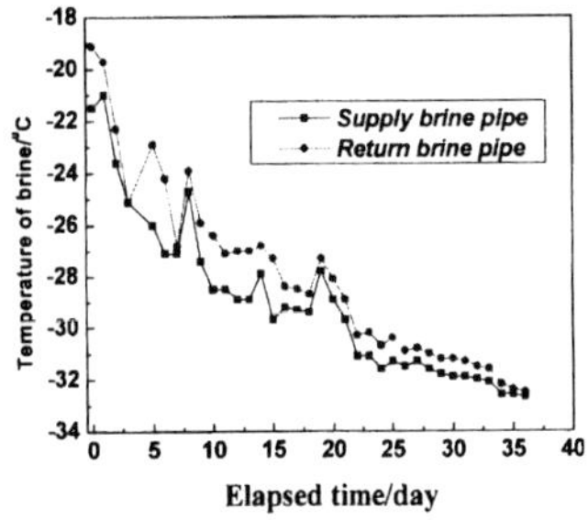

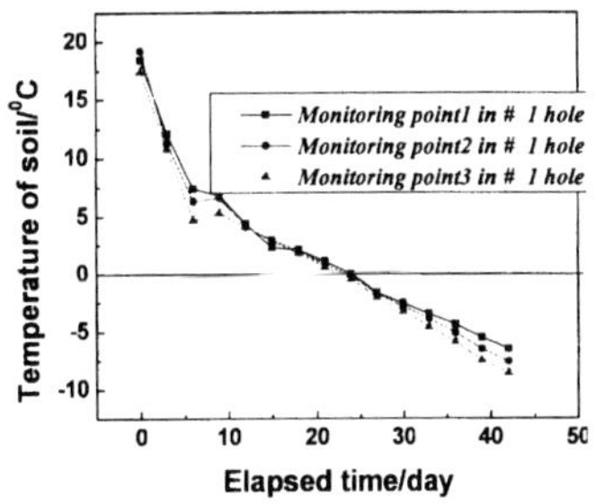

FIG.2. Brine temperature and elapsed time. FIG.3. Soil temperatures and elapsed time.

Temperatures of soil due to ground freezing

To monitor temperatures of the freezing soil, ten probe holes (See Fig.4) are set up near the two entrances to the cross passage, four on the side of the down tunnel, six on the side of the up tunnel. Three thermo-transducers are placed for each monitoring hole. The spacing between two neighboring monitoring points is 600mm, and each monitoring hole is 2m in depth. Soil in the cross passage is started to be excavated from down tunnel.

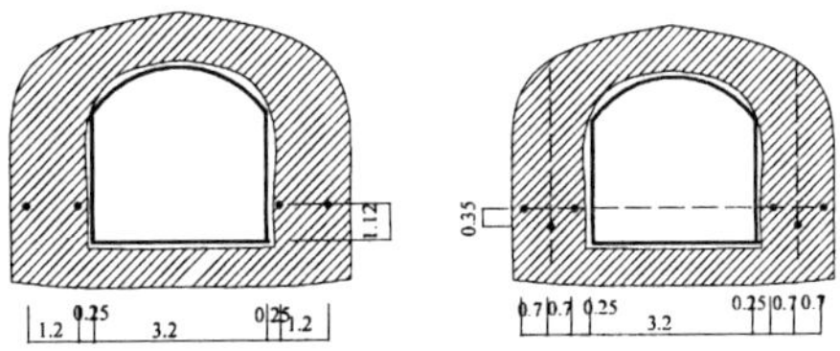

(a) In down tunnel (b) In up tunnel

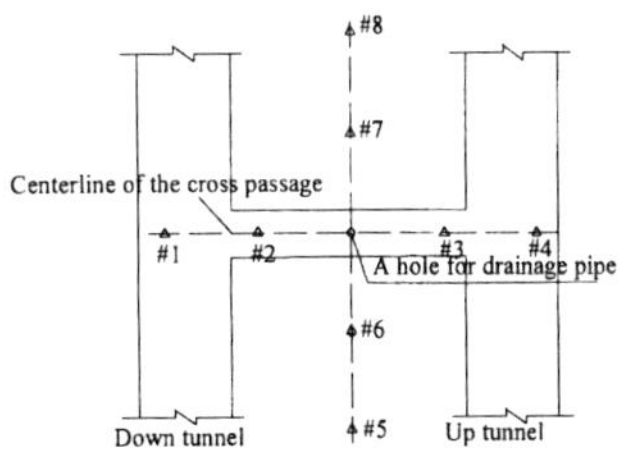

FIG.4. Temperature holes. (units: m) FIG.5. Survey points on the ground surface above the cross passage.

Data records of the frozen soil monitoring are given against elapsed time in figures 3,6 and 7. Taking # 1 probe hole as an example in Fig.3, although the differences of temperatures in the three monitoring points are not very apparent, we still may obtain a trend: The deeper the hole is the lager the decrease in temperature is. This reason results from the thermal interchanging between soil and the outer space.

The trends of the decrease in temperatures of soil are very similar, shown in Figs.3 and 6. When the freezing lasted 15 days, the development of soil freezing reached at an average speed of 26.7mm/d according to an evaluation. The temperature exploring holes are installed symmetrically in up and down tunnels. The results shown in Fig.5 to Fig.7 are almost similar; hence, it shows the freezing soil is homogeneous. Up to 8 January 2003, the day of soil excavation start, the freezing lasted 39days, and the thickness of frozen soil reached 2.06m in excess of the designed thickness, 0.46m.

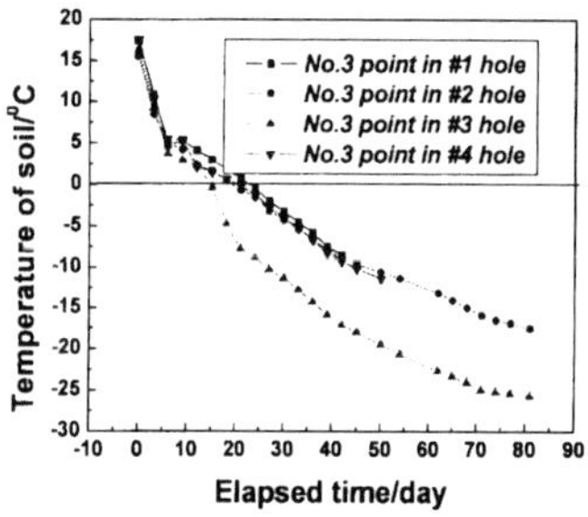

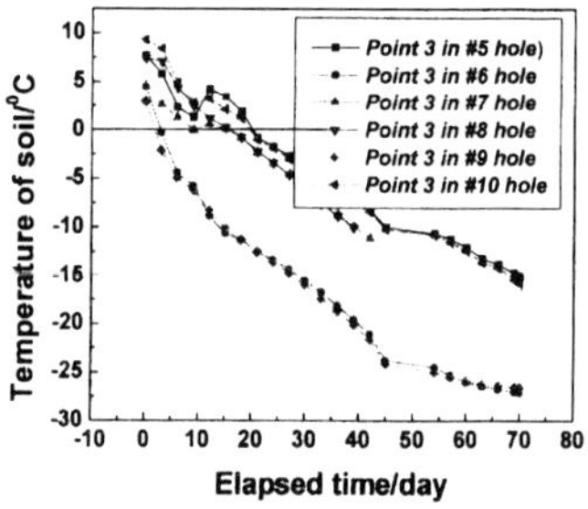

FIG.6. Soil temperature and time in down tunnel. FIG.7. Soil temperature and time in up tunnel.

Monitoring settlement/heave of ground surface

The survey points with spacing of 5m are illustrated in Fig.5. The Fig.8 shows: During phase I (boring holes), the ground surface increases to settles to a maximum value due to boring holes. Then it increase to heave gradually from settlement state to heave state due to gradual soil freezing developing. This phenomenon reflects features of ground freezing (See in phase II-active freezing). During phase III (retaining freezing), the deformation of ground surface tends towards an even state. It also shows in this phase that soil freezing develops slowly meaning to form a complete frozen overlap. That deformation of every survey point is not the same results from the uneven distributions of freeze tubes and in-situ soil properties. During phase IV (freezing complete), to prevent the ground from the excessive settlement due to frozen soil thawing, ground grouting was taken. The even curves in Fig.8 prove that grouting adopted is an effective approach to preventing excessive settlements.

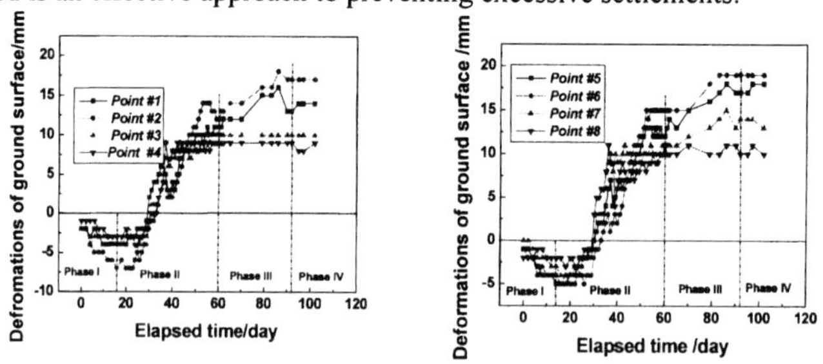

(a) Survey point in the longitudinal direction (b) Survey points in the transverse direction

FIG.8. Relations between settlement/heave of ground surface and elapsed time.

Deformation of the lining of a tunnel due to ground freezing

Ground freezing may have adverse impact on the lining of tunnels. It causes tunnels to move obliquely due to the increase in volume of clay with freezing. Excessive movement of tunnels may cause the water leakage. To explore this movement, some points with an interval of 3m are placed in the crown of the down tunnel. The movement is divided into two components: the horizontal part, and the vertical part shown in Figs.9 and 10 respectively. Due to the inflation of freezing soil, tunnel moves to a maximum value of 32mm. During frozen soil thawing, tunnels approach its initial positions. The movement of tunnels is so large to some extent that some water leakage occurred in few interfaces between precast concrete segments. But actually, in the project, these problems induced by ground freezing are not serious, and almost negligible. The ground freezing used in the project proved greatly successful.

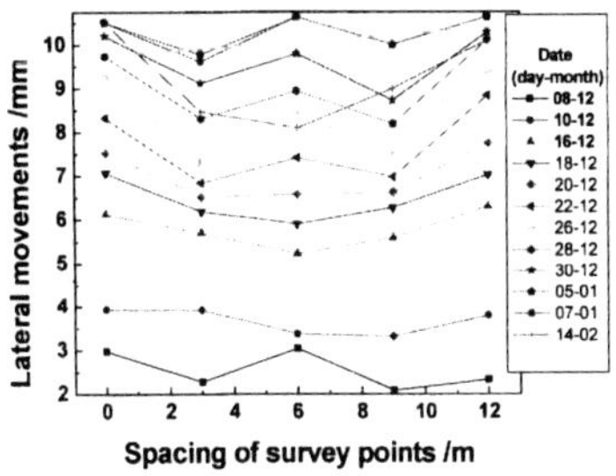

FIG. 9. Lateral movements.

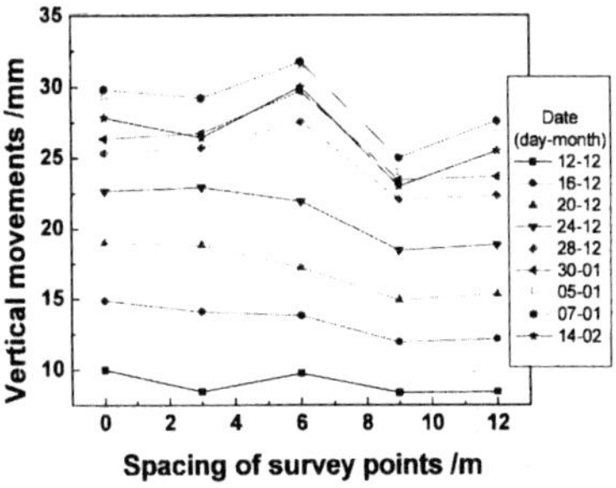

FIG. 10. Vertical movements.

CONCLUSIONS

The practical project shows that ground freezing is very suitable for sandy soils with a high water table and low permeability. However, to guarantee the success for the construction, many monitoring measures such as monitoring temperatures of brine and soil, deformations of ground surface and lining should be taken to feedback information to guide the construction all the time.

ACKNOWLEDGMENTS

The authors appreciate the support of Shangdong "Taishan Scholar Fund", and Shandong Province Natural Science Foundation (ZR2009FM003).

REFERENCES

Jessberger H. (1979). "Ground freezing." Elsevier Science Publishing Company.

Andersland O. and Ladany B. (2004). " Frozen ground engineering (second edition) ". John Wiely & Sons , Inc.

The Studies on Intelligent Construction Pre-control of a Foundation Pit in Shanghai

Ming Wang[1] Youliang Chen[1,2] Junhua Wu[1]

[1]School of Urban Construction and Environmental Engineering, University of Shanghai for Science and Technology, 516 Jungong Road, 200093 Shanghai, PR China;wangming03101@163.com
[2]Department of Engineering Geology and Hydrogeology, RWTH Aachen University, 52064 Aachen, Germany;chenyouliang2001@yahoo.com.cn

ABSTRACT: Controlling the displacements of foundation pit walls and supporting structures is one of the most effective measures for ensuring safety of foundation pit construction. Because of this, the accurate displacement computation of foundation pit walls and supporting structures is paid great attention by geotechnical scientists and engineers. In this paper, the three-dimensional displacement field of Deping Station foundation pit, which is part of Shanghai Track Traffic Line 6 project, is computed by means of ANSYS software. The "Artificial neural networks (ANNs)" have very excellent ability in simulating nonlinear and complicated problems. In the current paper, the authors try to combine FEM method with ANNs so as to improve the computational accuracy. Basing on the computed results, fuzzy control theory is applied to construction pre-control. The application effect is satisfactory. The research result of current paper is very helpful for geotechnical construction and the development of geotechnical theory.
Key words: Foundation pit, FEM method, Artificial neural networks, Fuzzy logic, Construction pre-control

INTRODUCTION

With the rapid growth of urban population and the lack of building land, more and more super high-rise buildings and city track traffic lines are built in Shanghai. And many buildings and track traffic lines must be constructed in very complicated environments, for instance, some buildings, municipal pipes and wires are situated closely around the construction site. Under these circumstances, the designing and construction ideas of foundation pits must be adjusted in order to fit in with the complicated environments.

Li Jia-Chuan (1992) adopted incoordinate finite element method with 8 nodes to compute three-dimensional stress and displacement fields of some foundation pits, in which continuous underground walls are employed as supporting structures. Wang Guang-Guo (1994) employed large-deformation theory to study the displacements of some foundation pits and summed up the applicable scope of large- and small-deformation theory separately. By utilizing ANNs, fuzzy logic and Matlab 5.2, Yuan

Jin-Rong and Sun Jun (2001) have developed a software packet for predicting and controlling foundation pit construction and applied it to analyzing Metro construction in Shanghai. Chua CG and Goh ATC (2003) have adopted the hybrid Bayesian back-propagation neural network methodology to conduct multivariate modelling. Basma Adnan A. and Kallas Nabil (2004) have successfully employed BP neural networks to model soil collapse. Goh ATC and Kulhawy FH (2005) have employed artificial neural networks to study the reliability assessment of serviceability performance of braced retaining walls.

In this paper, the displacements of continuous underground wall are computed and the pre-control of foundation pit construction is conducted by means three-dimensional FEM method, artificial neural networks and fuzzy logic.

PROJECT BACKGROUND

The main supporting structure of Deping station foundation pit is continuous underground walls with the thickness of 600mm. Round lock-mouth pipe is employed to connect the normal part of continuous underground wall and its end pit section. The end pit section of continuous underground walls is 27m in depth, and the normal part, 24m in depth. The foundation pit bottom of border between the normal part and end pit section is consolidated by means of grouting. The consolidated depth is 4.0 m. 67 drilling grouting piles are arranged below the foundation pit bottom for anti-floating. The pile diameter is 600mm and the depth of piles into soil layers is 20m. The pile top is at the same height as the lower end of bottom beams. The bottom soil of boundary between station and track lane should be consolidated by means of deep stirring piles before the digging of foundation pit.

THE BASIC THEORY OF ARTIFICIAL NEURAL NETWORKS AND FUZZY CONTROL

Artificial neural networks (ANNs) are a broad category of computer algorithms which have the ability to learn some target values (desired output) from a set of chosen input data that has been introduced to the network under both supervised and self-adjusted or unsupervised learning algorithms. The learning or training process is achieved by minimizing the error between the desired output and the output computed by the neural network using different learning algorithms. The predictive ability of the trained neural network can be tested by adding observations not included previously.

Fuzzy models describe input–output relationships by fuzzy if-then rules (fuzzy propositions). They make use of fuzzy sets and approximate reasoning to find an overall `good enough' solution to a particular problem domain without using detailed first-principle knowledge of that domain. Fuzzy rules may be formulated on the basis of expert knowledge of the system.

An interesting and attractive characteristic of fuzzy logic is that the training samples may be both linguistic (symbolic) and numeric ones. Fuzzy logic have reasoning and switch functions besides its prediction ability, they can produce both quantitative and qualitative results.

THREE-DIMENSIONAL FEM COMPUTATION

In this paper, 8-node three-dimensional solid element and Druck-prager ideal elastic-plastic model are adopted to compute the stress and displacement fields. 4-node plate element is employed to simulate the reinforced concrete continuous underground walls. Bar element is adopted to imitate the internal steel bar supports. It is assumed that the continuous underground walls are consisted of nonlinear elastic material and internal steel bar supports are composed of linear elastic material.

The control of displacement field of continuous underground walls is especially important for ensuring the safety of foundation pit construction. So, in this section, the displacement field of continuous underground walls is discussed. As the tentative work, only the displacements under following states are studied.

State 1 : Having finished the second digging step;

State 2 : Having finished the third digging step;

State 3 : Having finished the fourth digging step and the relative inner supports.

The horizontal displacements of continuous underground walls for the fourth vertical rank of nodes at different depth are computed by means of ANSYS and shown in Table 2. These nodes are situated in the normal section of continuous underground walls. The positive x-direction is perpendicular to the continuous underground wall and from outside to inside. The y-axis is parallel to continuous underground wall.

Table 1. The computed horizontal displacements of continuous underground walls at different depth (displacement unit: mm)

Depth(m)	State 1		State 2		State3	
	In x-axis direction	In y-axis direction	In x-axis direction	In y-axis direction	In x-axis direction	In y-axis direction
0	2.7472	-1.2513	4.0118	-1.7255	5.9862	-1.0332
2	5.1498	-2.4253	7.0368	-5.2124	9.8032	-7.8134
4	8.9744	-5.6021	12.246	-10.2131	16.658	-12.2543
6	8.6128	-5.6284	25.77	-10.7159	35.354	-21.6237
8	7.8235	-4.1002	27.455	-10.681	37.724	-23.3538
10	6.6032	-3.3245	25.238	-7.9419	34.667	-20.978
12	5.2134	-3.0235	18.741	-6.1586	25.669	-14.653
14	1.2761	-2.4913	4.3256	-3.8045	5.8249	-5.6149
16	0.226	-1.8167	0.7857	-2.2621	0.8957	-1.9961
18	0.0689	-0.8538	0.4052	-1.1298	0.5024	-1.1867

20	0.3976	-0.1037	0.7794	-0.2082	1.1647	-0.4267
22	0.9988	-0.3447	1.607	-0.352	2.4484	-0.0695
24	0.1797	-0.6904	0.2729	-0.8325	0.4122	-0.1823

BP-SIMULATION OF THE COMPUTED RESULTS

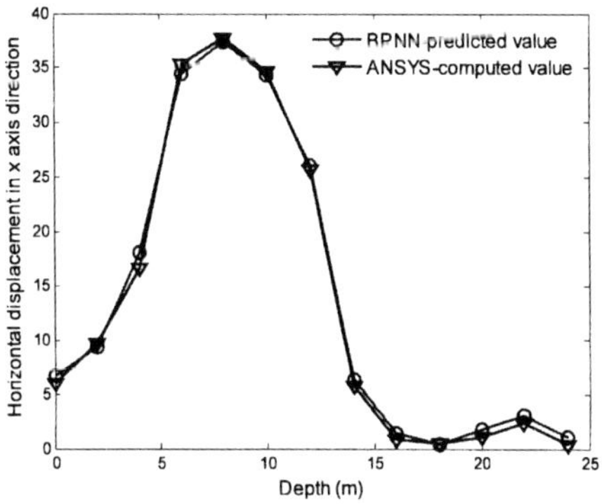

FIG.1. The contrast between ANSYS-computed and BPNN-predicted horizontal displacements in x-axis direction (Displacement unit: mm)

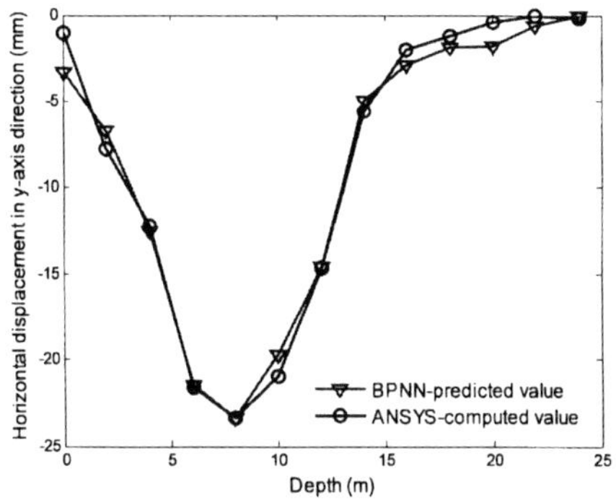

FIG.2. The contrast between ANSYS-computed and BPNN-predicted horizontal displacements in y-axis direction

Based on the computed data of States 1-2, a BP neural network model is established. The input layer is composed of digging depth of foundation pit, elastic modulus of soil, cohesion intercept and angle of shearing resistance. The output layer is composed of displacement of continuous underground walls. 3-layer BP neural network model is adopted. The sigmoid function is employed by BP neural network. The displacement of State 3 will be predicted by means of above-obtained BP neural network.

From Figs 1-2 it could be found that the error between ANSYS-computed and BPNN-predicted displacements is very small.

THE FUZZY PRE-CONTROL OF FOUNDATION PIT DIGGING

In this section, a fuzzy logic model is set up for pre-control of foundation pit digging. In the fuzzy logic model, the otal horizontal displacement and incremental horizontal displacement are adopted as input variables. The output variable is the pre-control measure. The purpose of fuzzy control is to ensure the safety of the foundation pit digging. Following *if-then* rules are adopted in this paper.

Table 2. "if-then" rules for fuzzy pre-control of Deping Station foundation pit digging of Shanghai Track Traffic Line 6 project

Rule No.	Definition of if-then rules
1	**If** *the total horizontal displacement is small* and *the incremental displacement is small* **then** *output is normal construction*
2	**If** *the total horizontal displacement is small* and *the incremental displacement is medium* **then** *output is slowdown construction*
3	**If** *the total horizontal displacement is medium* and *the incremental displacement is small* **then** *output is slowdown construction*
4	**If** *the total horizontal displacement is medium* and *the incremental displacement is medium* **then** *output is slowdown construction*
5	**If** *the total horizontal displacement is small* and *the incremental displacement is large* **then** *output is suspend construction*
6	**If** *the total horizontal displacement is medium* and *the incremental displacement is large* **then** *output is suspend construction*
7	**If** *the total horizontal displacement is large* and *the incremental displacement is small* **then** *output is suspend construction*
8	**If** *the total horizontal displacement is large* and *the incremental displacement is medium* **then** *output is suspend construction*
9	**If** *the total horizontal displacement is large* and *the incremental displacement is large* **then** *output is reinforce*

For the adopted *if-then* rules, the grade of total and incremental horizontal displacements is defined as follows.

Table 3. The grade of total and incremental horizontal displacements employed by above-mentioned "if-then" rules

	Small	**Medium**	**Large**
Total horizontal displacement	Below 30mm	30-40mm	Greater than 40mm
Incremental horizontal displacement	Below 10mm	10-20mm	Greater than 20mm

Table 4 can be obtained from fuzzy logic operation.

Table 4. The pre-control measures obtained through fuzzy logic operation (After integrating the results in x-axis direction and in y-axis direction)

Depth (m)	State 2	State 3
0	*normal construction*	*normal construction*
2	*normal construction*	*normal construction*
4	*normal construction*	*normal construction*
6	*slowdown construction*	*slowdown construction*
8	*slowdown construction*	*slowdown construction*
10	*slowdown construction*	*slowdown construction*
12	*slowdown construction*	*normal construction*
14	*normal construction*	*normal construction*
16	*normal construction*	*normal construction*
18	*normal construction*	*normal construction*
20	*normal construction*	*normal construction*
22	*normal construction*	*normal construction*
24	*normal construction*	*normal construction*

The result of Table 4 is the same as the measures that employed in practical construction.

CONCLUSIONS

Following conclusions could be summarized from these studies.

(1) BP neural networks have a very excellent ability in simulating the result and process of geotechnical FEM computation.

(2) Fuzzy logic has the unique superiority in simulating the pre-control measures of foundation pit digging.

(3) It should be noted that the proposed method is generic and can be easily extended to other engineering domains such as modeling of the dynamic interaction between the rock mass and the excavating machine, and controlling the functioning of the machine. The model can also be used for risk analysis of slopes, rock mass classification, hydrogeology and so forth. The coupled BP neural network-fuzzy

logic model is much more efficient in predicting and controlling a geotechnical engineering problem than either BP neural network or fuzzy logic alone.

(4) The traditional FEM method and intelligent technique should be closely integrated for better solving of geotechnical practical problems.

ACKNOWLEDGMENTS

The authors appreciate the support of the Innovation Fund Project for Graduate Student of Shanghai (JWCXSL0802), the National Natural Science Foundation of China (10872133) and the Pujiang Talent Program (09PJ1407700)

REFERENCES

[1] Basma Adnan A. and Kallas Nabil. Modeling soil collapse by artificial neural networks. *Geotechnical and Geological Engineering*, 22(3), 2004: 427-438.

[2] Chua CG and Goh ATC. A hybrid Bayesian back-propagation neural network approach to multivariate modeling. *International Journal for Numerical and Analytical Methods in Geomechanics*, 27(8), 2003: 651-667.

[3] Goh ATC and Kulhawy FH. Reliability assessment of serviceability performance of braced retaining walls using a neural network approach. *International Journal for Numerical and Analytical Methods in Geomechanics*, 29(6), 2005: 627-642.

[4] Li Jia-Chuan. Three-dimensional FEM computation and experimental study of foundation pit digging in soft soil areas. Thesis for applying doctoral degree, Tongji University, 1992. (in Chinese)

[5] Wang Guang-Guo. Large-scale deformation theory of foundation pit digging. Thesis for applying doctoral degree, Tongji University, 1994. (in Chinese)

[6] Yuan Jin-Rong. The intelligent prediction and control on displacement of underground engineering construction. Thesis for applying doctoral degree, Tongji University, 2001. (in Chinese)

Analytical Study on the Control of Ground Subsidence Arising from the Phenomenon of Accompanied Settlement Using Foot Reinforcement Side Pile

Ying Cui[1], Kiyoshi Kishida[2] and Makoto Kimura[3]

[1]Department of Urban & Environmental Eng., Kyoto University, Kyoto 615-8540, Japan; cuiying1981@gmail.com
[2]Department of Urban Management, Kyoto University, Kyoto 615-8540, Japan; kishida@mbox.kudpc.kyoto-u.ac.jp
[3]Inovative Collaboration Center, Kyoto University, Kyoto 615-8520, Japan; kimura@icc.kyoto-u.ac.jp

ABSTRACT: During the construction of new bullet train line where excavating shallow overburden tunnel using NATM, a phenomenon of equal settlement of the ground surface, crown and foot of tunnel called an accompanied settlement occurred. Foot reinforcement side pile (FRSP) was used to prevent the ground subsidence and it had been reported that it could prevent settlement of ground and tunnel effectively. Nevertheless the mechanism of preventing settlement by using FRSP is still not yet clear. In this study, 2-D elasto-plastic FEM analysis is carried out to clarify the mechanism of the effect of FRSP. Phenomenon of accompanied settlement was simulated pertinently and the results indicate that FRSP can prevent ground subsidence by increasing effect of shear reinforcement, load redistribution and internal pressure.

INTRODUCTION

Due to topographic and route conditions, shallow overburden tunnel on soft ground using NATM have been excavated in the construction of new bullet train line. In the mentioned construction field, the same settlements of ground surface, crown and foot of tunnel have been measured as shown in FIG. 1 (Kitagawa, et al., 2005) and its phenomena is called accompanied settlement. The mechanism of this phenomenon is

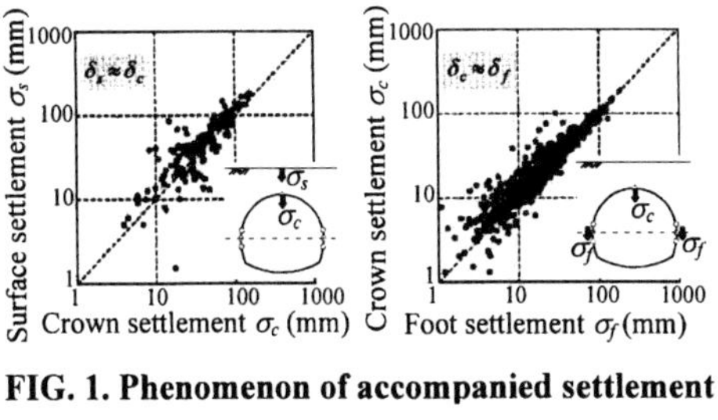

FIG. 1. Phenomenon of accompanied settlement

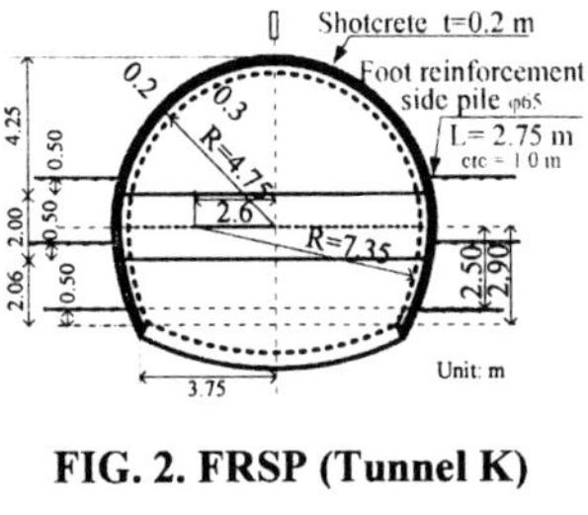

FIG. 2. FRSP (Tunnel K)

not clearly understood and preventing land subsidence has become a real problem.

It was assumed that if the settlement at the foot of the tunnel had been prevented, the ground subsidence can be controlled effectively. Based on this idea, FRSP (foot reinforcement side pile) have been utilized in the construction site as shown in FIG. 2 (Kitagawa, et al., 2005). FRSP is one of the foot reinforcement methods which insert steel pipe to ground from inside of tunnel. Kitagawa, et al. (2005) reported that FRSP can control the settlements of the ground surface and tunnel effectively. Nevertheless, the mechanics on how the FRSP control the settlement is still not clearly understood.

Cui, et al. (2007) carried out the simple 3-D trapdoor model test as shown in FIG. 3 and numerical simulation on the experiments concerning the effects of the FRSP. The result showed that the FRSP can prevent surface settlement by increasing the shear reinforcement effect and the load redistribution effect.

However, mechanical behaviors of ground and FRSP in model experiment are different with real field condition. Therefore, in this study 2-D elasto-plastic FEM analysis is carried out to simulate the phenomenon of accompanied settlement and clarify the mechanism of the effect of FRSP in real tunnel construction field.

OUTLINE OF NUMERICAL ANALYSIS

Modeling of ground, tunnel and FRSP

FIG. 4 shows the analysis area and boundary condition. This analytical object was determined based on construction filed data.

Subloading t_{ij} model (Nakai and Hinokio, 2004) was used to simulate the ground material. Properties of the model ground are as given in Table 1.

Ground improvement has been carried out before tunnel excavation because of the ground around tunnel was so soft with the N values smaller than 10. The ground in the improved area was modeled as an elastic material. Young's modulus was calculated from Eq. (1) based on compressive strength q_u as shown in FIG. 4.

$$E = 2800 \times N = 2800 \times \frac{q_u \times 8}{100} (kN/m^2) \quad \textbf{(1)}$$

The section of the tunnel is as shown in FIG. 2. For convenience, tunnel lining was modeled as composite elastic beam unifying the tunnel support and shotcrete. The young's modulus of the composite elastic beam was taken as 1.23×10^7 kN/m^2.

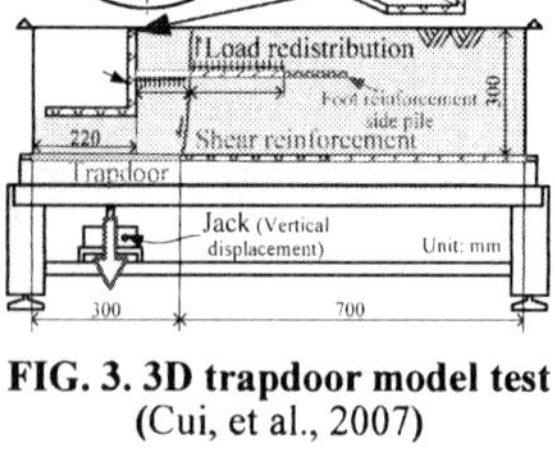

FIG. 3. 3D trapdoor model test
(Cui, et al., 2007)

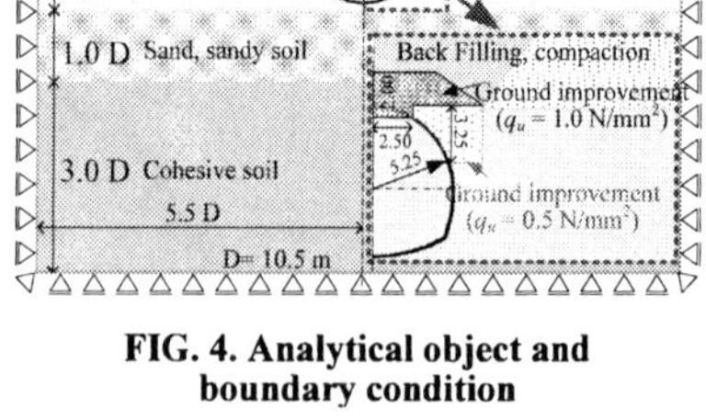

FIG. 4. Analytical object and boundary condition

Table 1. Format and Layout of Tables in GSP Papers

	Clay loam	Sandy soil	Sand, sandy soil	Cohesive soil	Back fiilling
Density (kg/m^3)	1.609	1.805	1.832	1.550	1.609
Poisson' s ratio (ν)	0.360	0.360	0.290	0.290	0.261
Void ratio (e_0)	1.704	1.071	1.012	0.613	0.852
Coefficient of earth pressure at rest k_0	0.562	0.562	0.409	0.409	0.353
Principal stress ratio at critical state R_{cs}	2.550	2.550	3.888	3.888	4.668
Compression index λ	0.137	0.070	0.070	0.082	0.094
Swelling index κ	0.030	0.005	0.005	0.018	0.021

Hybrid element model (Zhang et al., 2004) was used to modeling FRSP. The hybrid element model is made up of the elastic solid element and beam element, with the beam element surrounded by the solid element. The young's modulus of the solid element and beam element are respectively 1.20×10^7 kN/m^2 and 1.90×10^8 kN/m^2.

Simulation of tunnel excavation and phenomenon of accompanied settlement

In this study, tunnel excavation is simulated by release an equivalent in-situ stress due to excavation. Objective tunnel was constructed by alternate excavation of top heading and bottom section. Stress release rate of the top heading before installing tunnel support was determined as 40 % and the value of bottom section was determined as 15 %. Stress release rate was determined by setting the convergence of this analysis to be the same as the observed field data as shown in FIG. 5.

Variation of settlement of tunnel and ground surface with time from analysis and observed field data are as shown in FIG. 5. From the figure it can be seen that these analytical data are approximately equal to the observed filed data. In the other hand, the figure shows that ground surface, crown and foot of tunnel are sinking at the same value and it suggests that the phenomenon of accompanied settlement is simulated in this analytical condition. These results indicate that tunnel excavation process was simulated precisely and from the result it can be confirmed that these assumption mentioned above were pertinent.

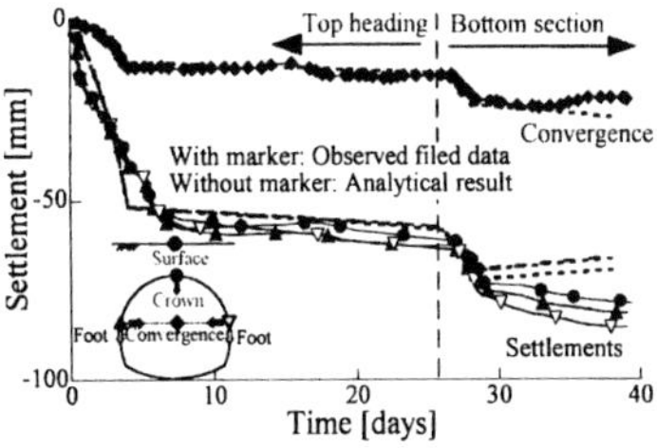

FIG. 5. Comparison of field data with analytical data

EFFECTS OF FRSP

Settlement of ground and tunnel

Variation of settlements with time for the case in which no pile is installed and the case where 2.75 m long FRSP is installed are as shown in FIG. 6. When FRSP are installed, the settlement value becomes smaller and it can be concluded that FRSP is effective on controlling the settlement of ground surface and tunnel. Additionally, ground surface, crown and foot of tunnel are sinking with same value in the case where FRSP is installed as same as the case where no pile installed.

Reduce ratio of settlement of ground surface, crown and foot of tunnel when tunnel excavation is finished for different length of FRSP are as shown in FIG. 7. From the figure it can be seen that FRSP can prevent settlement of ground and tunnel at all of the cases and it becomes more effective when the length of FRSP is increased. In addition, the effect of preventing settlement are increase rapidly when FRSP are longer than 2.5 m and it hit a peak when the length is longer than 4 m.

Mechanical behavior of ground

FIG. 8 shows the vertical earth pressure distribution on marked position when top heading excavation completed. The straight line shows the initial vertical load that before tunnel excavation. In the case in which 2.75 m long FRSP is installed, the vertical earth pressure acts on the upper part of tunnel become smaller when the vertical earth pressure act on the upper part of FRSP become larger. This result shows that FRSP can disperse the load of tunnel and ground above tunnel to ground next to the tunnel. This effect is named effect of load redistribution in this study.

FIG.9 shows the reduce ratio of vertical load acting on the marked position for different length of FRSP. From the figure it can be seen that FRSP demonstrate redistribution effect when it is longer than 2.5 m. This is the possible cause of the result

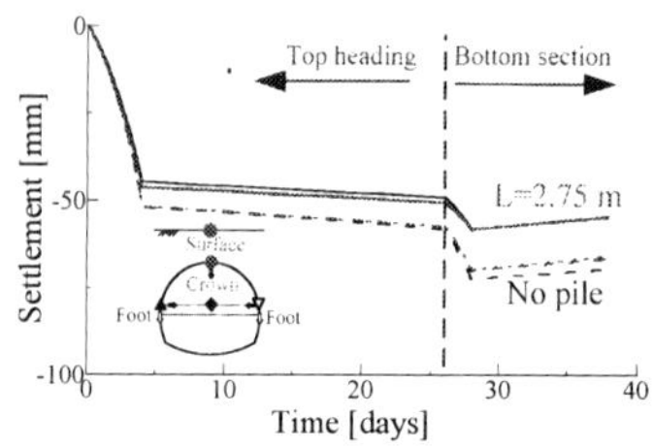

FIG. 6. Effect of foot reinforcement side pile

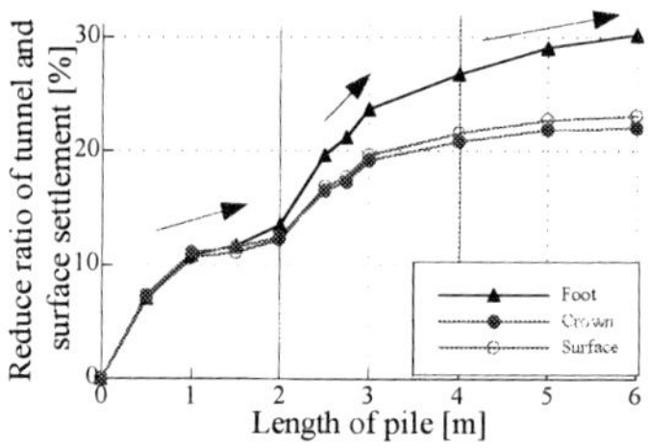

FIG. 7. Effect of preventing settlement of tunnel and ground at different length

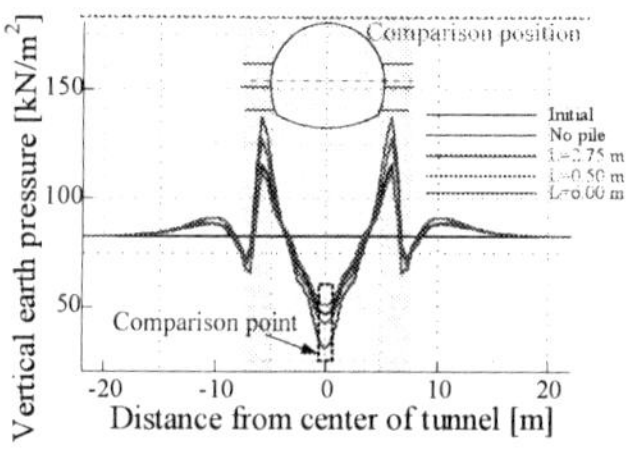

FIG. 8. Vertical load distribution at different length

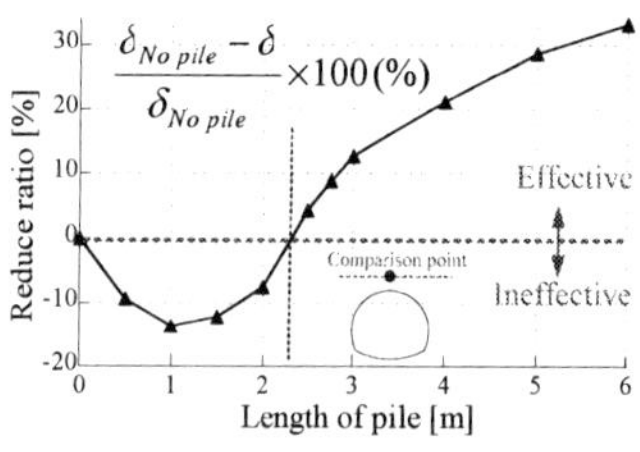

FIG. 9. Redistribution effect at different length (center line)

that the effect of preventing settlement of tunnel and ground increases at a great rate when FRSP is longer than 2.5 m. However, vertical load acting on the upper part of tunnel is bigger than the case where no pile is installed when the length is shorter than 2.0 m. The possible cause of this phenomenon is that when the FRSP is shorter than 2.0 m, it will interfere with the soil between the lining and the slip line giving rise to the load acting on the upper part of tunnel.

Shear strain distribution when top heading excavation completed is as shown in FIG. 10. It can be seen that when no pile installed, a large shear strain is generated at the foot of tunnel and it develops downward obliquely and another large shear strain is generated from the edge of improved ground and it develops upward vertically. When FRSP was installed, shear strain generated form the edge of improved ground is intercepted by the piles thereby making the shear strain around tunnel to become small. This effect is referred to as the effect of shear reinforcement. However, FRSP is no effect on the shear strain generated from the foot of tunnel.

Mechanical behavior of foot reinforcement side pile

Bending moment and axial force distribution of the FRSP for different length when top heading is excavated are shown in FIG. 11. The length of the curve in X direction is the same as the length of FRSP.

From the bending moment distribution, it can be seen that when the piles are shorter than 3.0 m, the upper side of the pile is tensioned and the maximum moment become bigger when the length of FRSP is increased. In addition, the distance from the lining to the position where the maximum moment occurs is about 2.0 m in all of the cases and it can be suggest that the slip line is passes through the mentioned position. When FRSP is longer than 3.0 m, the under side of the pile is tensioned in the part that is at a distance from tunnel and there is almost no change when FRSP is longer than 5.0 m.

The axial force distribution condition shows that the pile is tensioned in all of the cases independently of the length of FRSP. This is due to the friction between piles and ground and this effect is named internal pressure effect in this study. More over the maximum axial force increase when the FRSP become longer.

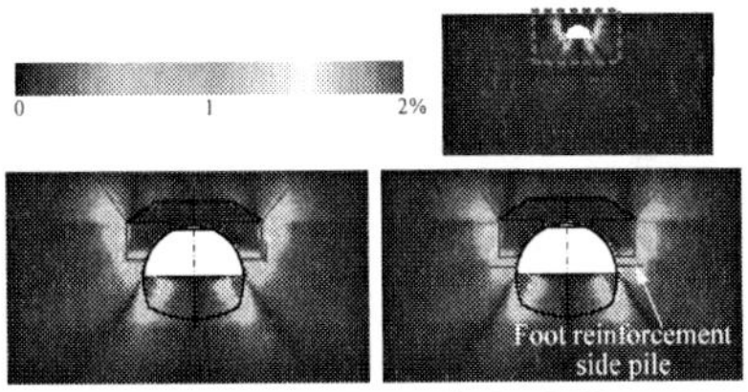

(a) No pile (b) L= 2.75 m

FIG. 10. Shear strain distribution

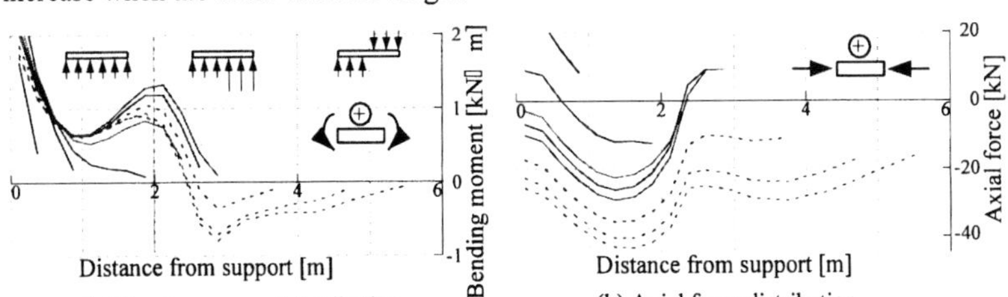

(a) Bending moment distribution (b) Axial force distribution

FIG. 11. Mechanical behavior of foot reinforcement side pile at different length

CONCLUSIONS

In this research work, the process of tunnel excavation and the phenomenon of accompanied settlement were simulated pertinently. Thereafter mechanical behavior of ground and FRSP are examined in this research work.

The analytical results indicated that FRSP can prevent ground subsidence and tunnel settlement effectively. The advantages of FRSP are presented as the three matters as shown in FIG. 12.

First one is the effect of shear reinforcement. Slip lines occurring in the ground by tunnel excavation are as shown in FIG. 12. The slip line generate from the corner of the improved ground is 2.0 meter away from tunnel in this research work and the distance is named as Q. FRSP demonstrate the effect of shear reinforcement to preventing ground subsidence when the piles goes past the slip line.

Second one is the effect of load redistribution. When the FRSP is longer than Q that enough to intersect with the slip line, it demonstrates bending stiffness as a beam to distribute earth pressure over tunnel to the ground next tunnel.

Third one is the effect of internal pressure. FRSP demonstrate effect of internal pressure to reinforce the ground around tunnel in all of the cases. Settlement of ground and tunnel is prevented as a result of the reinforcement of the ground.

In addition, these three kinds of effects are become more effective when the length of FRSP is increased. However, FRSP is no influence on the occurring of the phenomenon of accompanied settlement.

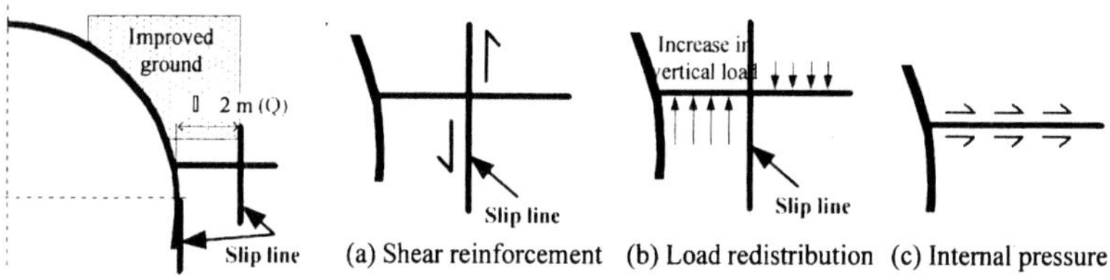

FIG. 12. Slip line and the mechanism of the effect of foot reinforcement side pile

REFERENCES

Kitagawa, T., Goto, M., Isogai, A., Yashiro, K. and Matsunaga, T. (2005), "Analysis on Behavior of Soft Ground Tunnels with Low Earth covering", *Proceedings of tunnel engineering, JSCE*, Vol. 15, 203-210. (in Japanese)

Cui, Y., Kishida, K. And Kimura, M. (2007), "Experimental study on effect of auxiliary methods for simultaneous settlement at subsurface and surface during shallow overburden tunnel excavation", *JGS Journal*, Vol.3, No.3, 261-272. (in Japanese)

Nakai, T. and Hinokio, M. (2004), "A Simple Elastoplastic Model for Normally and Over Consolidated Soils with Unified Material Parameters", *Soils and Foundations*, Vol. 44, No. 2, 53-70.

Zhang, F., Kimura, M., Nakai, T. and Hoshikawa, T. (2000), "Mechanical Behaviour of Pile Foundations Subjected to Cyclic Lateral Loading Up to the Ultimate State", *Soils and Foundations*, Vol. 40, No. 5, 1-17.

3D Numerical Simulation on the Failure Mechanism of Tunnel Working Face by Particle Flow Code

Chengbing Wang[1], Hehua Zhu[2], Wensheng Gao[1]

[1]Institute of Foundation Engineering, China Academy of Building Research, Beijing, 100013, China;
[2]Department of Geotechnical Engineering, Tongji University, Shanghai, 200092, China

ABSTRACT: Collapse of tunnel working face often occurred in tunnels constructed in cracked rock. The collapse has two forms, the one is arched collapse which has stable slumping arch before the tunnel working face, the other is caving collapse which the pit occurs in the ground surface before the tunnel working face. The effects of buried depth and surrounding rock strength on the failure mechanism of tunnel working face were studied through series numerical simulations by using PFC3D. The input parameters of PFC3D such as particle size, friction coefficient, bonded strength and so on were calibrated by numerical triaxial test. The calculating results indicate that the collapse of tunnel working face initiates from its upper part and progressively develop upward. And in the case of shallow buried, if the surrounding rock has high strength, the tunnel working face can keep self-stabilization, while the surrounding rock strength is low, collapse of tunnel working face occurs easily and it is likely to spread to the ground surface. Meanwhile, the smaller the buried depth is, the more easily the caving collapse of tunnel working face occurs.

1 INTRODUCTION

Collapse of tunnel working face often occurred in tunnels constructed in cracked rock. For example, at the site ZK1+426m along the left hole of Haichang Tunnel in Xiamen, China, just right the beginning of the concrete ejecting on the upper bench, 110m^3 clay soil fell down suddenly from the working face(Liu,2004). A large quantity of slimes mingled with boulders poured out from the working face when Tuojiashan Tunnel constructed at the section which was 70m deep. The mass of collapse was about 1000m^3 and a 4m deep round pit which was about 314m^2 occurred in the ground surface before the working face(Li, 2005).

The collapse of the working face has been an important potential safety hazards which causes construction delay, loss of life and riches, and tunnel operation. Finite Element Method and Finite Difference Method based on traditional continuum mechanics are suitable for forecasting the damage and failure region, but they can't be used to simulate the whole process of tunnel collapse directly(T.G.Sitharam, 2002; D.Sterpi, 2002). PFC can simulate the problem of separation, large deformation of particles, and also it can simulate the micro structure alternation and mechanical characters of particles(Cundall, 1999). Mitarashi(Mitarashi, 2006) studied the effect of different bolting separation on the failure mechanism of tunnel working face by using PFC3D. Fakhimi(Fakhimi, 2002) simulated the failure process of an opening and it was shown that the numerical model was able to reproduce the damage zone observed in the laboratory test. However, few studies focused on the failure mechanism of tunnel working face. In this paper the effects of buried depth and surrounding rock strength on the failure mechanism of tunnel working face were studied through series numerical simulations by using PFC3D.

2 NUMERICAL TESTS ON THE MICRO PARAMETERS OF THE SURROUNDING ROCK

Parameters assigned to particles in PFC3D are obtained through numerical triaxial compression tests. In order to give a quantitative simulation on engineering problems, numerical triaxial tests must be proceeded firstly to get the particle's micro parameters which can be used to simulate the actual engineering problems. For the limitation of computation, the method of combining model test and numerical simulation is used to study the problem of tunnel working face collapse. The geometric similarity ratio of the model is set as 50 and the gravitational similarity ratio is set as 1. According to the similarity theory, the similarity ratio of Poisson ratio and internal friction angle are equal to 1, and the similarity ratio of cohesive strength and Young's modulus are equal to 50.

The simulation of the compression test is composed of three steps, they are test sample generation, consolidation and loading. The test sample of the numerical test is 8.0cm high and its diameter is 3.91cm, which is the same as normal triaxial compression test. The test sample is built up by uniform size particles which the diameter is 0.16mm. The initial porosity is 0.3 and there are 3919 particles in the test sample of the numerical test.

Fig.1 shows the process of triaxial compression test simulated by using PFC3D. Fig.1 (a) presents the generated walls. The distribution of the generated particles is shown as Fig.1 (b). Unbalanced force occurred between the particles after the particle radius expanded. Fig.1 (c) shows the balanced state of the particles achieved automatically through circulate computing. The test sample after loading is shown as Fig.1 (d).

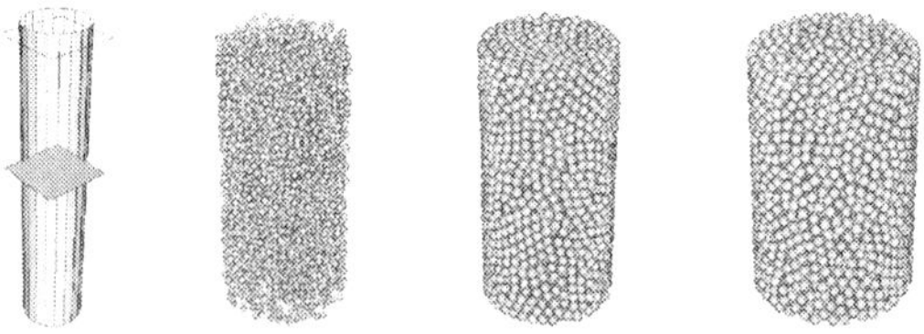

(a) (b) (c) (d)

Fig.1 Simulating process of triaixal test in PFC3D

By taking the physical and mechanical parameters of road tunnel surrounding rock (grades IV, V) as references, three kinds of materials are simulated in the numerical triaxial tests. Tab.1 and Tab.2 show the basic parameters and the calculated results of the model after a number of trials.

Tab.1 Basic parameters of PFC model

Material No.	Particle density (kg/m^3)	Normal contact stiffness (GPa)	Stiffness ratio	friction coefficient	Bonded strength(kPa)
No.1	2400	40	1	0.9	35
No.2	2000	30	1	0.5	15
No.3	2000	30	1	0.5	5

Tab.2 The calculated results of numerical triaxial test

Material No.	E(MPa)	μ	C(kPa)	φ(°)	surrounding rock grade
No.1	36.8	0.32	6.6	31.5	□
No.2	25.1	0.36	3.8	25.9	□
No.3	23.2	0.37	2.1	26.5	□

3 SCHEME OF NUMERICAL SIMULATIONS

The cross section of the tunnel adopted in the numerical simulations is shown as Fig.2. The tunnel is a two-lane road tunnel, and it is 175.4mm high and 220.0mm wide. According to symmetry principle, one half of the tunnel is considered in numerical simulations. Fig.3 illustrates the calculating range. The calculating model is 0.5m long along the tunnel axial orientation, and it is 0.3m wide along the cross section orientation. The distance between the tunnel and the bottom boundary is 0.1m in the model. The collapse status of tunnel working face is simulated after the tunnel excavated 0.2m long and was supported. The scheme of numerical simulations is shown as Tab.3, in which B means the span of tunnel.

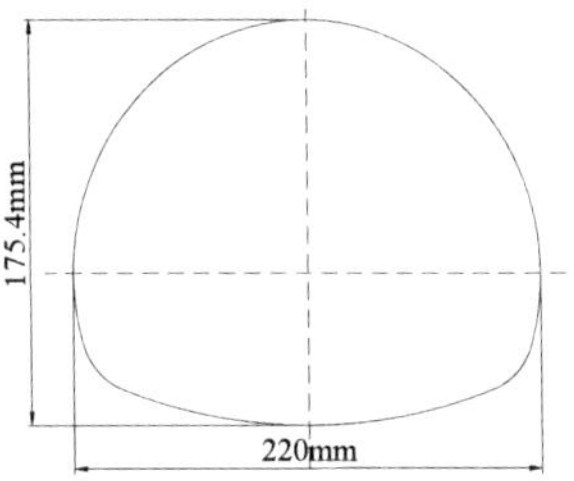

Fig.2 Tunnel cross section

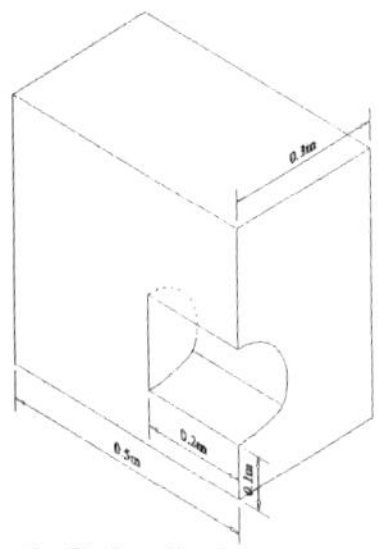

Fig.3 Calculating range

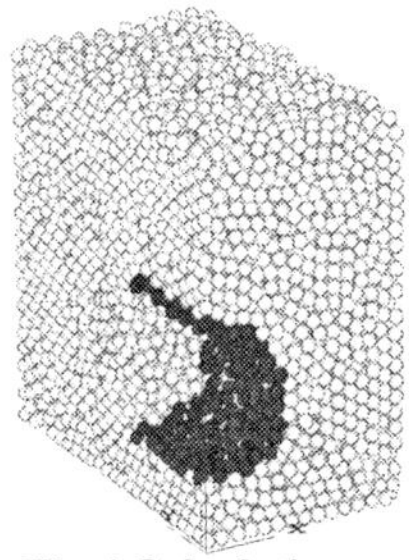

Fig.4 Calculating model of scheme No.3

The process of numerical calculation is as follow: (1) Generated particles and compacted them under the action of gravitational stress. (2) Assigned friction coefficient and bonded strength to particles, then excavated the tunnel and installed lining.

Tab.3 Scheme of numerical simulations

Scheme No.	Buried depth of tunnel	Material of surrounding rock
No.1	1B	Material No.1
No.2	1B	Material No.2
No.3	1B	Material No.3
No.4	2B	Material No.3
No.5	3B	Material No.3

For the inherent porosity in the particle assembly reduces the gravitational loading, a larger value of particle density must be used to compensate. After a number of trials, 2400 kg/m^3 and 2000 kg/m^3 are assigned to the particles to get the same gravitational stress of rock masses which densities are 2100 kg/m^3 and 1900 kg/m^3. Fig.4 shows the calculating model of Scheme No.3. Scheme No.1 to No.3 had 11611 particles, scheme No.4 had 16139 particles and scheme No.5 had 21101 particles.

4 RESULTS AND ANALYSIS OF NUMERICAL SIMULATIONS

Fig.5 shows the final failure shapes of each scheme, and the failure process of scheme No.3 is shown as Fig.6. From Fig.6 it can be seen that the tunnel working face cannot keep self-stabilization after excavation. The collapse of tunnel working face initiates from its upper part and progressively develops upward. A transient slumping arch occurred before the tunnel working face while the collapse develops to some extent. After the transient slumping arch broken, the collapse develops upward continually and finally it reaches the ground surface and a pit occurred in the ground surface.

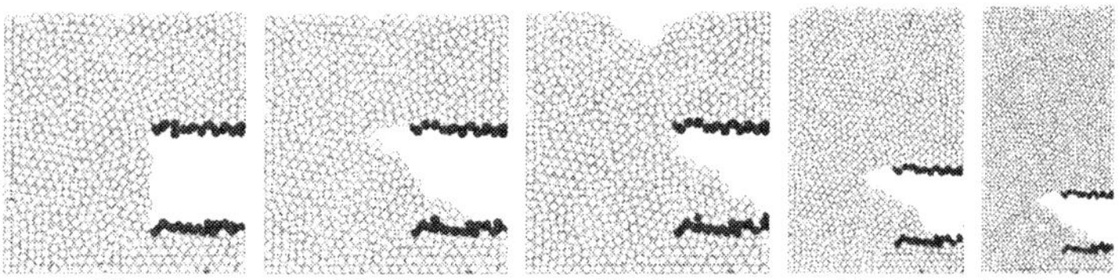

No.1 No.2 No.3 No.4 No.5

Fig.5 The final failure shapes of each scheme

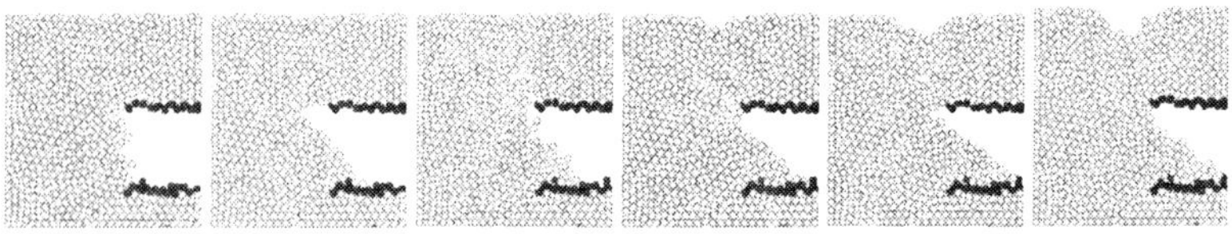

Fig.6 The failure process of scheme No.3

4.1 ANALYSIS ON THE EFFECT OF SURROUNDING ROCK STRENGTH

The final failure shape sketches of scheme No.2 and scheme No.3 are shown as Fig.7. Comparing the calculating results of scheme No.1, No.2 and No.3, the following conclusions can be drawn. The calculated result of scheme No.1 indicates that the tunnel working face can keep self-stabilization after excavation. The tunnel working face collapses immediately after excavation of scheme No.2, meanwhile the collapsing body swarms into the excavated part of the tunnel, and a slumping arch which is 94mm long formed before the working face, the dip angle of the collapsing body accumulates in the tunnel is 35°. The collapse condition of the tunnel working face of scheme No.3 is the most serious. The excavated range of tunnel is packed by collapsing body which dip angle is 33°. The collapse spreads to the ground surface and a pit which is 66mm deep and 186mm wide occurred before the tunnel working face in the ground surface. The distance between the after-edge of the pit and the tunnel working face is 40mm.

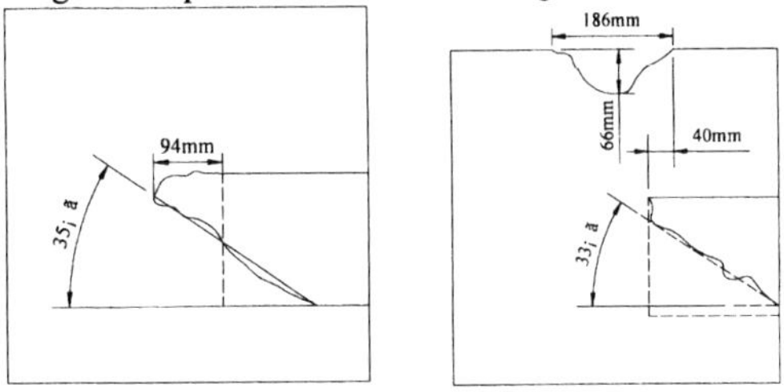

Scheme No.2 scheme No.3

Fig.7 The final failure shape sketches of scheme No.2 and No.3

The calculated results show that in the case of shallow buried, the tunnel working face can keep self-stabilization while the surrounding rock strength is high, but collapse of tunnel working face occurred easily and it is likely to spread to the ground surface while the surrounding rock strength is low. So pre-reinforcement should be adopted to prevent the accident of collapse when tunnel constructed in soft rock.

4.2 ANALYSIS ON THE EFFECT OF TUNNEL BURIED DEPTH

The final failure shape sketches of scheme No.4 and scheme No.5 are shown as Fig.8. The collapse of scheme No.3 spreads to the ground surface which formed a caving collapse. The tunnel working face of scheme No.4 collapses immediately after excavation, the slumping arch before the working face is 86mm long and the dip angle of the collapsing body accumulated in the tunnel is 34°. The slumping arch before the working face is 110mm long and the dip angle of the collapsing body accumulated in the tunnel is 32° of scheme No.5.

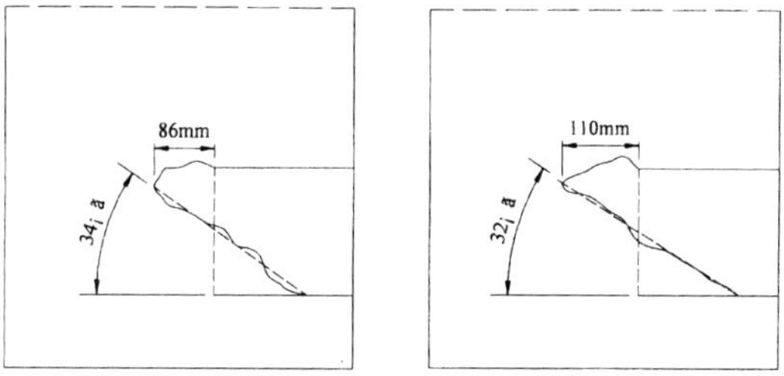

Scheme No.4 scheme No.5

Fig.8 The final failure shape sketches of scheme No.4 and No.5

From the calculated results it can be seen that the smaller the buried depth is, the more easily the caving collapse of the tunnel working face occurred. With the increase of buried depth, a slumping arch before the tunnel working face can be formed. And the slumping arch of scheme No.5 is longer than that of scheme No.4, that is to say the bigger the buried depth is, the more serious the collapse condition of the tunnel working face is.

5 CONCLUSIONS

Numerical simulations of the failure mechanism of tunnel working face are presented here. Some conclusions can be drawn from the study:

1. The collapse of tunnel working face initiates from its upper part and progressively develops upward, and finally a slumping arch or a caving collapse is formed.
2. In the case of shallow buried, if the surrounding rock has high strength, the tunnel working face can keep self-stabilization, while the surrounding rock strength is low, collapse of tunnel working face occurs easily and it is likely to spread to the ground surface.
3. The smaller the buried depth is, the more easily the caving collapse of tunnel working face occurs. With the increase of buried depth, a slumping arch before the tunnel working face can be formed, and the bigger the buried depth is, the more serious the collapse condition of the tunnel working face is.

ACKNOWLEDGMENTS

The authors appreciate the support of the Items of National Nature Science Fund, Grant No.40672184.

REFERENCES

[1] Lin Hui(2004)."The collapse and its treatment in the construction of Haicang Tunnel. " *Journal of Fujian Agriculture and Forestry University*, Vol.33(1): 126-128 (in

Chinese)

[2] Li Jian-jun, Jiang Shu-ping, Wu Ming-sheng(2005)."Pipe roofing construction technology for treatment of land slip of Tuojiashan Tunnel." *Technology of Highway and Transport*, No.4: 151-155 (in Chinese)

[3] T.G. Sitharam, G. Madhavi Latha(2002). "Simulation of excavation in jointed rock masses using a practical equivalent continuum approach. " *International Journal of Rock Mechanics & Mining Sciences*, No.39: 517-525

[4] D. Sterpi(2002). "Application of the FEM to the stability of shallow tunnels." *New Frontiers in Computational Geotechnics-Preceeding of 1st International Workshop on New Frontiers in Computational Geotechnics-banef*, September 14-15: 61-68

[5] Cundall P A(1999). "Particle Flow Code in 2 Dimensions Command and Fish reference summary". *Minnesota: Itasca Consulting Group Inc.*

[6] MITARASHI Y., TEZUKA H.(2006). "The evaluation of the effect of long face bolting by 3D distinct element method". *ISRM International symposium 2006, 4th Asian Rock Mechanics Symposium*

[7] Fakhimi, Carvalho F., Ishida T. & Labuz J.F(2002). "Simulation of failure around a circular opening in rock." *International Journal of Rock Mechanics & Mining Sciences* , No.39: 507–515

The Application of F&EI Method in Risk Assessment of Tunnel Gas Explosion

Jifei Wang[1], Hongwei Huang[2], Xiongyao Xie[3]

1 Key Laboratory of Geotechnical & Underground Engineering of Ministry of Education, Department of Geotechnical Engineering, Tongji University, Shanghai 200092, China; PH(86) 021-65982386; email: csu98013054@yahoo.com.cn
2 Key Laboratory of Geotechnical & Underground Engineering of Ministry of Education, Department of Geotechnical Engineering, Tongji University, Shanghai 200092, China; PH(86) 021 65989273; email: huanghw@tongji.edu.cn
3 Key Laboratory of Geotechnical & Underground Engineering of Ministry of Education, Department of Geotechnical Engineering, Tongji University, Shanghai 200092, China; PH(86) 021 65982986; email: xiexiongyao@tongji.edu.cn

ABSTRACT: Explosion may happen if the gas is rich in strata during tunnel construction, which may results in tremendous losses. Many methods have been used by engineers and scholars to evaluate the risk of gas explosion, but how to evaluate the losses accurately caused by gas explosion is still not solved. In this paper, the Dow Chemistry Method, namely F&EI method (Dow Chemical Company, 1994) is introduced to evaluate the losses, by using this method, the characteristics of gas can be taken into account, and so the results of risk assessment is much more reliable than those obtained by traditional method.

1. INTRODUCTION

Gas accidents seriously threaten the security of coalmine industry, it seems happens all the time, leading to heavy injuries, deaths and money loses (Zhou, etc.2006; Liu, etc. 2008). Gas is also can be found frequently during tunnel construction, if effective control measures are not taken, gas explosion also will happen and result in huge losses. After investigating the causes of gas explosion in the affected tunnels, it was found tha one of the most significant reasons is a complete lack of detailed evaluation of th potential gas explosion hazard before tunneling was initiated, so that no effective preventive measures were undertaken during tunnel design and construction. (Einstein, H.H.,1996), Therefore, it is very important to evaluate the risk involved in the tunneling process. Risk management is an effective way to control such kind of risk, to evaluate the risk of gas explosion, there are many ways, the most important but difficult thing is how to evaluate the risk accurately. There are many scholars use different methods to assess the gas explosion risks, but few of them can get the accurate results, because they do not take the chemical characteristics of the gas into account. The Dow fire and explosion index (F&EI) (Dow Chemical Company, 1994) has been widely used in assessing the risk within the chemical industry, and has been proven to be a reliable and effective methodology. However, after an exhaustive literature search there is no evidence of the Dow F&EI method being used to evaluate the gas explosion in tunnels. Since the gas explosion is a chemical reaction phenomenon similar to that of other chemicals, the authors believe that the principle of the Dow F&EI method can be effectively used in gas explosion situations.

2. DETERMINATION OF DOW INDICES APPLICABLE TO GAS EXPLOSION RISK

The gas explosion risk assessment of Xishan tunnel is based on the seventh edition of the F&EI hazard classification guide (Dow Chemical Company, 1994), the diagram of the evaluation program is shown in figure1, The F&EI is based on historical loss data, the potential energy of materials, and the extent to which loss prevention practices are currently applied. It uses numerical values of hazard factors associated with different material and process characteristics to determine the fire and explosion hazards in a step-by-step objective evaluation. The participation of personnel who have the appropriate working knowledge and experience is very important in evaluating the F&EI. The original aim of the Dow Chemical Company was to use the F&EI as a hazard classification guide; i.e. to identify process equipment that requires more detailed hazard analysis. 4 factors (material, general process hazards, special process hazards and process unit hazards) are used in calculating the F&EI. Among them, the material factor (MF) is dominant. The material factor is a measure of the inherent rate of potential energy release due to fire and explosion produced by combustion and chemical reaction. The analysis process is: selecting assessment unit, determining MF, calculating general process hazard factor F1, special process hazard factor F2, risk coefficient of the unit F3 ($F3 = F1 \times F2$), F&EI ($F\&EI = F3 \times MF$), compensation factor of safety measures F4. In the case of the gas explosion risk evaluation, consultation with personnel who have research experience of tunneling and gas treatment practices is very important in establishing a reliable risk evaluation result.

2.1. Select of the Assessment Unit

Usually gas explosion only happens in one tunnel, it is quite unusual that it happens in two or three tunnels at the same time. Thus the tunnel with a high risk of gas explosion can be defined as one evaluation unit. Experience has shown that a tunnel with the following features is suitable for evaluation purpose, e.g. strata rich in gas and fault fracture zone, area with poor ventilation and high ambient temperature.

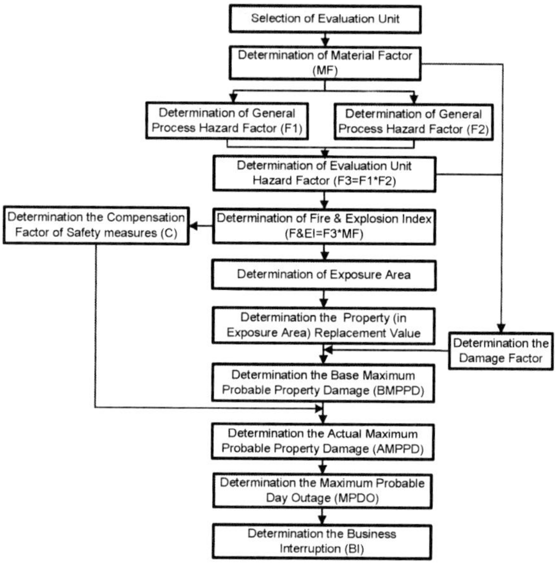

FIG 1. Diagram of Dow Chemistry Method.

2.2. Determination of Material Factor (MF)

Material factor is an inherent characteristic of the material which can measure the magnitude of energy released by the material when it explodes or catches on fire. It is determined by the N_F (combustion of the material) and N_R (chemical activity of the material), which are regulated by the National Fire Protection Association. According to previous investigations, the main component of gas is methane (CH_4), which makes up 95%, the other 5% is make up of carbon monoxide (CO), carbon dioxide (CO_2) and sulfur dioxide (SO_2), and methane (CH_4) is the dominant material that results in explosion. So the material factor can be obtained from the technical regulations of NFPA.

2.3. General Process Hazard Factor F_1

General process hazard factor F_1 determines the magnitude of the damage, which is mainly affected by the geological conditions, tunneling method, type of the chemical reaction, and state of the workspace. Experience has shown that a tunnel with a poorer ventilation and higher content of gas will result in a larger gas explosion, namely, a heavier loss.

2.4. Special Process Hazard Factor F_2

Special process hazard factor F_2 affects the occurring probability of gas explosion, and it is mainly determined by ventilation conditions, fire control measures and gas control measures. If the ventilation conditions are poor, gas will accumulate in the tunnel, if the gas content is between 5%~16%, it is extremely dangerous. Fire is one of the main causes of gas explosion, if the fire control measures are not implemented effectively, such as smoking in the tunnel, discharge spark or friction sparks of the electrical apparatus, the chance of gas explosion will increase dramatically. Gas control measures is the most important factor to reduce the probability of gas explosion, such as grouting for the reduction of gas emission, gas detecting and alarm meters, these measures can reduce the probability of gas explosion effectively. The value of these factors is recommended in the Dow Chemistry Method and also should consult the results of field investigation.

2.5. Compensation Factor of Safety Measures F_4

F_4 is determined by various fire preventive measures, such as gas control measures taken during tunnel design and construction, gas isolation measures taken during the construction and application of fire rescue equipments. The contribution of gas control measures if given between 0.8~1.0, the contribution of gas isolation measures is given between 0.9~1.0, the contribution of fire rescue equipments is between 0.7~1.0. All these values are given according to the field conditions and the guidance of handbooks (Safety Assessment, 2005).

3. VALUES OF ALL FACTORS IN GAS EXPLOSION RISK ASSESSMENT

The values of all factors in the gas explosion risk assessment system are shown in Table1. After completing a field investigation and obtaining the required values listed in Table1, *MF*, F_1 and F_2 can be calculated by summarizing all values with weighted coefficients in the corresponding item respectively. Then, F_3 ($F_3 = F_1 \times F_2$), F&EI (F&EI=$MF \times F_3$). After, the exposure area and the property value in the area can be determined according to the field investigation, here, we use *EA* and *PV* to stand for its value. And the damage factor of the evaluation unit also can be decided by the relationship of *MF* and F_3, which is shown in Figure2 (Safety Assessment, 2005). Then we can get base maximum probable property damage (BMPPD= $=EA \times PV$) and actual maximum probable property damage (AMPPD= $=BMPPD \times DF$)

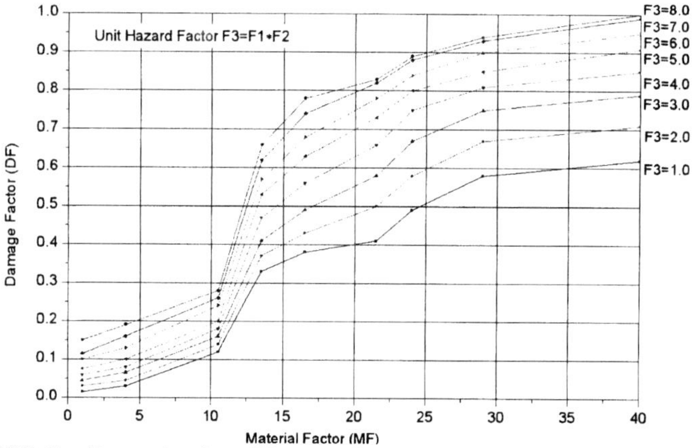

FIG. 2. Determination of Damage factor (DF) of the Evaluation Unit.

4. CASE STUDY OF GAS EXPLOSION RISK ASSESSMENT OF XISHAN TUNNEL

The Dow Chemistry Method is used to assess the gas explosion risk of Xishan tunnel, which is located in Taiyuan, Shanxi, China. There are many coal mines in this area, fault and gas is also very rich in the construction site. Gas explosion risk assessment is quite necessary before the construction of the tunnel. The classification of F&EI is shown in Table1. According to the field investigation and the guidance of handbooks, the value and calculation of all the factors are shown in Table2.

Table 1. Ranking of F&EI

F&EI	1~60	61~96	97~127	128~158	>159
Ranking	1	2	3	4	5
	Lightest	Lighter	Medium	Large	Very Large

Table 2. Calculation of F&EI

Material: CH4	**Operation Temperature: Normal Temperature**	**Material Factor: MF= 21**
1: General Process Hazard Factor	**Value Range**	**Coefficient Adopted**
Basic Factor	1.00	1.00
A: Exothermic Chemical Reaction	0.30~1.25	1.00
B: Gas reserves	0.25~1.05	0.50
C: Tunneling Method	0.25~0.9	0.50
General Process Hazard Factor: F1=3.00		
2: Special Process Hazard Factor		
Basic Factor	1.00	1.00
A: Failure of Ventilation System	0.30	0.30
B: Gas Emission	0.10~1.50	1.00
C: Machine erosion-corrosion	0.10~0.75	0.20
D: Use of Open Fire Equipments	0.15~1.15	0.70
E: Gas Control Measures	0.20~0.50	0.40
Special Process Hazard Factor: F2=3.60		
3: Hazard Factor of The Evaluation Unit :F3=F1×F2=10.80		
4: Fire & Explosion Index: F&EI=MF×F3=226.80		
5: Ranking of F&EI: 5 (Very Large)		
6: Exposure Area: EA= πR^2=113.04m^2 (Here R Is Tunnel radius)		
7: Property Values: PV=0.50 million (Based on Field Investigation)		
8: Damage factor: DF=0.84		
9: Base Maximum Probable Property Damage: BMPPD=EA×PV=57 million		
10: Compensation Factor of Safety Measures: C= $C_1 \times C_2 \times C_3$=0.57		
11: Actual Maximum Probable Property Damage: AMPPD=BMPPD×C=32.3 million		

5. CONCLUSIONS

This paper established a new risk assessment system for the gas explosion in tunnels based on the DOW F&EI methodology supported by field investigation and related laboratory test data. Since the system has considered all relevant factors concerning with gas explosion, it is considered to be competent and effective in application. From the results of the case study we can find out that if the gas explosion happens during the construction of Xishan tunnel, the actual probable maximum property damage may reach 32 million, and 84% of the exposure area will be damaged, so special attention must be paid on the control of gas explosion.

REFERENCES

Dow Chemical Company. Fire and Explosion Index Hazard Classification Guide (7th ed.). New York: American Institute of Chemical Engineers, 1994. 10

Einstein, H.H., (1996), "Risk and Risk Analysis in Rock Engineering." Tunneling and Underground Space Technology, Vol. 11(2): 141-155.

State Administration of Work Safety. Safety Assessment. Beijing: China Coal Industry Publishing House, 2005

Zhou Jiahong, Xu Kaili, (2006), "Assessment Methods of Consequences of Explosion Accidents." Industrial Safety and Environmental Protection, Vol. 32 (3): 52-53.

Liu Ying, Gu Yimin, Xuan Meiju. (2008), "Fire Explosion Index Application to Environmental Risk Assessment on Oil Storages." Environmental Science and Management, Vol. 33 (6): 165-168.

Analysis on Ground Deformation Caused by Tunnelling of Large-Diameter Tunnel Boring Machine

Zhiyong Yang [1], Hongwei Huang [2], Dongmei Zhang [3]

[1] PhD, Key Laboratory of Geotechnical & Underground Engineering of Ministry of Education, Department of Geotechnical Engineering, Tongji University, Shanghai, China, yzy621@163.com
[2] Professor, Key Laboratory of Geotechnical & Underground Engineering of Ministry of Education, Department of Geotechnical Engineering, Tongji University, Shanghai, China
[3] Associate professor, Key Laboratory of Geotechnical & Underground Engineering of Ministry of Education, Department of Geotechnical Engineering, Tongji University, Shanghai, China

ABSTRACT: The large-diameter shallow buried tunnel is not advantageous for ground surface deformation control. It's necessary to make study and research on the ground deformation caused by the advance of TBM. In this paper, the ground surface deformation curve is obtained based on the deformation measured data during the excavation of TBM in the soft ground of Shanghai, and the parameters of excavation, including slurry pressure and volume of synchronized grouting that affect the ground deformation are discussed. In addition, a fitting method to estimate the ground loss is put forward, which reduces the experience dependence and enhances the reliability of peck formula. It is also very simple for practical application to estimate the ground deformation.

INTRODUCTION

Shanghai Yangtze River Tunnel is located at the estuary of Yangtze River in the northeast of Shanghai. The large pressurized and mixed slurry TBM with a diameter of 15.43m is used for the tunnel construction. Comparing to its large diameter, the ratio of thickness of covering soil to diameter is small, which causes many difficult for deformation control. The ground deformation happens inevitably when the TBM drives through soft soil. If there is any large deformation, both the ground and underground buildings in the vicinity of the tunnel will be endangered. It is necessary to make some research on the ground deformation caused by the advance of large diameter TBM.

There are many methods to predict the ground deformation caused by tunnelling. J.

Ghaboussi et al. (1983) used FEM^{3D} to model the ground deformations caused by tunneling. K.M. Lee and R.K. Rowe (1990) described an elasto-plastic soil model suitable for a 3-D tunnelling analysis. Z.W. Liu et al. (2003) compared the result of a case study for a tunnel construction with that of FEM^{3D}, and the fitting curves agree with each other well. H. B. Zhang et al. (2005) proposed a three-dimensional nonlinear finite element model for simulating EPB shield tunneling. G. Wei et al. (2006) assumed that soil was undrained and proposed a computing formula of ground deformation induced by both bulkhead additive thrust and friction force between shield and soil from the Mindlin solution with elastic mechanics. Zhou and Pu (2002) investigated the ground settlements during tunnel exaction in sandy soil by centrifuge model test and discussed the relationship between support pressure and settlement. R.B. Peck (1969) put forward an experience formula to estimate the ground settlement. In Peck's opinion, the ground deformation caused by tunnelling happens under no drainage at the construction stage. As a result, the volume of the settlement tank was equal to the volume of the ground loss. Michael J. Kavvadas (2003) described the types of ground deformation measurements often used in tunnelling, the difficulties in obtaining ground measurements and the subsequent evaluation, and the application of these measurements in modeling tunnel excavation and support and in establishing early warning systems against incipient ground collapses or damage to structures on ground. Dimmock (2003) found during EPB tunnelling in dewatered dense silty sand at Southwark Park, London, as part of the Jubilee Line Extension Project, that all the volume loss occurred due to convergence of the soil on the overcut around the shield, and convergence around the tail void. During EPB tunnelling in mixed soft ground conditions for the recent Channel Tunnel Rail link (CTRL) project, Moss and Bowers (2005) reported that the proportions of volume loss at the face, body and tail skin were 25%, 25% and 50%. Zheng, Ding and Dai (2006) described change of ground/water pressure and ground movement under different condition, which provide useful experiences to similar project.

Among theses methods, numerical simulation and test method are more complex than experience formula method. Though experience formula method depends on area experience, it is easy for practical application.

In this paper, peck formula is used to estimate the ground deformation. The key factor of the formula is chosen not by experience but inversion result according to the measured data. This approach reduces the experience dependence while enhances the reliability. It is also very simple for practical application.

MEASURED DATA ANALYSIS

Deformation monitoring points (DMP) are set from the new dam to the retrieval shaft. There is one monitoring point (MP) every 5m and one monitoring section (MS) every 30m. There are 9 MPs in one section and the distances to longitudinal axis of tunnel are 14m , 10m , 6m , 3m , 0m , 3m , 6m , 10m , 14m in order of west to east.

Figure 1 shows the deformation curves of MP 7702 to 7712 corresponding to different distances to cutter head (CH).

As the figure 1 shows, when TBM approaches the MP, soil heaves or settles slightly and when TBM traverses the MP, the settlement of soil increases gradually. When the tail of TBM leaves the MP, the settlement of soil reaches the extremum.

When TBM approaches the MP, soil is subjected to the interaction of slurry pressure and disturbance of cutter head. If the slurry pressure is too low to support the soil, the ground will subside. On the contrary, the ground will heave if the slurry pressure is too high.

When TBM drives through the MP, the force situation of soil is similar as TBM approaches. The difference is that the distance to cutter head is smaller,and the settlement is more sensitive to slurry pressure when TBM traverses.

When the tail of TBM leaves, if the grout is not injected in time or the grout volume is not adequate, there will be large settlement caused by the gap created between the tail skin and the rings. But, if the grout volume is excessive, ground heave will occur.

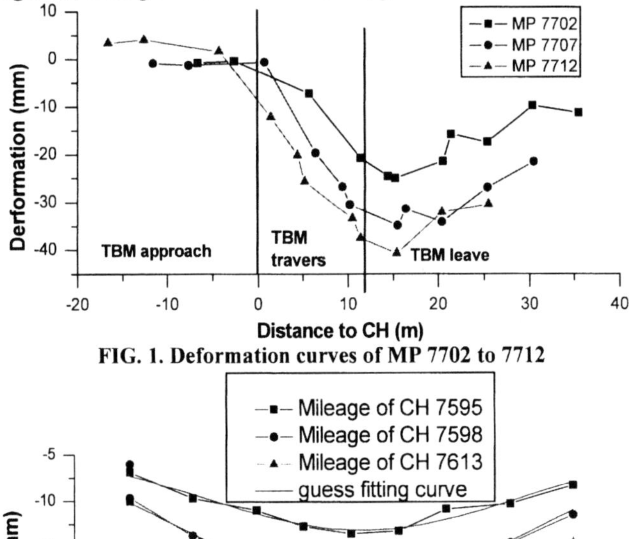

FIG. 1. Deformation curves of MP 7702 to 7712

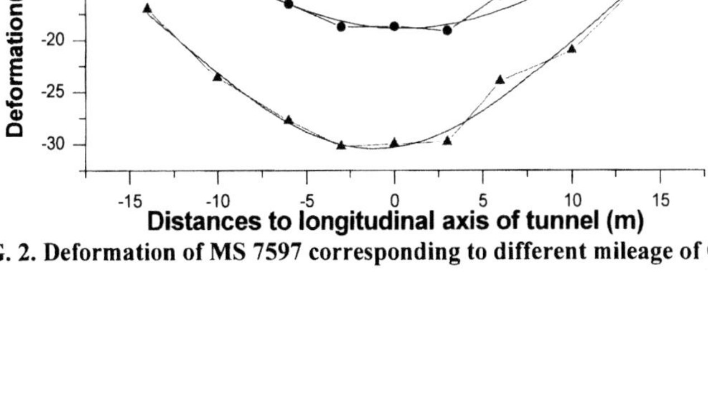

FIG. 2. Deformation of MS 7597 corresponding to different mileage of CH

The deformations of MPs on MS7597corresponding to different mileage of cutter head (CH) are shown in Figure 2.

Obviously, the deformations of MS 7597 are similar to Guess curve. It indicates the ground deformations caused by tunnelling accord with peck formula. It is practicable using the peck formula to estimate the ground deformation in this case.

GROUND LOSS RATE CALCULATION

The peck formula (equation 1) is used to estimate the transverse distribution of the settlement.

$$S(x) = \frac{\pi R^2 V_l}{\sqrt{2\pi} i} \exp\left(-\frac{x^2}{2i^2}\right) \quad \textbf{(1)}$$

Where V_l is the ground loss rate (%); R is the radius of the TBM (m); x is the distance to the longitudinal axis of tunnel (m); i is the width coefficient of settlement tank (m).

And the width coefficient i can be determined by equation (2).

$$\mathrm{i} = \left(\frac{Z}{2R}\right)^{0.8} R \quad \textbf{(2)}$$

Where Z is the distance between ground and centre of tunnel (m).

According to the peck formula, the ground loss is an important factor to estimate the settlement. How to determine the ground loss accurately is the key to estimate the settlement. According to the experience, the ground loss relates to geological condition, construction method, the level of construction technology and so on. Therefore, this coefficient is determined by area experience in many cases. If there is no similar experience to refer to, the ground loss would be difficult to judge. In this paper, an approach which uses the measured data to inverse the ground loss is introduced.

When TBM is advancing, the deformation is different at different stages. It means the ground loss at different stages is also varied. The plus deformation corresponds to plus ground loss, which means heave, while the minus deformation corresponds to minus ground loss, which means settlement. In this paper, ground loss is divided into three stages, which are TBM approaching, TBM traversing and TBM leaving.

In Table 1, it is listed that ground loss rates computed by measured data of MS 7597 based on peck formula.

The abnormal data of ground loss can be eliminated by σ -method with 2σ. And then the ground loss of MS 7597 is equal to the average value of the rest MP. The ground loss of the other MS could be got by this method. The result is listed in Table 2.

In Table 2, it is observed that the ground loss of different MS is varied even at the same stage. That is because there are many factors which affect the ground loss such as buried depth, slurry pressure, grout volume, operation factors and so on.

When TBM approaches, the main factors which affect ground loss are slurry pressure (P) and buried depth (H). When buried depth is constant, the ideal slurry pressure should support ground accurately which makes ground no settlement or heave. The ground will heave as the slurry pressure increases and subside as the slurry pressure decreases. Therefore, the ground loss is in direct proportion to the slurry pressure.

When slurry pressure is constant, ground would heave in a shallow tunnel and subside in a deep tunnel. Therefore, the ground loss is in inverse proportion to buried depth.

Table 1. The ground loss rates calculated by measured data of MS 7597

MP	x(m)	H(m)	R(m)	i(m)	TBM approach		TBM traverse		TBM leave	
					S(mm)	V_l (%)	S(mm)	V_l (%)	S(mm)	V_l (%)
E-4	-14	17.12	7.715	11.29	-6.78	-0.22	-9.67	-0.32	-17.03	-0.56
E-3	-10				-9.17	-0.21	-13.71	-0.31	-23.68	-0.53
E-2	-6				-11.04	-0.19	-16.56	-0.29	-27.74	-0.48
E-1	-3				-13.09	-0.21	-18.72	-0.29	-30.19	-0.47
7597	0				-13.26	-0.2	-18.7	-0.28	-29.97	-0.45
W-1	3				-13.78	-0.22	-19.12	-0.3	-29.8	-0.47
W-2	6				-10.68	-0.19	-15.77	-0.27	-24	-0.42
W-3	10				-9.93	-0.22	-14.52	-0.33	-21.01	-0.47
W-4	14				-7.79	-0.25	-11.5	-0.38	-14.5	-0.47

Table 2. The ground loss rates calculated by measured data

MS	TBM approach	TBM traverse	TBM leave
	V_l (%)	V_l (%)	V_l (%)
7238	-0.08	-0.17	-0.28
7241	-0.19	-0.28	-0.26
7567	-0.05	-0.12	-0.31
7572	-0.04	-0.09	-0.53
7577	0.05	-0.12	-0.51
7587	-0.15	-0.17	-0.38
7597	-0.21	-0.3	-0.48

At the stage of traverse, the ground loss is also in direct proportion to the slurry pressure and inverse proportion to buried depth.

At the stage of leave, the main factors which affect ground loss are buried depth and volume of grout. It is easy to know the ground loss increases as the buried depth decreases or the volume of grout increases.

In Table 3, the known parameters are listed and non-linear fitting is done to get the undetermined coefficients by the known parameters. And based on the analysis the ground loss rates could be estimated by equation 3 to 5.

$$V_a = -32.284 - 12.276 \times \left(\frac{P}{\gamma_w D}\right) + 3.006 \times \left(\frac{P}{\gamma_w D}\right)^2 + 105.066 \times \left(\frac{H}{D}\right)^{-1} - 61.67 \times \left(\frac{H}{D}\right)^{-2} \quad (3)$$

$$V_t = -32.48 - 20.713 \times (\frac{P}{\gamma_w D}) + 4.769 \times (\frac{P}{\gamma_w D})^2 + 130.47 \times (\frac{H}{D})^{-1} - 77.55(\frac{H}{D})^{-2} \quad \textbf{(4)}$$

$$V_l = 135.57 - 286.7 \times (\frac{V_g}{V_r}) + 119.58 \times (\frac{V_g}{V_r})^2 + 80.69 \times (\frac{H}{D})^{-1} - 45.5 \times (\frac{H}{D})^{-2} \quad \textbf{(5)}$$

Where V_i is the ground loss rate (%) and subscript *a, b, l* means three stages of TBM driving; V_g is the volume of grout (m^3) and V_r is the volume of gap between tail and segment of one ring (m^3); *P* is the slurry pressure (m); *H* is the buried depth (m); *D* is the diameter of TBM (m); *L* is the unit length (m); γ_w is the unit weight of water (N/m^3).

Table 3. The ground loss rates calculated by measured data

MS	*D* (m)	*L* (m)	*H* (m)	TBM approach		TBM traverse		TBM leave	
				V_a (%)	*P*(bar)	V_t (%)	*P*(bar)	V_l (%)	V_g(m^3)
7238	15.43	1	22.26	-0.08	4.24	-0.17	4.24	-0.28	27.24
7241			22.30	-0.19	4.24	-0.28	4.24	-0.26	27.28
7567			18.03	-0.05	3.28	-0.12	3.28	-0.31	23.74
7572			17.92	-0.04	3.28	-0.09	3.28	-0.53	24.60
7577			17.80	0.05	3.48	-0.12	3.48	-0.51	24.70
7587			17.45	-0.15	3.17	-0.17	3.17	-0.38	25.22
7597			17.12	-0.21	3.16	-0.30	3.17	-0.48	25.00

VALIDATION

In order to validate the equations 3 to 5, the compares between the estimated values by the equations and the measured data are shown in Figure 3 and Figure 4.

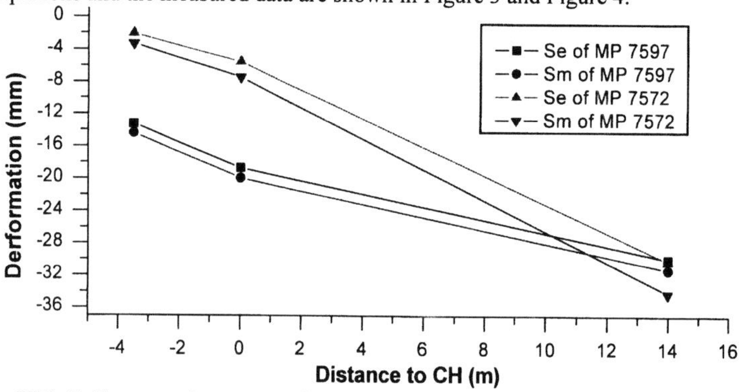

FIG. 3. Compare between estimated value (Se) and measured data (Sm)

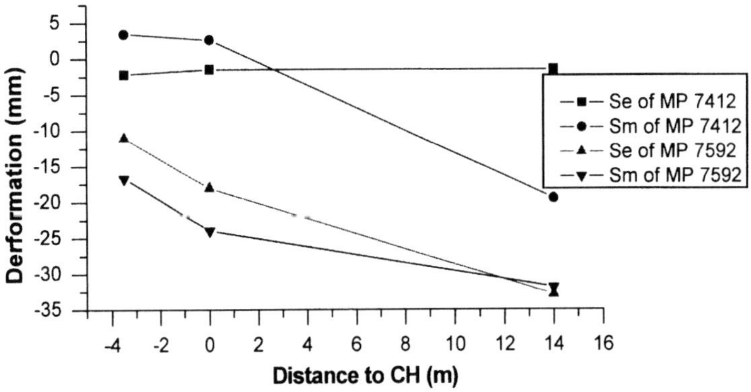

FIG. 4. Compare between estimated value (Se) and measured data (Sm)

In Figure 3 and Figure 4, it is observed that the estimated values are close to the measured data except MS 7412. That is because MS 7412 is on the dam and the dam affects the ground deformation. It shows that the equations 3 to 5 are applicable to where there is no building on the ground.

CONCLUSIONS

In tunnelling, ground deformation is affected by many factors including buried depth, slurry pressure, grout pressure, advance rate, and other operation factors. The research proves that the ground deformation is in direct proportion to the slurry pressure and volume of grout and in inverse proportion to buried depth and the ground loss rates at different stage are different. The equations for ground loss rates estimation given in this paper are simple for practical application but they are also immature which need to be improved so that the extent of application can be enlarged.

REFERENCES

Peck R B (1969). "Deep excavations and tunnelling in soft ground" Proceedings of the 7th International Conference on Soil Mechanics and Foundation Engineering. Mexico City 1969: 225-290.

Ghaboussi J, Hansmire W H, Parker H W, et al (1983). "Finite element simulation of tunnelling over subways". *Journal of Geotechnical Engineering*, 109(3):318-334

Lee, K.M and Rowe (1990), R. K "Finite element modeling of three-dimensional ground deformation due to tunnelling in soft cohesive soils." *Computers and Geotechnics*, (10):87-109

Liu Zhaowei, Wang Mengshu, Dong Xinping (2003). "Analysis Of Ground surface subsidence of metro tunnel induced by shield construction". *Chinese Journal of Rock Mechanics and Engineering*, 22(8): 1297-1301.

Michael J. Kavvadas (2003). "Monitoring and modeling ground deformations during Tunnelling", Proceedings, 11th FIG Symposium on Deformation Measurements, Santorini, Greece.

Dimmock, P. S. (2003). "Tunnelling-induced ground and building movement on the Jubilee Line Extension". PhD thesis, Cambridge University.

Moss, N. A. and Bowers, K. H. (2005). "The effect of new tunnel construction under existing metro tunnels". Proc. 5th International Symposium: Geotechnical Aspects of Underground Construction in Soft Ground, Amsterdam, 45-50.

Zhang Haibo, Yin Zongze， Zhu Jungao (2005). "3D finite element simulation on deformation of soil mass during shield tunneling". *Chinese Journal of Rock Mechanics and Engineering*, 24(5):755-760.

Wei Gang, Zhang Shimin, Qi Jingjing, et al (2006). "Study on calculation method of ground deformation induced by shield tunnel construction". *Chinese Journal of Rock Mechanics and Engineering*, 25(Supp.1): 3317-3323.

Zheng Yifeng, Ding Zhicheng, Dai Shimin (2006). "An analysis on surrounding ground deformation and ground/water pressure fluctuations caused by a super large diameter TBM", *Chinese Journal of Underground Space and Engineering* Vol.2: 1349-1353

Han Xuan， Li Ning (2007). "Comparative analysis of strata prediction models for ground movement induced by tunnel construction" *J. of Rock Mechanics and Engrg*, 26 (3):594-600.

The Technological Issues of Health Monitoring in Wuhan Yangtze River Tunnel Operations

Haitao Dou[1], Hongwei Huang[2], Yadong Xue[3] and Qunfang Hu[4]

[1, 2,3] Key Laboratory of Geotechnical and Underground Engineering of Ministry of Education, Department of Geotechnical engineering, Tongji University, Shanghai, China;

[1] douhaitao888@yahoo.com.cn [2] huanghw@tongji.edu.cn [3] yadongxue@126.com

[4] Shanghai Institute of Disaster Prevention & Relief, Tongji University, Shanghai, China; huqunf@tongji.edu.cn

ABSTRACT: The first highway tunnel (Wuhan Yangtze River Tunnel) in Yangtze River was built to operate in 2009 through about 5 years of construction. There are large-scale investment, heavy traffic and the complexity of operating conditions of the tunnel. To ensure that the tunnel is in the safety during its operations, the health monitoring of the tunnel will be carried out. It will not only escort the normal operation of the project, but also provide reliable early warning information. Besides, it will also accumulate basis and valuable information for large-scale river shield tunnel health operation of China.

INTRODUCTION

At present, the importance of structural safety monitoring in operational phase has been generally recognized in the field of engineering and technology, and th long-term stability of the monitoring have been added to growing number of larg tunnels, bridges and other public structures in the world, which is called "health monitoring", Such as Haibei XIONG (2006), Jie SU (2007) and Zhenggen LIU (2008). This has a decisive significance to ensure the design life of the tunnel.

Wuhan Yangtze River Tunnel project was the key construction project which was included in "Tenth Five-Years Plan" in Wuhan. The objective of health monitoring is accurate evaluation of the tunnel structural health and working condition timing. It can provide direct technical basis for timely maintenance of the tunnel structure. There are large-scale investment, heavy traffic and the complexity of operating conditions of the tunnel. To ensure that the tunnel is in the safety during its operations and always in a safe, reliable, controlled state in the design of 100 years, the health monitoring of the tunnel work is essential. It will not only escort the normal operation of the project, but also provide reliable early warning information. Besides, it will also accumulate basis and valuable information for large-scale river shield tunnel health operation of China.

PROJECT OVERVIEW

Wuhan Yangtze River Highway Tunnel is the first river-crossing highway tunnel project in the Yangtze River. The total length of the tunnel is 3630m, in which the north approach road is 200m, tunnel is 3270m, and south approach road is 160m. This is a twin-tube tunnel for 4 lanes. The external diameter of the tunnel is 11m, and the design traffic speed is 50 km/h. The horizontal alignment and longitudinal profiles of Wuhan Yangtze River Tunnel Project are shown in Figure 1, while its cross-sectional diagram is shown in Figure 2.

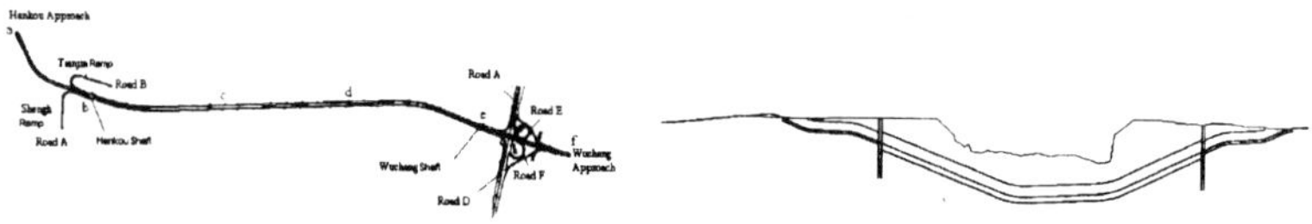

FIG. 1. Horizontal alignment and longitudinal profile of Wuhan Yangtze River Tunnel

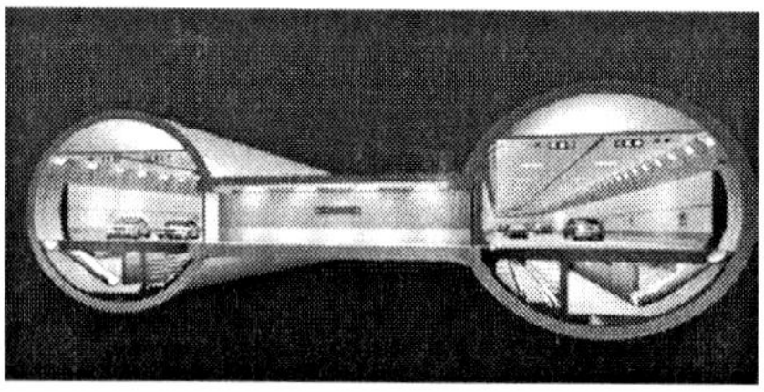

FIG. 2. Cross-sectional view of Wuhan Yangtze River Tunnel

Wuhan Yangtze River Tunnel is a new traffic trunk which solves the problem of traffic crossing the river in downtown. After it was built to operate in 2009, the time of trip from Wuchang to Hankou can be shortened to about seven minutes, greatly reducing the transit time of vehicles through Wuhan. There will be more than 23 percent of vehicles crossing the river by the way of the tunnel, and the pressure of traffic through Yangtze River Bridge and second Yangtze River Bridge will be greatly alleviated.

Through comprehensive analysis of the safety in operation and the potential disaster risks of the Wuhan Yangtze River Highway Tunnel, the main purpose of health monitoring is as follow.

(1) Real-time settlement, deformation and stress state observation of tunnel lining structure to ensure the health operations.

(2) Prediction of the deformation trend of the tunnel structure, in accordance with the development of deformation to determine whether there is a need to adopt corresponding protection measures, and to provide a reasonable decision making basis for maintenance programs.

(3) The establishment of a multi-level safety early warning system and the improvement of tunnel health protection mechanism in operation to ensure the overall and local safety during the operation.

(4) To provide reliable data, accumulate shield tunnel engineering experience and lay the foundation for the future research on shield tunnel health monitoring and operational management.

HEALTH MONITORING AND DIAGNOSIS OF TUNNEL STRUCTURE

Health monitoring program

The indicators of Tunnel Structural Health Monitoring included the usability evaluation index system, the strength evaluation index system and the durability evaluation index system. Usability evaluation indicator is the index system to evaluate the overall vertical along the tunnel from the perspective of the overall usability of the tunnel structure and function. Strength evaluation indicator is the index system to evaluate the vertical and horizontal performance from the perspective of tunnel structural safety. Durability evaluation indicator is the index system to evaluate the tunnel current situation of structural durability and the remaining service life. Shown in the tables below:

Table 1. The Usability Evaluation Index System and Inspection Program

Content	Specific Requirements	Inspection Methods
Water Leakage	Classify the leakage level of the whole tunnel. The site of serious leakage: to investigate the amount of leakage and leakage paths, where there is mud leak, to detect hole behind lining, and to plan leakage drawing.	Visual, supplemented by infrared thermal imager, radar, etc.
Cracks	The pattern cracks of main structural elements, the distribution of crack width, length, depth, direction and other	Visual, with cracks ruler, crack width measurement and

	information, and to plan cracks drawing.	tape.
Breakage	To determine the location of damage, the size of the area or volume, etc., and plans structural damaged drawing.	Visual, tape, etc.
Joints	Investigation of the tunnel joints, including the deformation of joints and the wrong units, such as segment opening.	Visual, vernier caliper, etc.
Vertical Deformation	Measurement of the tunnel center deviation.	Water Level, Total Station, GPS, etc.
Convergence of Tunnel	Key section.	Laser rangefinders

Table 2. The Strength Evaluation Index System and Inspection Program

Content	Specific Requirements	Inspection Method
Concrete Strength (on site)	Key section	Ultrasonic rebound method
Concrete Strength (sampling)	Key section	Core sampling strength test
Elastic Modulus of Concrete (sampling)	Key section	Core sampling strength test
connecting Bolts Strength	In accordance with the actual conditions at the scene to collect bolt and test it	Experimental tensile or Richter hardness tester
Strength of Reinforced Bar	Key section	Experimental tensile or Richter hardness tester
Joint Strength(rubber gasket)	Possible sampling of the joint material and the tensile, compressive strength test.	compressive strength test

Table 3. The Durability Evaluation Index System and Inspection Program

Content	Specific Requirements	Inspection Method
Air Quality	In addition to the key section of the tunnel, sampling should include the entrances of tunnel and air vents.	Spectrophotometer, Air sampler, analytical instrument.
Water Quality	To sample based on the key section or the location of the actual water leakage, and determine chloride ion / sulfate ion.	PH testing equipment

Content	Specific Requirements	Inspection Method
Temperature and Humidity	In addition to the key section of the tunnel, sampling should include the entrances of tunnel and air vents.	temperature and humidity testing equipment
Depth of Carbonation	Key section	Phenolphthalein reagent
Thickness of Concrete Protective Layer		Radar
Concrete Permeability		Water and gas permeability equipment
Anti-corrosion Properties of Chloride ions	Select a cross-section in Each key section, each taking a set of core samples.	RCM method, flux method
Chloride Ion / Sulfate Ion Concentration	Select a cross-section in Each key section.	Silver sulfate solution titration
Concrete Ingredients		Scanning Electron Microscope , energy spectrum analysis.
Steel Corrosion	Select two cross-sections in each key section, steel corrosion included the scope and depth of corrosion.	Visual, supplemented by tape and vernier caliper, etc.
Corrosion of Steel Components	Inspection of bolt corrosion depth, corrosion current.	Field measurements and laboratory test.
Joints Material Aging	Tensile, compressive strength test for joints material.	Indoor mechanics experiment
Impermeability of Joint Material	Indoor impermeability test.	Indoor impermeability test.

With Wuhan Yangtze River Tunnel design, construction and operational management and other relevant information, considering the importance and special nature of the project characteristics, the main indicator of health monitoring program of Wuhan Yangtze River shield Tunnel project can be the index system of strength and durability, including: load-bearing capacity, stress, deformation, structural damage, cracks in concrete, concrete protection layer, such as injury, as well as the contents of steel corrosion.

Key sections of health monitoring

Considering the characteristics of vertical section of the Wuhan Yangtze River Tunnel, a few key sections are selected as follows.

The first key section, LK2 +970, features: Hankou shallow silty clay layer.

The second key section, LK3 +955, features: the maximum depth of the river.

The third key section: LK4 +220, features: differentiation layer at the bottom of the river.

The fourth key section, LK4 +500, features: the maximum depth above the shore.

The fifth key section, LK5 +000, features: Wuchang shallow silty clay layer.

The content, frequency of monitoring and data acquisition

In accordance with the monitoring needs of Wuhan Yangtze River Tunnel, manual or automatic monitoring can be used in data collection. In view of 24-hour operations, automatic monitoring data collector is recommended. DATA GeoLogger DT515 automatic data acquisition is proposed, shown below.

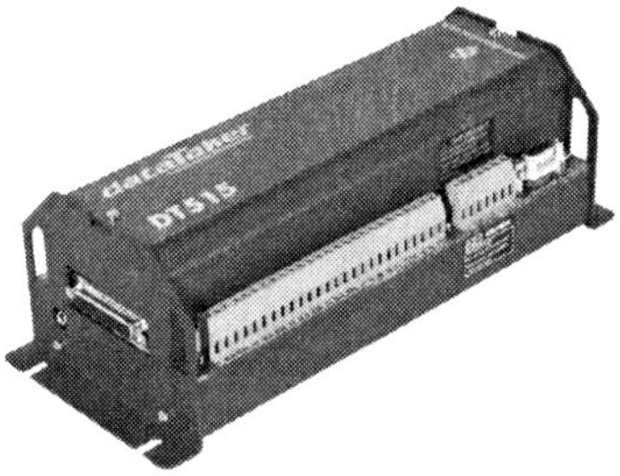

FIG. 3. DATA GeoLogger DT515 automatic data acquisition

To sum up, the specific content of health monitoring, measuring points and monitoring frequency are included in the table below:

Table 4. The Contents of Health Monitoring, Measuring Points and Frequency

Monitoring Content	Measuring Points	Monitoring Frequency
Tunnel Settlement	Key section	Daily
Open Degrees and Deflection of Segment Joint	Segment Joint of Key Section	Daily
Concrete Cracking	Potential cracks occurrence department of Key parts	Daily
Water Leakage	Seen in Table 5	
Stress-strain Monitoring of Reinforced Bar	The same to the crack monitoring	Daily
Corrosion of Bar	The same to the crack monitoring	Daily
Integrated Detection System	Automatic data collection	

Table 5. The Frequency of Leakage Check

Inspection Type	A Type of Work Status	B Type of Work Status
General Inspection	2-4 times / year (regular)	1-2 times / year (regular)
Focus Inspection	1-2 times / year (regular)	1 times / year (regular)
Comprehensive Inspection	By the need	By the need

The alarm value of monitoring

The alarm value of corresponding monitoring of Wuhan Yangtze River Tunnel project, which is based on the actual situation, surrounding environment and design calculations, should be pre-determined whether or not the strength and displacement of the tunnel structure is more than the extent permitted to determine the safety of tunnel construction, then to determine whether or not to adjust the design and construction methods and take effective and timely measures. Specific alarm values are as follows.

(1) The accumulated changes value of tunnel settlement is over 3cm, and the deformation rate is over 0.1mm / d.

(2) The stress of reinforced bar in tunnel lining is over 80% of the design value.

(3) The open volume of segment joint is over 5mm.

(4) The crack width of segment is over 0.2mm.

(5) The deformation of tunnel lining is over 3 ‰ D (D is the external diameter of tunnel).

Alert grading can be divided into four levels, and the specific criteria is shown in the table below.

Table 6. Alert Level for Health Monitoring and Response Measures

Grading	Security Standards	Response Measures
Safety	$U<0.5U_0$	No measures needed.
Third Grading (concerns)	$0.5U_0 \le U<0.7U_0$	Concerns needed, and should keep a close watch on the data and analyze data trends in the next stage.
Second Grading (Warning)	$0.7U_0 \le U<0.9U_0$	Oral report, or meeting, even to write a written report and recommendations.
First Grading (Alarm)	$U \ge 0.9U_0$	Conduct on-site investigations, site meetings, to study the emergency measures.

Note: U_0—Alarm value, U—Measured values.

CONCLUSION

(1) Through the analysis of health monitoring, the health monitoring framework is provided for the operation management of Wuhan Yangtze River Tunnel.

(2) Access to health testing data, the key issues of engineering can be grasped. Through the health status evaluation of the tunnel structure, credible evidence is provided for tunnel maintenance and environment protection.

(3) Through the implementation of the health monitoring program, tunnel maintenance management model will be developed from the previous visual analysis of appearance survey and with some simple testing equipment to judgment with modern technology. A full range of health monitoring management and evaluation with computer is carried out.

(4) In accordance with alert principles of the health monitoring, the specific alert value was determined. Alert grading was divided into four levels, which is easy for operation and management.

ACKNOWLEDGMENTS

The authors appreciate the support of 863 national high-tech research and development program (2006AA04Z442) and Action Plan for Innovation project of Shanghai Science and Technology Committee (07DZ12065).

REFERENCES

Guixiang Chen, Hongwei Huang, Jianxin You. (2006). "Study on Life Cycle Risk Management of Metro." *J. Chinese Journal of Underground Space and Engineering*, Vol. 2 (1): 47-51.

Zhenggen LIU, etc. (2008). "Immersed tube tunnel real-time health monitoring system." *J. Chinese Journal of Underground Space and Engineering*, Vol. 4 (6): 1110-1115.

Tao LIU. (2008). "Evaluation of existing shield tunnel." (Doctoral Thesis), *D. Shanghai, tongji Univrtsity.*

Shanghai Design and Research Institute of Tunnel Engineering Rail Transit. (2006). "Health monitoring program of Chongming Yangtze River Tunnel Channel Tunnel project in Shanghai."

Jie SU, etc. (2007). "Research on design of subsea tunnel structural health monitoring." *J. Chinese Journal of Rock Mechanics and Engineering*, Vol. 26 (2): 3785-3792.

Mingqing Xiao. (2005). "General Situation of Yangtze River Tunnel Engineering in Wuhan." *J. Soil Engineering and Foundation*, Vol. 19 (1): 2-4.

Haibei XIONG, Zhiqiang LI. (2006). "State of the art of structural health monitoring." *J. Structural Engineers*, Vol. 22 (5): 86-90.

SAFETY INFLUENCED BY COMBINED ACTION OF SULPHATE AND CHLORIDE TO SHALLOW HIGHWAY TUNNEL

Rong-rong Yin[1,2], He-hua Zhu[1]

1. Dept of Geotechnical Engineering, Tongji Univ, Shanghai, 200092,China; zhuhehua@tongji.edu.cn
2. School of Naval Architectuer and Ocean Engineering, Jiangsu Univ of Science and Technology, Zhenjiang, Jiangsu 212003,China; ecsi_tangjian@163.com)

ABSTRACT: This study, using GeoFBA, generates a load-structure method tensile-compressional model for analyzing and calculating the tunnel that is 10m deep embedded by rock □ and is eroded by combined action of sulphate and chloride. The calculation indicates that first of all, the tunnel safety falls down gradually with the increase of circumferential crack width (the degree of corrosion of reinforcement) as well as longitudinal crack depth; furthermore, when lining suffers the same corrosion degree of reinforcement and the same depth of longitudinal crack, the problem occurs on the arch crown which would bring up bigger effect to the tunnel safety than that occurs in the haunch; In addition, the change of circumferential crack width, which is smaller than 0.2mm, would make less influence to the tunnel lining safety.

Key words: shallow; highway tunnel; sulphate; chloride; safety

0 INTRODUCTION

When the highway tunnel is located in a relative worse area both of geographical and geological conditions, aggressive medium in the environmental water always doesn't merely impact a tunnel's lining. But a comprehensive action causes the tunnel material worse, then fairly lowers and shortens working functions and ages of the tunnel, and then brings greatly snags or hidden danger to the transportation. There are plenty of aggressive ions, like chloride, SO_4^{2-}, so for cross-river tunnel, cross-sea tunnel as well as those ones near sea or relevant environment, it is not only should consider chloride corrosion, but also the comprehensive influence from chloride, SO_4^{2-}, etc[1]. to the tunnel concrete. In the south-western area and western salt lake area [2-4], the tunnel concrete structure is faced with the aggressive environment both of single sulphate corrosive liquor/solution and a high concentration of salt. Therefore, it is very meaningful to develop a study to find out changing rules of concrete under the combined action of sulphate and chloride in such tunnel subjects.

1 MUTUAL INFLUENCES BETWEEN SUPLPHATE AND CHLORIDE CORRODING CONCRETE SUBJECTS

When the combined action of sulphate and chloride happens, there is a mutual influence between them in the corrosion of concrete. These influences include:

(1) Sulphate weakens concrete's power of fighting against the corrosion of Cl^- . In the combined solution, Cl^- and SO_4^{2-} both spread inside concrete, and react on aluminum in the concrete, then separately produce Frield salt and Aft salt. The more aluminum reacts with SO_4^{2-}, the less aluminum reacts with Cl^-, which causes free Cl^- concentration increased. Furthermore, sulphate salt would cause concrete's mechanics impact and supply more Cl^- into the inner place of concrete. Study shows that in the concrete, ettringite is usually a phase more stable than Friedel salt that will transfer into ettringite and produce Cl^- with enough SO_4^{2}.

(2) Cl^- can delay sulphate salt's aggression to concrete[5]. It exists a competitive relation between Cl^- and SO_4^{2-} while they react with aluminum in concrete. Although Frield salt is not stable than ettringite, spreading coefficient of Cl^- is usually two ranks of quantity higher than that of SO_4^{2-}. In terms of it, Frield salt is easier to react with aluminum, producing a more stable Frield salt; thus, it can reduce such an opportunity that SO_4^{2-} and aluminum react to produce ettringite and gypsum, meanwhile, delay the damage speed of concrete. Ettringite's solubility in chloride salt is three times than its solubility in the water, which also lowers formation of ettringite in the corrosion course.

2 STEEL REINFORCED CONCRET'S MECHANICAL PROPERTY AFTER THE COMBINED ACTION OF SUPLPHATE AND CHLORIDE

2.1 strength of reinforcement concrete with the combined action of suplphate and chloride

When there exsits the combined action of sulphate and chloride, chloride would not just corrode concrete, but also corrode steel. This paper quotes Hui Yunlin's[6] rules and conclusion achieved by experiments: when steel in the concrete structure corrodes to a certain degree, the most usual appearance is that longitudinal corrosive cracks occur on concrete along the steel reinforcement. Its corrosive extent is mainly relevant to reinforcement diameter and corrosive cracks' width. Reinforcement corrosive extent can be indicated by ρ,reinforcement's loss of weight percentage, which can calculated by the formula below:

$$\rho = (32.43 + 0.303 f_{cu} + 0.65c + 27.45W) / d \quad (1)$$

ρ —— loss of weight percentage, %; d ——reinforcement diameter, mm; f_{cu} ——concrete cube compression strength, MPa; W ——longitudinal rusting crack width, mm; c ——thickness of concrete for protection, mm.

In the calculation of tunnel lining, loss of reinforcement cross section ratio can be inferred from the loss of weight ratio of reinforcement. Meanwhile, when the steel reinforcement is corroded, the lining's cross sectional area can be inferred from formula (2):

$$E_c A = E_c A_c + E_s A_s \quad (2)$$

E_c ——modulus of elasticity of concrete, kpa; A ——cross section area of lining after steel reinforcement is equivalent with concrete, m^2; A_c ——cross section area of concrete in the lining, m^2; E_s ——modulus of elasticity of steel reinforcement, kpa; A_s ——cross section area of steel reinforcement that is corroded, m^2, in formula (3):

$$A_s = (1 - \beta) A_{so} \quad (3)$$

β ——the ratio of the loss of reinforcement weight, %; A_{s0} ——cross section area of steel reinforcement that is not corroded, m^2

2.2 rigidity of reinforcement concrete after the combined action of sulphate and chloride

On the influence of combined action of sulphate and chloride to reinforcement concrete rigidity, this paper denotes Zhao Xin's[7] experimental results: The reinforcement concrete beam would both produce lateral cracks and longitudinal steel rust expansion destruction, when the steel reinforcement is corrosive under the external load action. And these lateral cracks and longitudinal steel rust expansion destruction would low down reinforcement concrete beam's bending rigidity. This bending rigidity extent depends on the deepness of lateral cracks as well as the width of longitudinal cracks formed by rust's expansion. If the reducing extent of corrosive reinforcement concrete beam's rigidity can be expressed by α, rigidity reduction coefficient, then:

$$\alpha = \alpha_1 \cdot \alpha_2 \quad (4)$$

α ——corrosive steel reinforcement concrete beam rigidity reduction coefficient; α_1 ——lateral crack reduction coefficient to steel reinforcement concrete beam initial rigidity; α_2 ——longitudinal steel reinforcement rust reduction coefficient to steel reinforcement concrete beam initial rigidity. α_1 、 α_2 calculation method in literature[8].

The rigidity of lining with the influence of combined action of sulphate and chloride can be indicated by this formula below:

$$EI' = \alpha \cdot EI = \alpha_1 \cdot \alpha_2 \cdot EI \quad (5)$$

E ——modulus of elasticity of steel reinforcement, kpa; I ——moment of inertia of lining after the steel reinforcement is corroded, m^4; α ——rigidity reduction coefficient of rust steel reinforcement concrete beam, %; I_0 ——lining's moment of inertia before the steel reinforcement is corroded, m^4 。

3 CHECKING CALCULATION METHOD OF LINING STRUCTURE SAFETY

Safety factor K is defined as a ratio of structure max loading ability N_u and axial force of real structure N by the tunnel design standard.

$$K = \frac{N_u}{N} \quad (6)$$

If the safety factor K is bigger than or equal to the safety factor K_u that is stipulated by the tunnel design standard, the structure is considered safe; otherwise, the structure is deemed dangerous.

4 MODEL CALCULATION

4.1 model situation

This study chooses a single-arch standard two lanes tunnel that is 10m-depth embedded by rock Ⅴ as the study object, using GeoFBA, generates a load-structure method tensile-compression mode[8] (setting pulling springs in the range of 140° of the arch crown and setting compress springs in the left part) for analyzing and calculating the tunnel after its lining is impacted by the combined action from sulphate and chloride. Initial lining in the model uses sprayed concrete, thickness 20cm, label C30; steel reinforcement concrete lining 2 inches, thickness of concrete 45cm, labelC30; diameter of steel is 22mm; coefficient of elastic resistance is chosen according to JTGD70—2004; $k_n = 200000 KN/m^3$, $k_s = 8000 KN/m^3$; the ratio of pulling spring coefficient of elastic resistance and that of compress spring is 1/2. Modulus of elasticity of concrete influenced by the combined action of sulphate and chloride decreases by 15%[9].

According to defect quantitative methods, there are many influence factors: defect tunnel model calculation consequences, defect position ,defect working range, defect extent（crack width, crack deepness）,defect of lining material property （modulus of elasticity、rigidity）. This paper mainly discusses that the tunnel lining impacted by combined action of sulphate and chloride causes cracks in the crown and haunch of arch ($30^{\circ}-50^{\circ}$) and such cracks bring what kind or degree of influence to the tunnel's working functions. The width of circumferential crack (representing steel corrosive extent) is selected by ten characteristic ranks：0.02m、0.04m、0.06m、0.08m、0.1m、0.2m、0.6m、1m、1.5m、2m while the deepness of longitudinal crack is selected by four characteristic rank：0.06m、0.08m、0.1m、0.2m; the length of steel corrosion in the crown and haunch of arch is 4m. In the range of steel corrosion，there is a longitudinal crack every 0.5m. When it exists cracks both in the crown and haunch of arch, in the range of 1m, there is a longitudinal crack every 0.5m, meanwhile, it is considered of full steel that has corrosion.

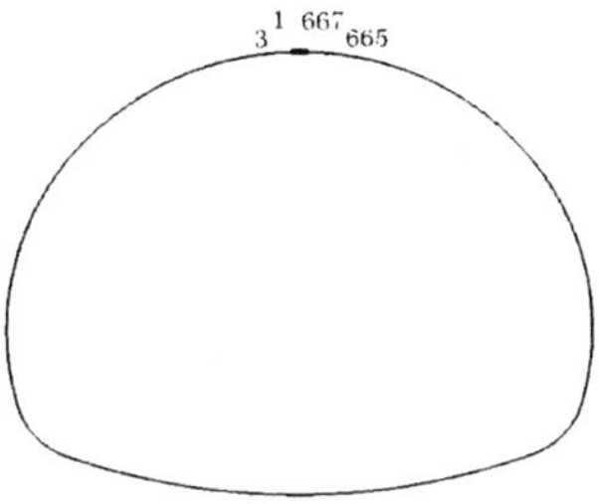

Figure 1 units' positions in the cross sectional figure when defeats exist at the arch crown

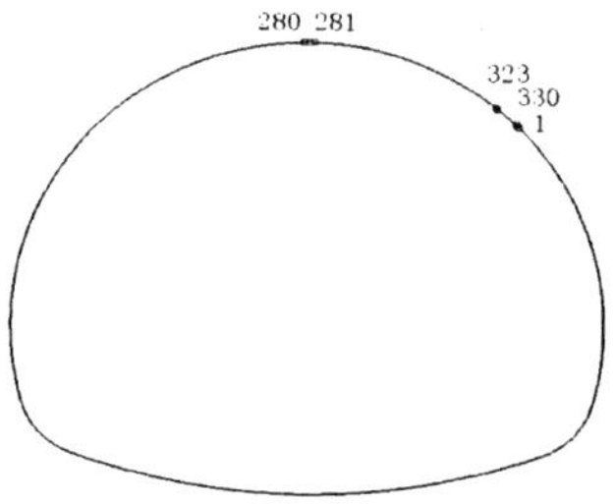

Figure 2 unit's positions in the cross sectional figure when defeats exist both at arch crown and haunch

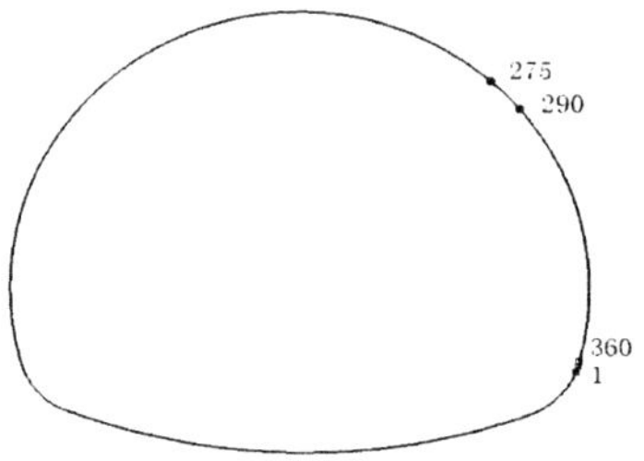

Figure 1 units' positions in the cross sectional figure when defeats exist at the haunch

4.2 calculation consequences

The combined action of sulphate and chloride corrodes the tunnel lining deeply embedded by 10m-depth rock Ⅴ. Safety factors of a tunnel full units calculated from tunnel models should be seen in figure 1 to 3. Figure 1 shows the safety factor when there are cracks in the crown of arch of tunnel lining whose defect unit number is 1—53 and 615—667;Figure 2 shows the safety factor of tunnel lining whose haunch has crack, and its defect unit number is 243-322. Figure 3 shows the safety factor of tunnel lining whose haunch and crown both have crack, and their defect unit numbers are 1—13、281—299、318—330. When the tunnel lining is in the state of press, its safety factor is supposed as "+" in the safety factor table and when the tunnel lining is in the state of pull, its safety factor is premised as "-". In the figure, thread 2 and 3 respectively represent boundary value of given concrete pressed or pulled safety factor. 0.06,0.08,0.1 and 0.2 represent the tunnel safety factor when longitudinal crack deepness is 0.06m,0.08m,0.1m and 0.2m. According to safety factor's calculating consequences, table 1,2,3 conclude the safety condition of tunnel after it has been influenced by the combined action of sulphate and chloride, and " √ " stands for safety while " × " stands for unsafety.

(1) **crack existing at arch crown**

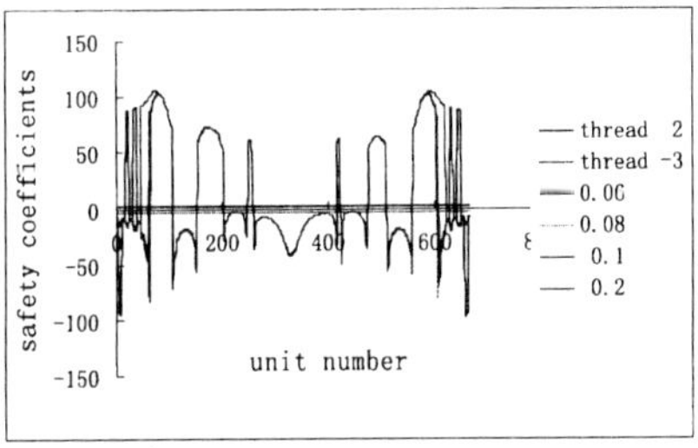

（a） circumferential crack width 0.02mm

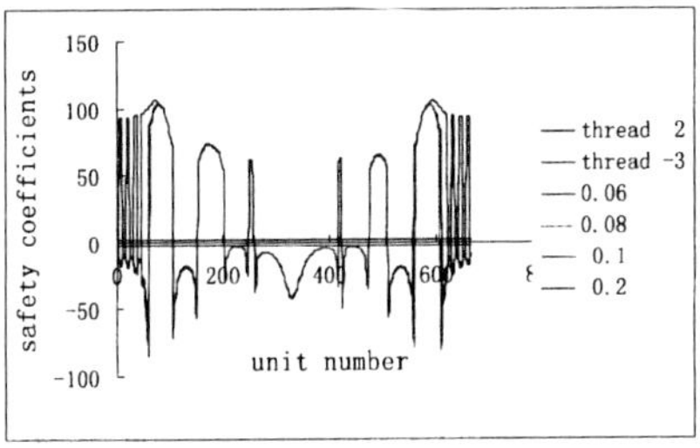

（b） circumferential crack width 0.2mm

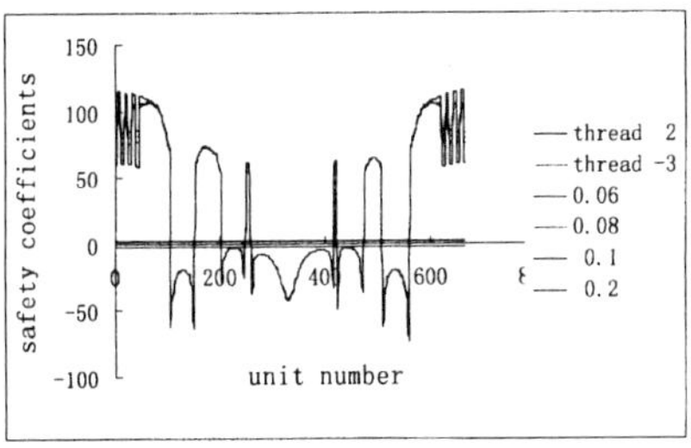

（c） circumferential crack width 2mm

Fig.4 Safety coefficients when crack exists at arch crown

Tab.1 Safety condition when crack exists at arch crown

Circumferential Crack Width (mm) \ Longitudinal Crack Depth(m)	0.06	0.08	0.1	0.2
0.02	√	√	√	×
0.04	√	√	√	×
0.05	√	√	√	×
0.08	√	√	√	×
0.1	√	√	√	×
0.2	√	√	√	×
0.6	√	√	×	×
1.0	×	×	×	×
1.5	×	×	×	×
2.0	×	×	×	×

(2) **crack existing at haunch**

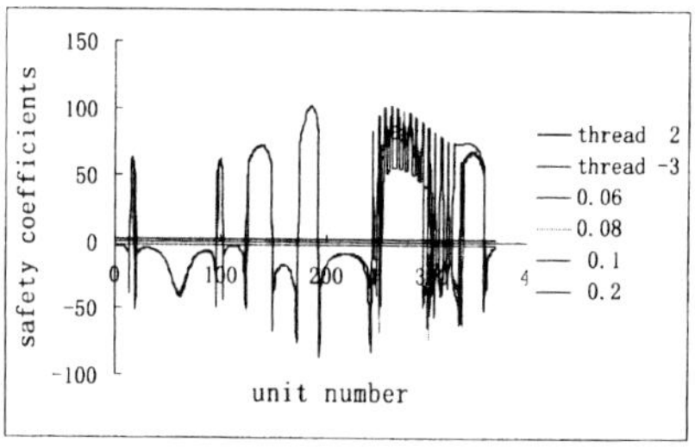

(a) circumferential crack width 0.02mm

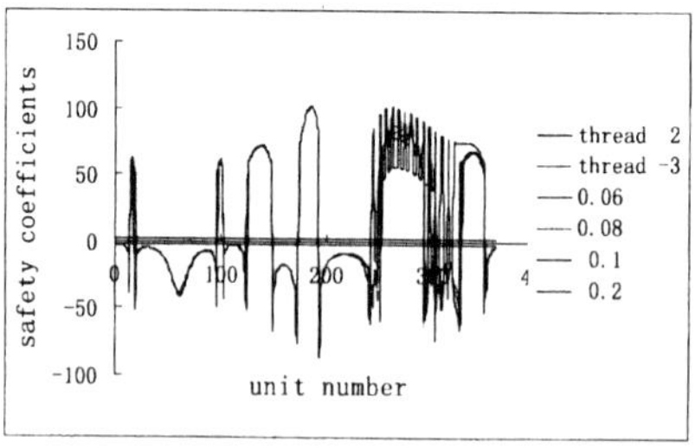

(b) circumferential crack width 0.2mm

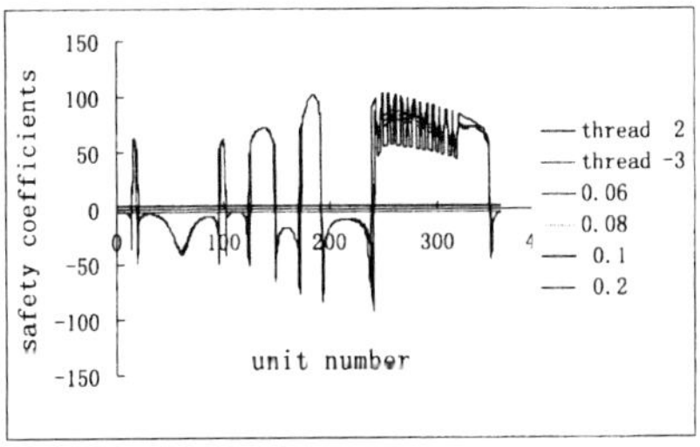

(c) circumferential crack width 2mm

Fig.5 Safety coefficients when crack exists at haunch

Tab.2 Safety condition when crack exists at haunch

Longitudinal Crack Depth(m) / Circum-ferential Crack Width (mm)	0.06	0.08	0.1	0.2
0.02	√	√	√	√
0.04	√	√	√	√
0.05	√	√	√	√
0.08	√	√	√	√
0.1	√	√	√	√
0.2	√	√	√	√
0.6	√	√	√	×
1.0	√	√	√	×
1.5	√	√	√	×
2.0	√	√	×	×

(3) crack existing at arch crown、haunch

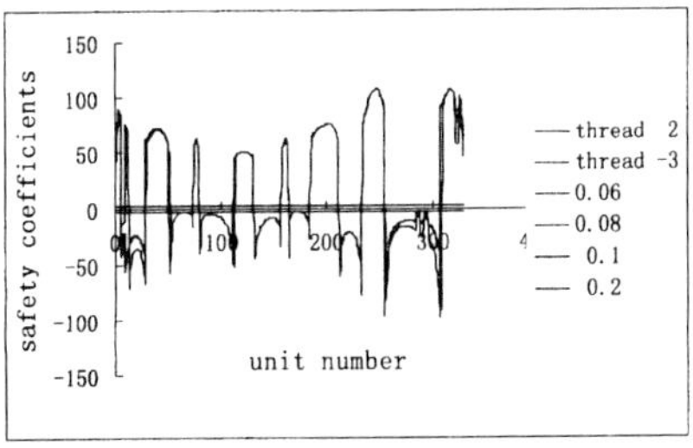

(a) circumferential crack width 0.02mm

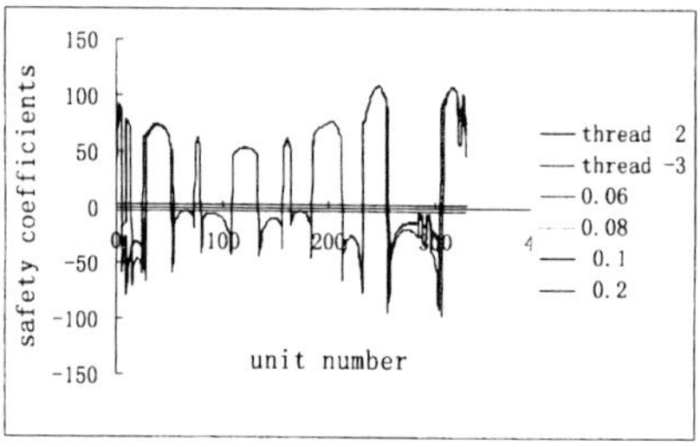

(b) circumferential crack width 0.2mm

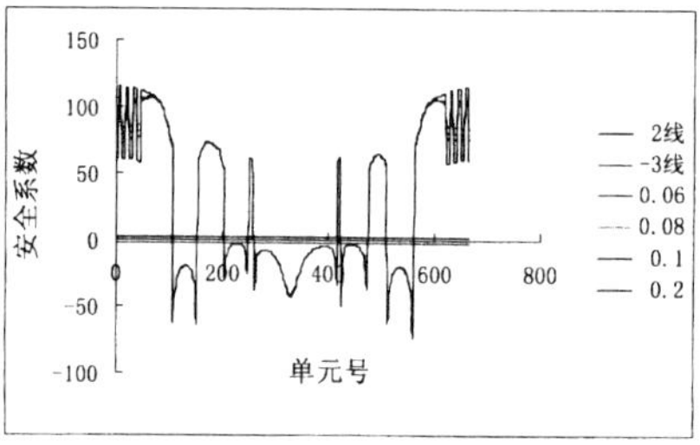

c) circumferential crack width 2mm

Fig.6 Safety coefficients when crack exists at arch crown、haunch

Tab.3 Safety condition when crack exists at arch crown、haunch

Longitudinal Crack Depth(m) / Circum-ferential Crack Width (mm)	0.06	0.08	0.1	0.2
0.02	√	√	√	√
0.04	√	√	√	√
0.05	√	√	√	√
0.08	√	√	√	√
0.1	√	√	√	√
0.2	√	√	√	×
0.6	√	√	√	×
1.0	×	×	×	×
1.5	×	×	×	×
2.0	×	×	×	×

CONCLUSIONS

(1) From the safety factor table, the combined action of sulphate and chloride makes a tunnel lining degraded, which brings more impact to mechanical property of the lining which has defects, yet brings less impact to mechanical property of the lining without defects.

(2) With the increase of circumferential cracks' width (increase of steel corrosive degree) as well as that of longitudinal cracks' deepness, a tunnel's safety conditions lower down gradually and the tunnel becomes not safe.

(3) If a highway tunnel lining has crack defects because of freezing-melting circulation, this tunnel starts to be in an unsafe situation when circumferential crack width in the arch crown is 1mm, 1mm, 0.6mm, 0.02mm separately and longitudinal crack deepness is 1mm, 1mm, 0.6mm, 0.02mm separately; meanwhile, if circumferential crack width in the arch haunch is 2mm, 0.6mm respectively and longitudinal crack deepness is 0.1m, 0.2m respectively, the tunnel begins to be in an unsafe condition. Hence, defects in the arch crown would make a larger influence to the tunnel safety than that existing in the arch haunch. In addition, the situation of crack distribution and steel corrosive length is different, if defects exist both in the arch crown and arch haunch or if they exist only in one place of the two. As a result, the calculating consequence separately from such two situations can not be compared or contrasted.

(4) As the calculation consequence analyzed, if circumferential crack width is smaller than 0.2mm, its change would brings little influence to the tunnel lining.

(5) It is can be seen from the analysis of lining mechanical situation that lining structure property changes a lot if steel corrosive degree is worse or longitudinal crack deepness is bigger. Then the lining looses its stability at this time. The value of safety factor calculated from the formal safety factor formula doesn't decrease but increase, which is mainly due to the formal safety factor formula that is not fit for the calculation of lining under the unstable situation. Therefore, this paper judges a tunnel's safety within the comprehensive consideration of safety factor and lining mechanical situation.

ACKNOWLEDGEMENTS

Thanks for the support of the International Engineering Foundation.

REFERENCES

1 Lei Shengxiang,Li Shuyuan. Discussion on the groundwater in undersea tunnels[J]. Modern Tunnelling Technology, 2004,41(5):6-11

2 Kang Yong,Li Xiaohong,Xu Mou,Yang Guang,Xia Xiaoquan. The application of

water quality analysis in corrosion prevention design of Bijiashan tunnel, Chongqing City[J]. The Chinese Journal of Geological Hazard and Control,2005,16(2):80-83

3 Zhou Junlong,Yang Debing. Problems of durability of concrete used in tunnel[J].Highway, 2002,12(12):136-139

4 Shi Changsheng,Gao Feng. Analyse diseases of Dong Gang tunnel on Bao Lan railway and deal with landslip[J]. Lanzhou Jiaotong University, 2004,23(1):53-56

5 Yu Zhenyu.Soil of Zhejiang [M].Hangzhou:Zhe Jiang science and technology press,1994.

6 Hui Yunling.Experimental study on assessment and prediction of corrosive degrees of reinforcements in concrete structures[J]. Industrial Construction: 1997,27(6)

7 Zhao Xin.Experimental study on the performance of corroded reinforced concrete beams[D].Changsha:College of Civil Engineering of Hunan University,2006

8 Yin Rongrong.Calculation method of road tunnel lining based on life cycle[D].Shanghai:Doctoral Dissertation, College of Civil Engineering of Tongji University,2008

9 Jin Zuquan,Sunwei,Zhang Yunsheng,Lai Jianzhong. Effect of chloride on damage of concrete attacked by sulfate[J]. Journal of Wuhan University of Technology, 2006,28(3):43-46

Numerical simulation of EPB tunnel face instability in dry sand by Discrete Element Method

Lvjun Tang[1,2], Renpeng Chen[1,2,*], Yunmin Chen[1,2], Daosheng Ling[1,2]

[1] MOE Key Laboratory of Soft Soil and Geoenvironmental Engineering, Hangzhou 310058, China
[2] Department of Civil Engineering, Zhejiang University, Hangzhou 310058, China
[*]Corresponding author, Email: chenrp@zju.edu.cn

ABSTRACT: Tunnel Face instability is greatly concerned in tunnel engineering. This paper presents the failure mechanism of EPB tunnel face in dry sand by using discrete element method (DEM). Laboratory direct shear tests on the dry sand were conducted for calibrating the meso-parameters of DEM model. Two stages of tunnel face failure associated with the development of soil arching were proposed. A new method for estimating the face failure zone based on sudden change of displacement was also presented. The results show that the maximum variation of vertical stress and porosity occurs at tunnel crown, width of the failure zone is about 0.5D~0.6D in front of the tunnel face.

KEYWORDS: DEM, Tunnel Face failure, Soil arching

INTRODUCTION

Face stability control of Earth Pressure Balance (EPB) shield is greatly concerned during tunnel construction. A variety of factors may induce face losing stability, such as adverse geology in the construction region, improper control of EPB shield, etc. Losing face stability will affect safety assurance of surrounding buildings; even bring a threat to human life. In order to design the cutter head and properly control the EPB shield during constructing, it is meaningful to understand the mode for face failure.

Much work has been done on the face stability. Anagnostou et al. (1994, 1996) gave empirical formula for limit supporting force by wedge stability model. Their model is suitable for EPB shield or slurry shield advancing in saturated uniform stratum. Broere (2001) improved traditional wedge stability model to the saturated layered soils. Chambon and Corte (1994) coducted centrifuge model test to simulate face failure when shield is stopping for lining at dry sand. A bulb shape failure envelope was proposed, and strain localization appeared at the envelope surface. With different ratio of tunnel depth to diameter C/D=0.5, 1, 2, the scope of envelope was also obtained. In-Mo Lee et al. (2003) simulated the distribution of seepage field in front of the tunnel face by using commercial software PENTAGON-3D (Emerald Soft Consulting 1998). The influence of seepage force on tunnel face stability was taken into account.

As sand is a kind of granular material, discrete element method based on mechanics of discrete media reflects sand behavior better than finite element method. Commercial software PFC^{2D} or PFC^{3D} based on DEM (Itasca Consulting Group) is widely used in numerical analysis for meso-mechanics. Masud Karim (2007) simulated the process of tunnel face failure in dry sand by PFC^{3D}. The failure zone was obtained by variation of displacement gradient for particles. Comparing with centrifuge model test (Chambon and Corte, 1994), the failure zone was larger in horizontal direction and smaller in vertical direction at longitudinal symmetry plane. According to relations between particles' volume into the ballast and time, Manuel J.Melis Maynar et al. (2005) investigated face stability in condition of lining and advancing separately. It was found that friction coefficient and bond strength between particles were two important factors affecting stability of the excavation face. T.Funatsu et al. (2008) simulated face stability of single tunnel and two tunnels facing each other separately by PFC^{2D}. Three face conditions for tunnels including unlined, fore-poling, and face bolts were considered. Failure zones and soil arching effect were reflected by the force chain between particles.

This paper studies the failure process of tunnel face in dry sand by DEM. Variation of the supporting force versus the tunnel displacement was investigated, the distribution of vertical earth pressure and variation of porosity were also considered. Finally, the failure zone in front of the tunnel face was discussed in this paper.

DISCRETE ELEMENT METHOD (DEM)

Discrete element method was first proposed by Peter Cundall (1971, 1979). It is based on force-displacement law between blocks and Newton's second law for block. Blocks achieves a relatively stable state by relaxation iteration. Block moving, rotating even separating between blocks are allowed in DEM, thus it has advantages in large deformation problem analysis for non-continuous geotechnical medium induced by force. In commercial DEM software PFC^{3D}, in order to transfer force in normal and tangent direction between particles, particles are connected by spring, visco-pot, slide and couplers. Four basic contact models between particles including linear contact model, simplified Hertz-Mindlin model, Coulomb slip model and bonding model are in PFC^{3D}. For dry sand with little cohesion, bonding model is seldom used. Considering simplicity and high-performance computing, linear contact model is applied in later sections to analyses meso- mechanism of tunnel face failure.

CALIBRATING SOIL MESO-PARAMETERS

Choosing suitable meso-parameters is important for accurate simulation. PFC^{3D} contains two kinds of parameters for particles and walls separately. Parameters for particles include k_n (stiffness in normal direction), k_s (stiffness in tangential direction), μ (friction coefficient). However, stiffness in normal and tangential directions are two parameters for walls. Meso-parameters are ordinarily obtained by two kinds of methods. For one kind of methods, meso-parameters are acquired by theoretical method (D.O. Potyondy et al., 2004; H. Lan,dry et al., 2006). For another kind of methods, parameters are achieved by calibration of test results (Manuel et al., 2005; Itasca, 2008). Meso-parameters used in this studies are obtained by calibrating results of the direct shear test (see Fig.1).

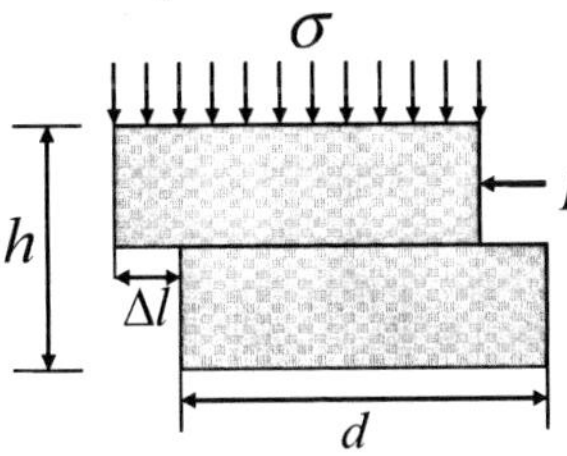

Fig. 1 Sketch of direct shear test

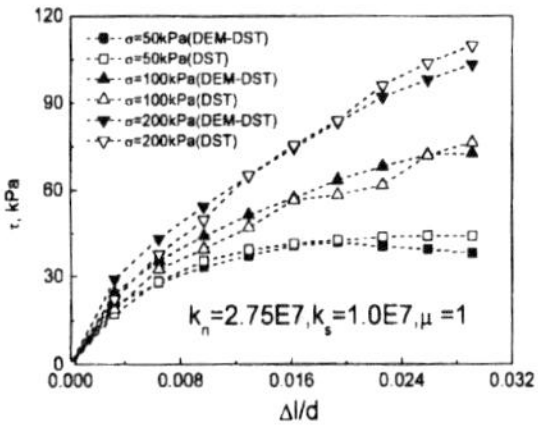

Fig.2 Results of DEM_DST and DST

Dry sand used in direct shear test (DST) is from Yangtze River. Table 1 presents basic physical parameters of the sand.

Table 1 Basic physical parameters of dry Yangtze Rive sand

Max. dry density (g/cm^3)	Min. dry density (g/cm^3)	Controlled dry density (g/cm^3)	Specific gravity	D_{50} (mm)	Friction angle (°)
1.71	1.45	1.61	2.65	0.33	37

Two key problems should be considered in the calibrations. one is the shape of particles, and the other is matching between particle size and shear box size. Particle shape is one of important factors that influence properties of soil cell. Spherical particles and special-shaped particles are included in PFC^{3D}. Dry sands are actually not spherical particles, using spherical particles instead of sands may overestimate rotation ability and underestimate shear dilatancy (X.Lin et al., 1997; Tang-Tat Ng et al., 2004) to some extent, yet spherical particles are still chosen in this paper as two following reasons. One is that shear properties of dry sand can be accurately reflected by spherical particles (L.Zhang&C.Thornton, 2007), the other is that contacts between spherical particles are easy to estimate which will bring high computation efficiency. As the computation capacity of computer is limited, enlarging particle size is a usual way to solve the real problem by DEM. It should be noted that enlarging particle size will lose computational accuracy to some extent. However, some studies (Manuel et al., 2005; German code DIN 18137) have shown credible results when minimum size of shearing box is no less than ten times of particle diameter. Considering diameter of particles used in this study is 0.66 meters, therefore the same size is used for DST simulation. Table 2 presents the geometric size of particles for laboratory DST and DEM simulation of DST (DEM-DST) separately.

In order to obtain meso-parameters matching well with the dry sands, repeated adjustments are conducted which are based on above conclusions, thus a group of proper meso-parameters (k_n=2.75E7, k_s=1.0E7, μ=1) is obtained via agreements of curves for average shear force versus normalized displacement by two methods (see Fig.2).

Table 2 Geometry size for DST and DEM-DST

Type	Radius of shear box (m)	Height of shear box (m)	D_{50} (mm)	Magnification of particles
DST	0.03	0.02	3.3E-04	1
DEM_DST	12.36	8	6.6E-01	2000

DISCRIPTION OF MODEL FOR TUNNEL FACE FAIURE

Wall and particle are basic components in DEM. The model box and shield which contains cylindrical shell and supporting plate are simulated with walls, while dry sands are simulated with sphere particles. In the process of tunnel face failure, variations of stress and porosity in front of the tunnel face are concerned both in practice and on theory. In DEM, stress or porosity is a mean value in the measure zone, thus two columns of measure spheres are adopted in the model (see Fig.3), which is based on the failure zone in front of the tunnel face (Chambon and Corte, 1994; Masud Karim, 2007).

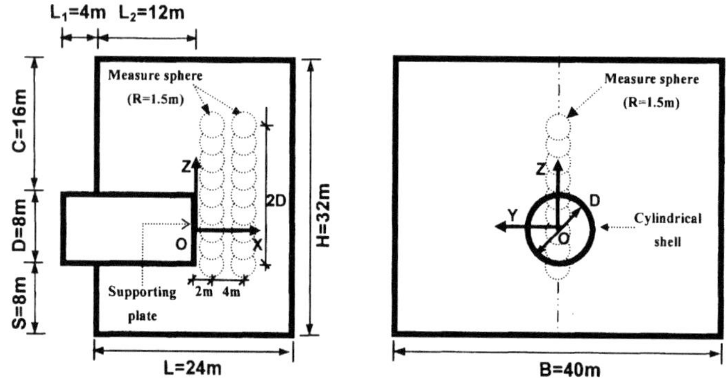

(a) Longitudinal profile of model (b) Transverse profile of model

Fig. 3 Model and arrangement of measure sphere with (a) longitudinal profile of model and (b) transverse profile of model

Simulating tunnel face instability by DEM contains three steps. Firstly, model box and particles are generated, besides that gravitational acceleration is also applied to particles to obtain the initial stress field. Then, the particles in excavated regions are deleted and the shield is installed at the same place simultaneously. Iterating is proposed subsequently to make particles in a state of equilibrium. Lastly, supporting plate is retreated at a speed of 0.01m/s until the displacement of the supporting plate is more than 0.53m.

RESULTS AND DISCUSSION

Supporting force on face plate

The force on the supporting plate is recorded during the process of face failure. Fig. 4 presents the curve of the normalized supporting force versus the displacement of supporting plate. The initial supporting force corresponding to zero displacement is about 6416kN.

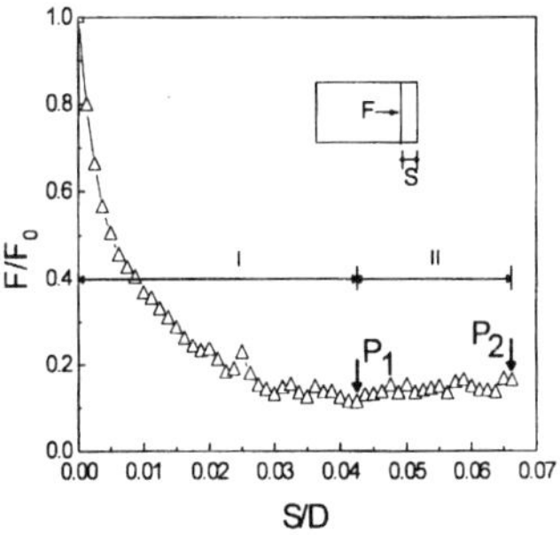

Fig. 4 Normalized supporting force versus displacement of tunnel face

Failure of tunnel face can be divided into two stages (see Fig. 4). Stage I is local failure stage. In stage I, supporting force decreases rapidly at the beginning as the displacement increases, then slowly until it reaches the minimum value. The minimum supporting force can be defined as limited supporting force (LSF). A local failure zone has gradually generated, while the original soil arching (see Fig. 5b) is formed completely under condition of LSF. During stage□, the original soil arching is collapsed and the new soil arching (see Fig.5c) is continuously formed, thus supporting force increases gradually with the increase of displacement.

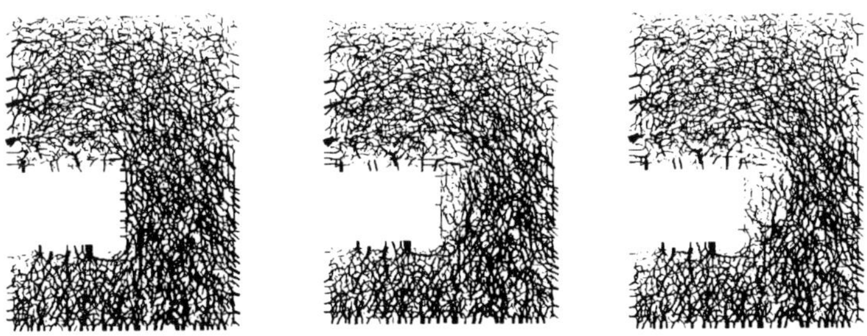

(a) Initial force chain (f_{max}=4.727E+005) **(b) Force chain of P1 (f_{max}=4.812E+005)** **(c) Force chain of P2 (f_{max}=4.834E+005)**

Fig.5 Force chain under different supporting force at longitudinal profile with (a) initial force chain, (b) force chain of P1, and (c) force chain of P2

Vertical stress and variation of porosity

Traditional stress based on continuum mechanics can not be obtained directly from DEM. The contact force between the particles and displacements of particles are concerned. Average field theory (Oda M, Iwashita K,1999) is usually adopted for definition of stress in a given volume. The analogous formula is given in PFC^{3D} (Itasca, 2008). Since horizontal stress in front of the tunnel face has been reflected by supporting force to some extent, thus vertical stress is mainly concerned in this study. Two column measure spheres are arranged along longitudinal profile (y=0). The changes of vertical stress in two regions during face failure are also presented (see Fig. 6). It is obvious that vertical stress of the near region (x=2) changes greater than that of the far region (x=6). The near region is in the failure zone while the far region is not. In the near region, vertical stress decreases with increase of displacement of the tunnel face. Maximum vertical stress reduction occurs at tunnel crown, while vertical stress reduction becomes smaller and smaller away from tunnel crown. At about 0.5D above tunnel crown, change of vertical stress is very small. It may be considered to be the crown of soil arching.

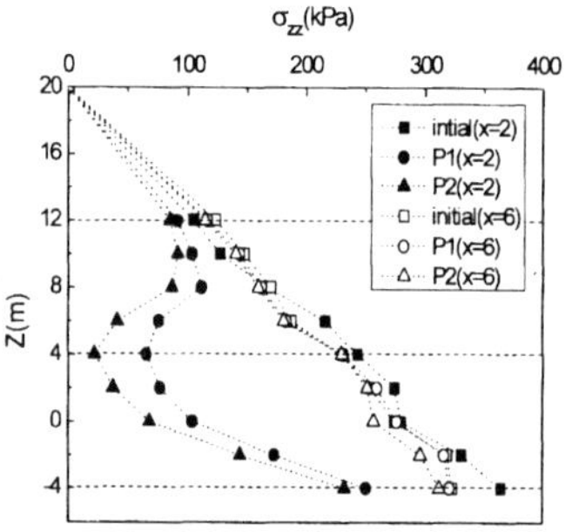

Fig.6 Vertical stress in longitudinal profile

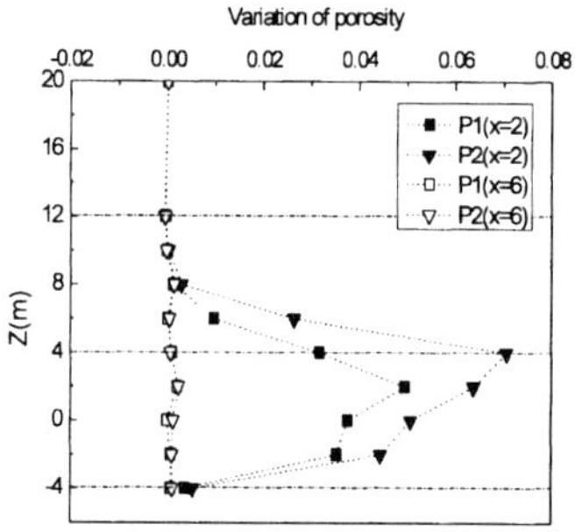

Fig.7 variation of porosity in longitudinal profile

Porosity is an important parameter reflecting soil characteristic. The variation of porosity in two regions (see Fig. 7) has been invested during the process of tunnel face failure. Obviously, porosity changes at failure zone while keeps invariant at stability zone. At near region, maximum change of porosity occurs at tunnel crown, away from tunnel crown, variation of porosity is gradually reduced analogous to variation of vertical stress.

Failure zone

Failure zone is closely related to LSF, while determination of failure zone in front of the tunnel face is difficult. To solve this problem, an applicable method is analyzing particles displacement at various planes. Taking final failure zone of P2 as example, vertical displacement for different horizontal planes at longitudinal profile (see Fig. 8) has been recorded, a sudden displacement changing point which is near the point of stable displacement is thought to be in the envelop of failure zone. Liking these points, envelop of failure zone at longitudinal profile is proposed (see Fig. 9), thus width of failure zone is about 0.5D~0.6D in front of the tunnel face.

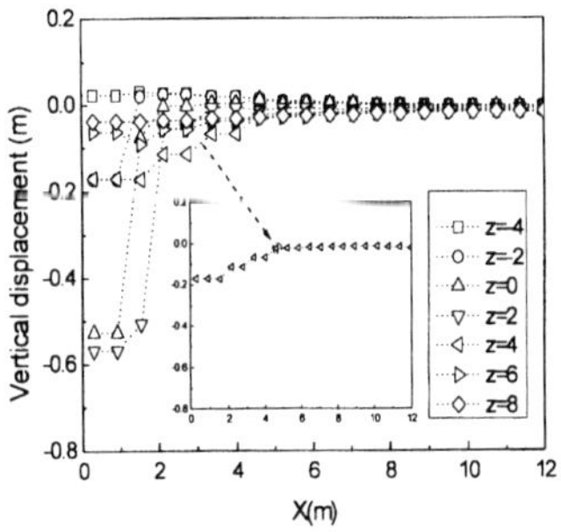

Fig.8 Final vertical displacement for various horizontal planes at longitudinal profile

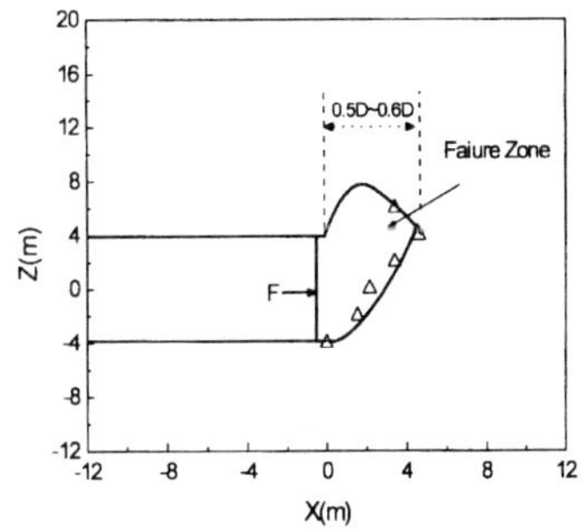

Fig.9 Final failure zone at longitudinal profile

CONCLUSIONS

A 3D numerical DEM model to analyze mechanism of tunnel face failure has been developed. The following results have been found:

(1) Tunnel face failure contains two stages. Firstly, as increases of displacement in tunnel face, a local failure happened and original soil arching is formed until limited supporting force is applied. Then, the original soil arching is collapsed and a new soil arching is formed as displacement of tunnel face, thus supporting force increase gradually from limited supporting force.

(2) In the regions which belong to failure zone, significant variation of vertical stress and porosity occurs at tunnel crown, while variation gradually reduces away from tunnel crown.

(3) A sudden displacement changing point near the stable displacement point is considered in the envelop of failure zone, width of failure zone at longitudinal profile is about 0.5D~0.6D in front of the tunnel face

ACKNOWLEDGMENTS

The authors wish to thank National Basic Research Program of China (research grant: 2007CB714001), National Natural Science Foundation of China (research grant: 50629802), Program for New Century Excellent Talents in University (NCET) for financial supports. The authors also greatly appreciate Doctor candidate Xiao-min Xu, and Doctor candidate Chao Han for their kind supports on DEM analysis.

REFERENCES

G.Anagnostou, K.kovári. (1994). "The face stability of Slurry Shield Driven tunnels. " Journal of Tunneling and Underground Space Technology, Vol.9(2):165-174.

G.Anagnostou, K.kovári. (1996). "Face stability condition with Earth Pressure Balanced shields." Journal of Tunneling and Underground Space Technology,

Vol.11 (2):165-173.

Broere W. (2001). "Tunnel Face Stability & New CPT Application[D]. " Netherlands: Delft University Press.

Pierre Chambon , Jean-Francois Corté. (1994). "Sallow tunnels in cohesionless soil: stability of tunnel face." Journal of Geotechnical Engineering, Vol.120(7): 1148-1164.

In-Mo Lee, Seok-Woo Nam, Jae-Hun Ahn. (2003). "Effect of seepage forces on tunnel face stability." Canadian Geotechnical Journal, Vol.40(2):342-350.

ASM Masud Karim. (2007). "Three-Dimensional Discrete Element Modeling of Tunneling in Sand." Cannada: Alberta University.

Manuel J.Melis Maynar, Luis E.Medina Rodríguez. (2005). "Discrete Numerical Model for Analysis of Earth Pressure Balance Tunnel Excavation." Journal of Geotechnical and Geoenvironmental Engineering, Vol.131(10): 1234-1242.

T.Funatsu, T.Hoshino, H.Sawae, N.Shimizu. (2008). "Numerical analysis to better understand the mechanism of the effects of ground supports and reinforcement on the stability of tunnels using the distinct element method." Journal of Tunneling and Underground Space Technology, Vol.23(5): 561-573.

Cundall P A. (1971). "A computer model for simulating progressive large scale movements in blocky system." Proc. Symp. On Rock Fracture(ISRM) Nancy, France.

Cundall, P., Strake, O.D. (1979). "A discrete numerical model for granular assemblies." Géotechnique, Vol.29(1): 47-65.

Itasca Consulting Group. (2008). "PFC3D (Particle Flow Code in 3 Dimensions), Version 4.0, User's Manual, Minneapolis. "

D.O. Potyondy, P.A. Cundall. (2004). "A bonded-particle model for rock." International Journal of Rock Mechanics & Mining Sciences, Vol.41: 1329-1364.

H. Landry, C. Lagu¨e, M. Roberge. (2006). "Discrete element representation of manure products." Computers and Electronics in Agriculture, Vol.51: 17-34.

X.Lin, T.-T. NG. (1997). "A three-dimensional discrete element model using arrays of ellipsoids." Géotechnique, Vol.47(2): 319-329.

Tang-Tat Ng. (2004). "Behavior of Ellipsoids of two Sizes." Journal of Geotechnical and Geoenvironmental Engineering, Vol.130(10):1077-1083.

L.Zhang, C.Thornton. (2007). "A numerical examination of the direct shear test." Géotechnique, Vol.57(4): 343-354.

Oda M, Iwashita K. (1999). "Mechanics of granular materials." Rotterdam: Balkema A A.

An Improved Pseudo-Static Seismic Analysis for Underground Frame Structures

Huiling Zhao and Yong Yuan

Key Laboratory of Geotechnical and Underground Engineering, Tongji University, Shanghai, 200092, China

ABSTRACT: There are great needs of simple but reliable engineering design method for underground frame structures subjected seismic loads. The paper describes how ground deformations are estimated and how they are transmitted to an underground structure, presenting methods used in the computation of deformation in the structure with the consideration of the structural stiffness variation with the different magnitude of applied loads. This paper provides an example of the application of the method by a typical two-story three-bay frame structures analysis for a subway station in Shanghai.

INTRODUCTION

Frame structural system has becoming one of the most predominate support systems of underground facilities such as subway stations, underground parking garage, and underground shopping mall. Compared to tunnel structures, the vulnerability of frame structures to seismic damage is great. Therefore, current engineering seismic design of underground frame structures gets into great needs.

There are two basic approaches in present seismic design of underground structures. One approach is to carry out dynamic, time-history analysis using finite element or difference methods. Sweet (1997) employed detailed three-dimensional model to analyses the seismic behaviors of Los Angeles Metro system. The second approach assumes the seismic ground motions to induce a pseudo-static loading condition on the structure. This approach allows the development of analytical relationships to evaluate the magnitude of seismic-induced deformations or strains in underground structures (Merritt el al., 1985; Penzien and Wu,1998; Hashash et al. 2001). These relationships are developed based on the ground deformation and the interaction between soil and structures. Generally, an interaction coefficient is introduced and depends on the relative flexibility of structures to soil. This coefficient is assumed as constant in most static analysis. Actually, it varies with the stiffness of structures when the structures, subjected to severe earthquake, fall into nonlinear state. Obviously, most previous static approaches could not be used in seismic analysis of the underground frame structures when in the elastic-plastic state. Therefore effects of the stiffness nonlinearity on structural responses are investigated in this paper.

This paper aims to produce a soil-structure interaction static analysis method that can be used by practicing engineers for the design of frame structures in seismic areas. The solution herein incorporates the push-over curve of structures to account for the stiffness variation with the earthquake loads.

ENGINEERING APPROACH TO SEISMIC ANALYSIS AND DESIGN

Earthquake effects on underground structures can be grouped into two categories (Hashasha,2001): ground shaking and ground failure such as liquefaction, fault displacement, and slope instability, which are out of the scope of this paper. Ground shaking refers to the deformation of the ground produced by seismic waves propagating through the ground. When the ground is deformed by the traveling waves deformations affect any underground structure. The most damaging to underground structures are shear waves, which tends to result in racking of the structure, shown as FIG.1. It is the most critical deformation, and also the scope of this research.

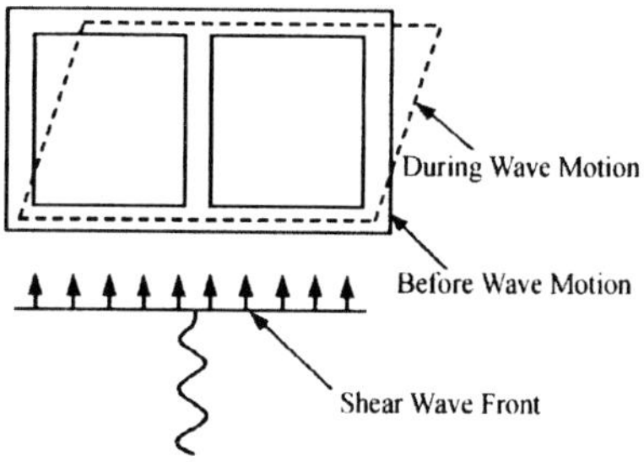

FIG.1 Shear deformation of underground structure

The evaluation of racking deformation for frame structures is conducted by means of the pseudo-static analysis which is based on the following assumptions:

1. The dynamic amplification effect associated with a wave impinging on the opening is negligible. This assumption is correct if the wave length of peak velocities is at least 8 times larger than the width of the opening (Hendron, 1983) ;
2. The surrounding ground is elastic;
3. The structure deforms with the surrounding soil.

The major factors influencing racking deformation of structures include: the shape; dimensions and depth of the structure; the properties of the surrounding soil or rock; the properties of the structure; and the severity of the ground shaking. With the considerations of these factors, the approach is given in detail as follows.

Free field deformation

Free-field deformation describes ground strains caused by seismic waves in the absence of structures or excavations. These deformations ignore the interaction between the underground structure and the surrounding ground, but can provide a first-order estimate of the anticipated deformation of the structure. A designer may choose to impose these deformations directly on the structure. This approach may overestimate or underestimate structure deformations depending on the rigidity of the structure relative to the ground.

Newmark (1968) and Kuesel (1969) proposed a simplified method for calculating free-field ground strains caused by a harmonic wave propagating at a given angle of incidence ϕ in a homogeneous, isotropic, elastic medium. St. John and Zahrah 1987 used Newmark's approach to develop solutions for free-field axial and curvature strains due to compression, shear and Rayleigh waves. Shear strains caused by shear waves is given by the following equation:

$$\gamma = V_S / C_S \cos^2 \phi \qquad \text{eq. (1)}$$

where C_S is the shear wave velocity propagating in soil, V_S is peak particle velocity associated with shear waves.

When incidence angle $\phi = 0^\circ$, the shear strain reaches the greatest value V_S / C_S.

Pushover $P-\Delta$ curve of frame structure

Lateral force-deformation curve can be obtained by pushover analysis of the frame structure which can determine the stiffness variation. Due to nonlinearities stemming from inelasticity of the materials including steel and concrete, or from changes in the geometry of the frame structure, nonlinear analysis models are necessary for the pushover analysis. Herein, the simplified models based fibered beam-column elements and connection regions enable nonlinear characteristics are used for the pushover analysis. The section of beam-column elements is subdivided into n fibers and the stresses are integrated over the cross-sectional area to obtain stress resultants such as force or moment. Each fiber in the section can be assigned concrete or steel stress-strain curve which are defined to express the whole behaviors of materials including linear-elasticity and nonlinear elasto-plasticity. According to the ''plane sections remain plane'' assumption and from the defined constitutive models, fiber stresses are calculated from the fiber strains. This kind of simplified models can arrive at a reliable solution efficiently, especially for a overall frame analysis, and is amenable to routine application in a design office environment.

The curve of structure stiffness varying with the deformation, $S_1(\Delta_{structure})$, can be calculated based on $P-\Delta$ curve.

Interaction between soil and structure

The presence of an underground structure modifies the free field ground deformations. Interaction effects exist between soil and structure. The interaction depends on the relative stiffness, structure geometry, input earthquake motions, and structure depth, among which the flexibility ratio R, i.e. the stiffness in simple shear of the soil relative to the structure that replaces it, is the most important factor (Wang, 1993).

Consider a rectangular soil element under simple shear condition, as shown in FIG 2. Based on the assumption 3, the structure deforms with the surrounding soil, the equation is given as follows (Wang, 1993):

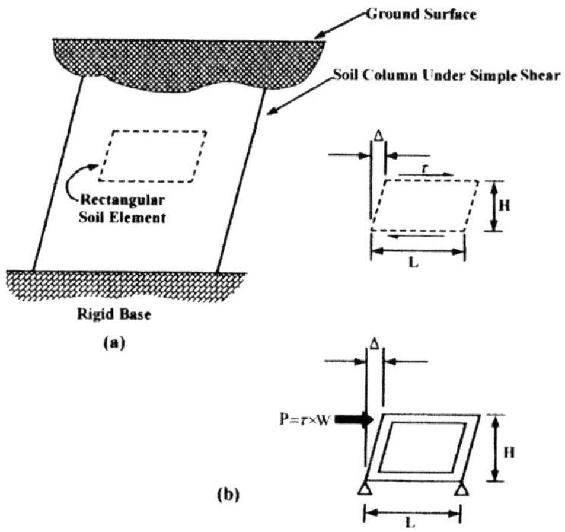

FIG.2. Relative stiffness between soil and a rectangular frame (after Wang, 1993) a) Flexural shear distortion of free-field soil medium. b) Flexural racking distortion of a rectangular frame

$$\frac{\tau}{G_m} = \gamma_s = \frac{\Delta_{structure}}{H} = \frac{P}{HS_1(\Delta_{structure})} = \frac{\tau W}{HS_1(\Delta_{structure})} \qquad \text{eq. (2)}$$

where W and H are the width and height of frame structure respectively, τ is the shear stress distributed on the bottom and top surface of the frame, $S_1(\Delta_{structure})$ is the stiffness of structure, which may varies with $\Delta_{structure}$ when the structure falls into nonlinear state, G_m is the shear modulus of soil.

The flexibility ratio $R(\Delta_{structure})$ can be given by:

$$R(\Delta_{structure}) = \frac{\Delta_{structure}}{\Delta_{free-field}} = \frac{\gamma_s / \tau}{1/G_m} = \frac{WG_m}{HS_1(\Delta_{structure})} \qquad \text{eq. (3)}$$

Lastly, $\Delta_{free-field}$ can be expressed as a function of $\Delta_{structure}$, given as follows, and the related $\Delta_{free-field}$ and $\Delta_{structure}$ can be determined. This curve is used to obtain the actual racking deformation if the free-field deformation is known by step 2.

$$\Delta_{free-field} = \frac{\Delta_{structure}}{R(\Delta_{structure})} = \frac{\Delta_{structure} HS_1(\Delta_{structure})}{WG_m} \qquad \text{eq. (4)}$$

The shear modulus of the soil G_m can be calculated from the relation between shear modulus and shear wave velocity is $G_m = \rho_m C_S^{\ 2}$, where ρ_m is the mass density of the soil and C_S is the shear wave velocity, measured by geophysical tests such as cross-hole tests, down-hole tests.

CASE STUDY

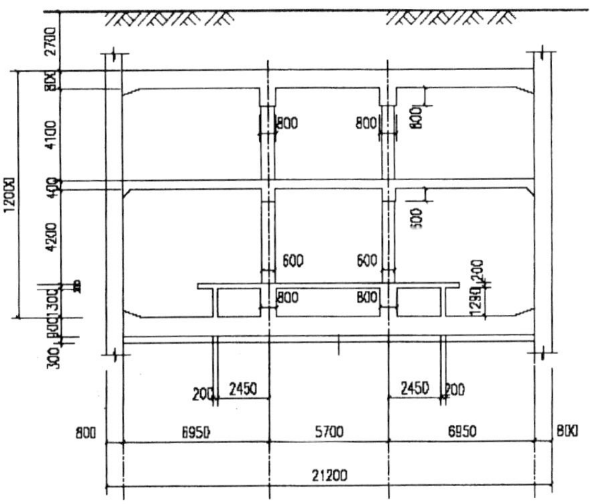

FIG.3. Cross section of the frame structure

A typical subway station in Shanghai is analyzed as an example. It is assumed to be built in a soft soil site, and its support structure is a two-story three-bay reinforced concrete frame structure, shown as FIG.3. The width and height of the structure is 21.20m and 12.02m respectively. The overburden depth is 3.5m. The geotechnical and earthquake parameters are listed as follows:

Soft soil density, ρ_m=1800 kg/m3; Apparent velocity of S-wave propagation in soil C_S=180 m/s; peak particle velocity V_S=0.25m/s.

Shear modulus $G_m = \rho_m C_S^{\,2}$=1800/1000*1802=58320 kPa.

The free-field deformation

$\Delta_{free-field} = \gamma_{\max} H = V_S / C_S * H$ =0.25/180*12.02=0.016m

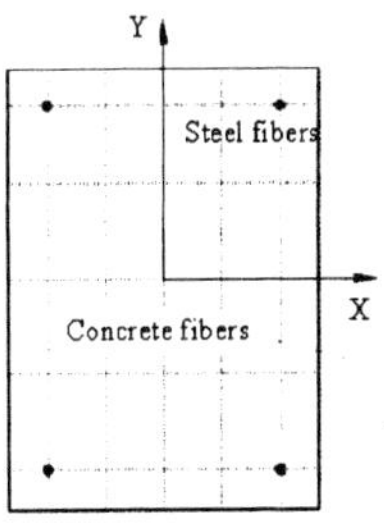

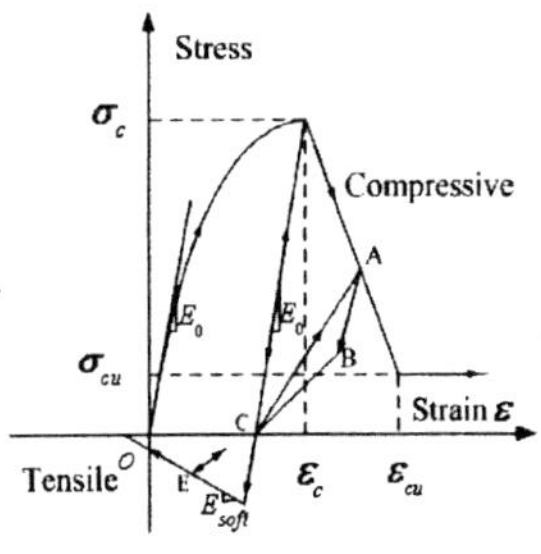

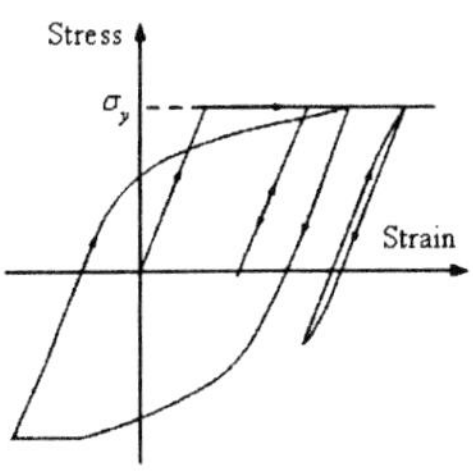

FIG.4. Fiber section **FIG.5. Concrete stress-strain curve** **FIG.6. Steel stress-strain curve**

FIG.4, 5 and 6 show the fiber section of beam-column elements, the stress-strain curves for concrete and steel fibers respectively.

FIG.7(a) shows the $P-\Delta$ curve by pushover analysis, and FIG.7(b) shows the curve of stiffness varying with the structure deformation. FIG.8(a) shows the curve of structural flexibility varying with the structure deformation, i.e. $R-\Delta_{structure}$ curve, and FIG. 8(b) shows the related free-field deformation and structure deformation curve, i.e. $\Delta_{freefeild}-\Delta_{structure}$ curve.

The free-field deformation

$\Delta_{free-field}=\gamma_{\max}H=V_S/C_S*H=0.016\text{m}$.

Therefore, according to FIG.8(b), the actual structure racking deformation is 0.017m.

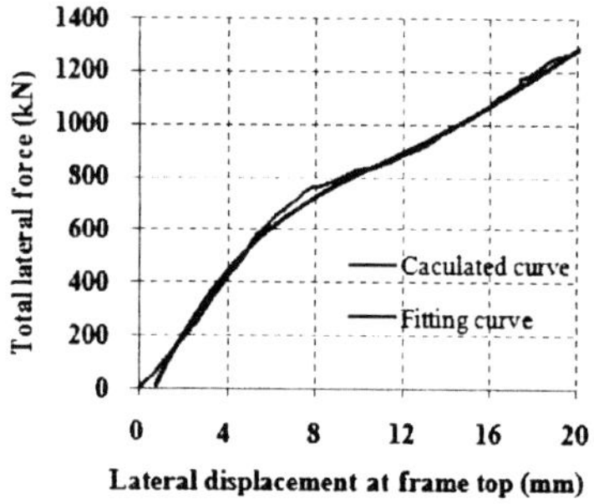

FIG.7(a) $P-\Delta$ curve

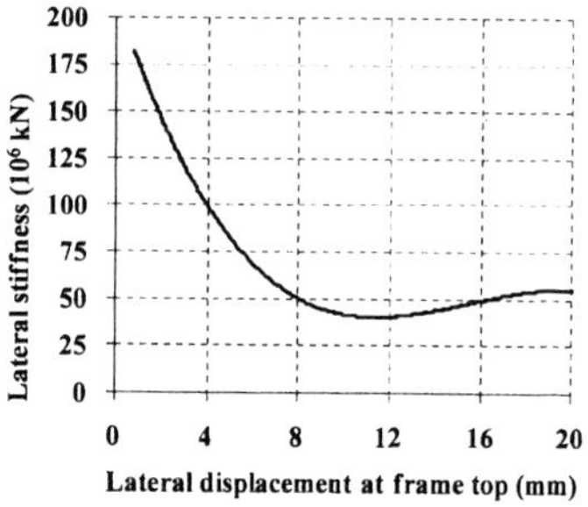

FIG.7(b) Stiffness-deformation $S-\Delta_{structure}$ curve

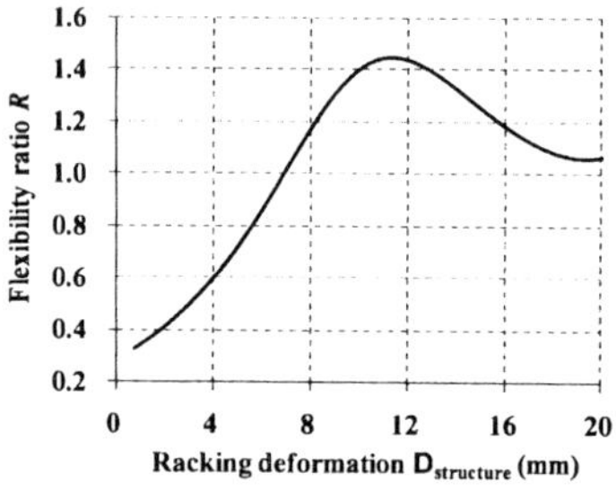

FIG.8(a) $R-\Delta_{structure}$ curve

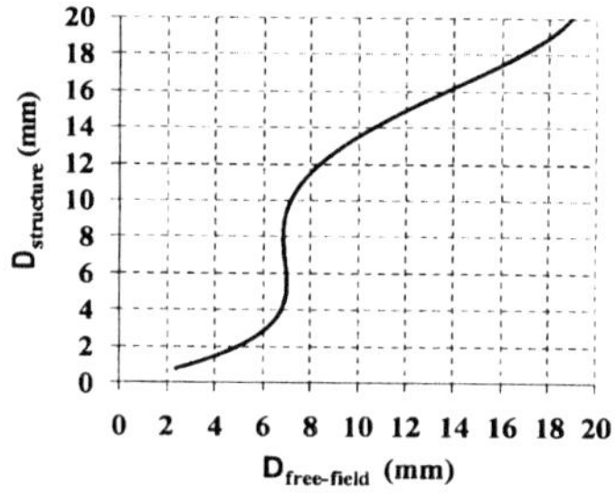

FIG.8(b) $\Delta_{freefeild}-\Delta_{structure}$ curve

CONCLUSION

The assessment of underground structure seismic deformation, therefore, requires an understanding of the anticipated ground shaking as well as an evaluation of the response of the ground and the structure to such shaking. The approach described in this paper consists of three major steps:

1. Estimate the free-field shear deformations, $\Delta_{free-field}$, of the ground at the depth of free-field interest using vertically propagating horizontal shear wave.

2. Conduct Pushover analysis of the structure to determine the lateral $P-\Delta$ curve and the stiffness variation curve with the deformation.

3. Calculate the flexibility ratio between the free-field medium and the structure to determine the flexibility ratio variation curve with the deformation; determine the $\Delta_{free-field} - \Delta_{structure}$ curve; Obtain the actual racking deformation of the structure based on the known $\Delta_{free-field}$, the seismic induced later load can be determined by means of the $P-\Delta$ curve from step 2.

The calculation of the seismic-induced racking deformation for a subway station frame structure shows that the approach herein accounting for the structure stiffness variation is simple but more reliable, and could be used for the engineering design of underground frame structure subjected seismic loads.

References

[1] Hashasha, M.A. et al, 2001. Seismic design and analysis of underground structures. Tunnelling and Underground Space Technology, 16: 247-293.

[2] Hendron AJ, Fernandez G. (1983). Dynamic and Static Design Considerations for Underground Chambers. Seismic Design of Embankments and Caverns, ASCE, New York, 157-197

[3] John ST, Zahrah CM,1987. Aseismic design of underground structures. Tunneling Underground Space Technol. 2 (2): 165-197.

[4] Kuesel, T.R., 1969. Earthquake Design Criteria for Subways. J.Struct. Div., ASCE ST6, 1213-1231.

[5] Merritt JL, Monsees JE, Hendron AJ, 1985. Seismic Design of Underground Structures. Proceedings of the 1985 Rapid Excavation Tunneling Conference, Vol. 1, 104-131

[6] Newmark, NM, 1968. Problems in wave propagation in soil and rock. Proceedings of the International Symposium on Wave Propagation and Dynamic Properties of Earth Materials.

[7] Penzien J, Wu CL, 1998. Stress in Linings of Bored Tunnels. Earthquake Engineering and Structure Dynamics 27, 283-300

[8] Sweet, J., 1997. Los Angeles Metro Red Line project: seismic analysis of the Little Tokyo Subway Station. Report no. CAI-097-100.Engineering Management Consultants.

[9] Wang, J.N., 1993. Seismic Design of Tunnels: A State-of-the-art Approach. Parsons Brinckerhoff Quade & Douglas, Inc., New York, NY, Monograph 7.

Author Index

Page number refers to the first page of paper

Subject Index

Page number refers to the first page of paper